applications index (cont'

(Applications Index is continued on inside back cover.)

Basic
Calculus
with applications

Burton Rodin
University of California
San Diego

Goodyear Publishing Company, Inc., Santa Monica, California

To Mari and Joey

Library of Congress Cataloging in Publication Data

Rodin, Burton.
Basic calculus with applications.

Includes index.
  1. Calculus.   I. Title.
QA303.R668   515   77–16093
ISBN: 0–87620–097–8

Current printing (last digit):
10 9 8 7 6 5 4 3 2

Y–0978–0

Printed in the United States of America

Cover illustration, $z = 3\sin(x)e^{y/5}$, from *Calculus and Analytic Geometry*,
by Al Shenk, copyright © 1977, Goodyear Publishing Company, Inc.

Contents

Preface

In questions of science the authority
of a thousand is not worth the humble
reasoning of a single individual.

Galileo Galilei (1564–1642)

You are probably majoring in the social sciences, life sciences, or business. Since you are not a mathematics major, you are likely to ask the important question, "What good will this calculus course do me?" I want to answer that question.

You may feel that calculus is remote from your area of interest. If so, you will be surprised at the variety of exciting applications that exist. In Chapter 8, for example, differential equations are derived for the Weber-Fechner law in psychology, for a complicated compound interest problem of finance, for diffusion through cell membranes of living cells, for the reaction rate of a chemical process, and for the motion of falling bodies. The Applications Index on the end pages gives many other examples; so do the footnotes scattered throughout the text.

You will not need to be a mathematical expert at calculus to work in your field. You will, however, need a basic understanding of calculus. For example, business executives do not perform sophisticated mathematical analyses in making decisions. They assign those tasks to actuaries or other mathematical consultants. To do this, however, they need enough mathematical knowledge to foresee the value of such analyses and to communicate effectively with their consultants. Here is how a noted biologist describes a similar situation in his field: "Biologists . . . increasingly will need to read papers containing mathematical reasoning, and . . . will occasionally need to use mathematics in their own work. Indeed, if a biologist can learn enough mathematics to know when the problem he is working on could usefully be discussed with a mathematical colleague, that alone will be worth considerable effort." (from *Mathematical Ideas in Biology* by J. Maynard Smith, Cambridge University Press, 1971).

The problems in this book are of two kinds: drill problems (*computational problems*) and word problems (*story problems*). It is best to resist the common tendency to concentrate on drill problems and neglect word problems. By struggling with word problems you will acquire the kind of understanding that will benefit you most. I hope you enjoy this course. I welcome any comments you care to send me.

November 1977   BURTON RODIN
La Jolla, California

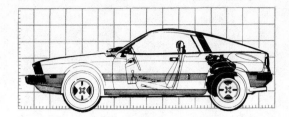

"R & T Road Test LANCIA BETA SCORPION"   Scale: 10" divisions

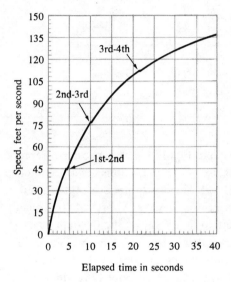

Elapsed time in seconds   **Figure 2**

Figure 2 shows the speed of a Lancia Beta Scorpion as it accelerated from rest.[2] It is apparent that the car lost speed when shifting gears from first to second, second to third, and third to fourth. It is not so apparent that the car traveled a quarter of a mile during the first 19 seconds. The method for finding this latter fact will also be studied in Chapter 6.

Example 2.   Figure 3 shows the Consumer Price Index for cost of living.[3] The graph shows clearly that prices are rising. Later in this chapter you will learn how to use such a graph to calculate the rate at which prices are rising, that is, inflation.

[2]Adapted from *Road and Track*, September 1976, p. 51.
[3]Data taken from *The World Almanac,* 1976.

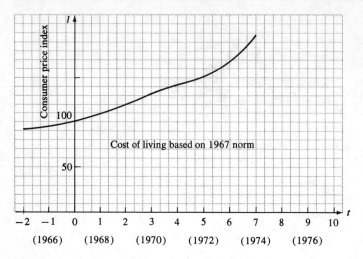

Figure 3

Reading graphs.   Consider the graph in Figure 1. The vertical axis represents the rate at which water was flowing, measured in millions of cubic meters per day. Abbreviate this quantity *rate* by *R*. The horizontal axis represents the time of the year, beginning January 1, measured in days. Abbreviate this quantity *time* by *t*. To find the rate *R* at a certain time *t*, say on the thirtieth day, we first locate the point on the horizontal axis that represents that time. Then we measure the height of the curve above that point using the scale on the vertical axis. We find that on the thirtieth day the rate was about 125 million cubic meters per day. This is abbreviated by writing

   $R = 125$ (million cubic meters per day) when $t = 30$ (days)

or, when there is no doubt about the units, simply by

   $R = 125$ when $t = 30$.

The following example illustrates how to read other information from this graph.

Example 3.   Use Figure 1 to find:

a.   the value of *R* when $t = 180$;

b.   the maximum value of *R* during the year and the time when it occurs;

c.   the minimum rate of flow and when it occurs.

SOLUTION.

a. From the point $t = 180$ on the horizontal axis, move your pencil upward until it hits the curve. Then move the pencil horizontally to the left until it hits the vertical axis; this point is $R = 100$. Thus when $t = 180$, $R = 100$.

b. The maximum value of R is about 790; it occurs when t is approximately equal to 243.

c. The minimum rate of flow occurs on about the 120th day (about May 1) and is approximately 60 million cubic meters per day.

Drawing graphs. Example 4 shows how to construct a graph from a given table of values. The following basic steps are used.

Step 1. Determine which of the given quantities will be represented by the vertical axis and which by the horizontal axis.

Step 2. Examine the range of values needed for each axis. Then select a convenient scale on each axis so these values can all be represented. The scale need not be the same for the two axes.

Step 3. Draw the horizontal and vertical axes, and mark off the scale for each one.

Step 4. Plot the values given in the table.

Step 5. Draw a smooth curve through the points you have plotted.

Graph paper is very helpful for making accurate and easy-to-read graphs.

In Step 1, explicit instructions will often be given telling which quantity is to be represented on each axis. The basis for making this choice is usually the following: To indicate the dependence of quantity B on quantity A (that is, if B depends on A), the graph is drawn with B on the vertical axis and A on the horizontal axis.

Example 4. If $1 is deposited in a bank at 6% annual interest compounded daily, the amount A in dollars in the account after t years is given by the following table.

Table 1

t (years)	A (dollars)	t (years)	A (dollars)
0	1.00	25	4.48
5	1.35	30	6.05
10	1.82	35	8.17
15	2.46	40	11.02
20	3.32	45	14.88

Draw a graph to show how A depends on t. Discuss the *investor's rule of thumb:* At 6% interest, money doubles in 12 years.[4]

SOLUTION.

Since A depends on t, we represent A on the vertical axis and t on the horizontal axis. The range of values for t is 0 to 45. We can represent these values conveniently by marking off 9 equally spaced points on the t- or horizontal axis, and letting each interval represent 5 years. The values for A on the vertical axis can be conveniently represented on a scale with 15 equal intervals, each representing $1 (see Figure 4).

[4]Actually, it takes 11 years, 201 days for a given sum of money to double.

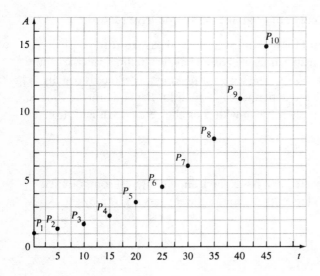

Figure 4

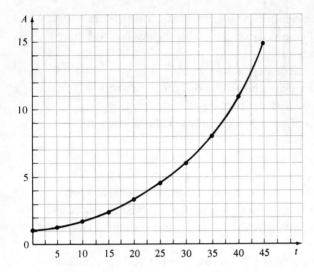

Figure 5

Now we plot the values in the table. The first pair, $A = 1.00$ when $t = 0$, is plotted as the point labeled P_1 in Figure 4; the second pair, $A = 1.35$ when $t = 5$, is plotted as the point P_2; and so on. Figure 5 shows the completed graph after these eight points are connected by a smooth curve.

From Figure 5, one can see that \$1 grows to \$2 in about 12 years; \$2 grows to \$4 in 12 years more, and so on. Since \$1 grows to \$2 in 12 years, \$50 will grow to \$100 in 12 years and, in general, x dollars will grow to $2x$ dollars in 12 years. Compound interest will be discussed in more detail in Chapter 5.

Problems

Graph paper is recommended for all problems in this section.
1. Psychologists have found that appetite decreases under a moderate exercise program. Rats were given access to an unlimited food supply.

Their daily food intake F in calories and weight W in grams were measured under various durations of enforced exercise. Table 2 gives F and W for an exercise program of H hours.[5]

(a) Draw a graph to show how F depends on H.

(b) How much exercise produces a minimum daily food intake?

(c) Draw a graph to show how W depends on H.

(d) What daily exercise program of 5 hours or less will produce the smallest body weight?

2. The *law of downward-sloping demand* in economics states that the demand q for a commodity will decrease if the price p increases. Table 3 gives values of p in dollars per bushel and q in millions of bushels per month for wheat. Graph this data with q on the vertical axis. Economists call such a graph the *demand curve for wheat*.[6]

(a) What would be the demand for wheat priced at $1.50 per bushel?

(b) How much does q decrease when p changes from 1.5 to 2.5?

Table 2			Table 3		Table 4		Table 5	
H	F	W	p	q	p	q	x	y
0	60	280	1	18	1	0	0	0
1	52	260	2	13	2	8	.2	.45
2	63	252	3	10	3	14	.4	.63
3	73	258	4	8	4	16	.6	.77
4	82	258	5	7	5	17	.8	.89
5	88	258					1.0	1.00
6	94	245						
7	93	237						
8	80	230						

3. Table 4 illustrates the *upward-sloping supply curve* for wheat. Here, q is the quantity of wheat in millions of bushels per month that sellers will supply

[5]Adapted from *Psychology Today*, January, 1970.

[6]At the turn of the century, the English economist Alfred Marshall introduced supply and demand curves. He drew them with p on the vertical axis. Although some economists today still follow that tradition, in this book we use the modern convention and represent q on the vertical axis.

when the market price is p, in dollars per bushel. Graph this supply curve with q on the vertical axis.

(a) How much wheat would sellers supply when the market price is $1.50 per bushel?

(b) How much does q increase when p changes from 1.5 to 2.5?

4. Draw the supply and demand curves in Problems 2 and 3 on a single graph. Their point of intersection is called the *equilibrium point* in economic theory. What are the values of p and q at the equilibrium point?

5. In Table 5, y is the square root of x. Draw a graph to show how y depends on x. Then find the amount by which y increases as x changes from 0 to .5 and as x changes from .5 to 1.

Table 6

t	F	t	F
0	50	210	400
30	30	240	550
60	20	270	360
90	15	300	140
120	15	330	60
150	50	360	50
180	80		

Table 7

t	T	t	T
0	24	210	74
30	27	240	66
60	37	270	55
90	50	300	40
120	60	330	29
150	71	360	24
180	75		

6. Table 6 gives the flow F in millions of cubic meters per day of the Blue Nile at Khartoum measured at various days of the year.[7] Here, t is the time in days from January 1. Draw a graph with t on the horizontal axis. What are the maximum and minimum values of F and when do they occur?

7. Table 7 shows temperature T in degrees Fahrenheit in Chicago, where t is the time in days from January 1. Draw a graph. Approximately how much did the temperature increase from January 15 to February 15? How much did it decrease from November 15 to December 15?

8. Let $y = 3x - 2$. Make a table showing the values of y when x has the values $-3, -1, 0, 1$, and 2. Draw a graph with x on the horizontal axis. What kind of curve do you obtain? Check your answer by using some additional values of x.

[7] *Encyclopaedia Britannica*, 1957, Vol. 16, p. 453.

9. Temperatures measured in degrees Fahrenheit, denoted by F, and in degrees Celsius or centigrade, denoted by C, are related by the formula $F = 1.8\ C + 32$. Make a table showing the values of F when C has the values 0, 50, and 100. Draw a graph with F on the vertical axis; the curve is actually a straight line. Use the graph to find F when $C = 25$ and when $C = 75$. Check your results by calculation.

Section 2 □ Rectangular Coordinates

The method of graphing discussed earlier leads to the idea of a coordinate system for the plane. Construct horizontal and vertical number axes in the plane (see Figure 6). Consider any point P in the plane. Each point determines an ordered pair of numbers (p_1, p_2) as follows: p_1 is the number on the horizontal axis which is directly above or below P and p_2 is the number on the vertical axis which is directly to the left or right of P. For example, point A in Figure 6 determines the ordered pair (6,4). Point B determines $(-7,5)$, C determines $(-5,-3)$ and D determines $(4,-6)$.

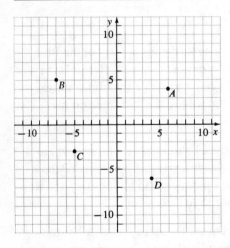

Figure 6

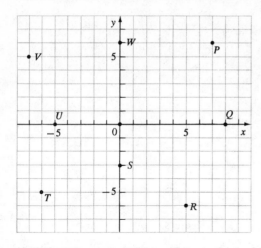

Figure 7

If the point P determines (p_1, p_2), we say that p_1 and p_2 are the **rectangular coordinates** of P. The number p_1 is called the **abscissa** of P and p_2 is called the **ordinate.** Often each axis is labeled with a letter. Suppose, for example, that the horizontal axis is called the **x-axis** and the vertical axis is called the **y-axis.** Then we also say that p_1 is the **x-coordinate** of P and p_2 is the **y-coordinate.**

Every point in the plane can be described unambiguously by its rectangular coordinates. Conversely, any ordered pair of numbers are the rectangular coordinates of a unique point in the plane. The point of intersection of the axes is called the **origin;** its rectangular coordinates are $(0,0)$.

There are other coordinate systems in use in addition to rectangular systems. In an *oblique coordinate system,* the axes are not perpendicular. In a *polar coordinate system,* angles are used to locate points. In this book, we shall always use rectangular coordinate systems. Therefore we often omit the word *rectangular* and say simply "the coordinates of P" rather than "the rectangular coordinates of P."

Example 1. Find the coordinates of the points O, P, Q, R, S, T, U, V, and W in Figure 7.

SOLUTION.
O: $(0,0)$; P: $(7,6)$; Q: $(8,0)$; R: $(5,-6)$; S: $(0,-3)$; T: $(-6,-5)$; U: $(-5,0)$; V: $(-7,5)$; W: $(0,6)$.

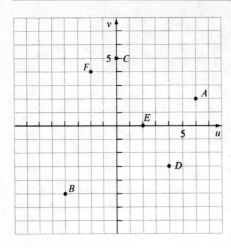

Figure 8

Example 2. Draw a u,v-coordinate system (that is, label the horizontal axis with u and the vertical axis with v). Then plot the points A, B, and C with coordinates A: (6,2); B: $(-4,-5)$; and C: (0,5). Also plot the points D, E, and F whose coordinates are D: $u = 4$, $v = -3$; E: $u = 2$, $v = 0$; and F: $u = -2$, $v = 4$.

SOLUTION.
The solution is shown in Figure 8.

Problems

1. Draw a coordinate system and plot the points A: (8,5); B: (1,6); C: (8,0); D: $(-6,3)$; E: (0,7); F: $(7,-5)$; G: $(-5,-4)$; and H: $(0,-4)$.

2. Draw an xy-coordinate system and plot the points P: $x = 2, y = 3$ and Q: $x = 8, y = 9$.
   Draw the line segment PQ and mark the midpoint M. Find the coordinates of M.

3. Draw a *uv*-coordinate system (*u* is the horizontal axis) and plot the points
   A: (3,8) and *B*: (6,1). Draw an extended line *L* through *A* and *B*.
   (a) Find the *v*-coordinate of the point on *L* which has 12 as its
       u-coordinate.
   (b) Find the *u*-coordinate of the point on *L* which has −6 as its
       v-coordinate.

4. The points *A*, *B*, *C*, and *D* are vertices of a square. The coordinates of
   A, *B*, and *C* are *A*: (4,1); *B*: (6,4); and *C*: (1,3). Find the coordinates of *D*.

5. Draw an *xy*-coordinate system, using equal scales on each axis. Draw the
   right triangle with vertices *A*: (1,2), *B*: (4,2), and *C*: (4,6). Find the lengths
   of the hypotenuse and the two legs.

6. Find the coordinates of the midpoint of the hypotenuse and of the
   midpoints of each leg of the triangle in Problem 5.

7. Find the area of triangle *ABC* of Problem 5.

8. The two axes in a coordinate system divide the plane into four regions
   called **quadrants**. Group the following points so that all the points in a
   group are located in the same quadrant.

A: (3,−4)	*E*: (5,7)	*I*: (−4,8)	*M*: (1,1)
B: (3,5)	*F*: (−2,−3)	*J*: (1,−1)	*N*: (−2,3)
C: (−2,7)	*G*: (−3,4)	*K*: (3,4)	*O*: (−8,−3)
D: (−3,−4)	*H*: (1,−5)	*L*: (−7,−1)	*P*: (8,−3)

Section 3 □ Functions

In Section 1, we saw that graphs provide an excellent way to depict how one
quantity depends on another. The graph in Figure 1 (page 2), for example,
can be used to determine the flow *F* of the Nile River at any time during the
year 1955. In other words, the graph shows a rule for finding *F* whenever *t* is
given, if *t* is between 0 and 365.

Let us be more abstract. Suppose a quantity *y* depends uniquely and
unambiguously on another quantity *x*. This means there is a rule for finding *y*
whenever *x* is given. In such a situation we say that *y* is a **function** of *x*. Since
the value of *y* depends on the value of *x*, often *y* is called the **dependent variable**
and *x* is called the **independent variable.** The **domain** of this function is the set
of numbers that the independent variable is allowed to assume.

In the above illustration concerning the Nile River, F is the dependent variable, t is the independent variable, F is a function of t, and the domain of this function is all numbers from 0 through 365.

The terminology of functions is used in all areas where mathematics is applied. To summarize, *a function is a rule which assigns a unique value to every number in its domain.* In calculus, various functions are introduced and methods are developed to study their properties. These methods are then applied in areas of science and business where functions are important.

Functions Determined by Graphs. Sometimes a function is determined from a graph. Consider Figure 9, for example. For any value of x from 1 through 5 inclusive, the curve determines a single value of y. Thus the graph determines y as a function of x. The independent variable is x, the dependent variable is y, and the domain is all numbers from 1 through 5, inclusive.

Functions determined by graphs frequently occur in applications. Experiments are performed or observations are made to determine how a quantity y depends on x. A table of values is thereby obtained. Then the graph is drawn, just as in Section 1. Functions obtained in this manner might be called *empirically-determined functions.*

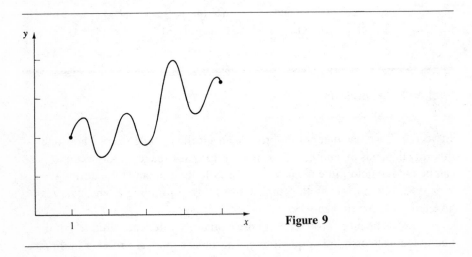

Figure 9

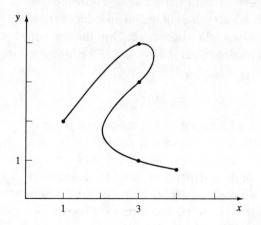

Figure 10

Figure 10 illustrates a curve which does not determine y as a function of x. One reason is that when x has the value 3, for example, the graph shows that y can be 1, 3, or 4. Since the definition of function requires that there is a unique value of y for each value of x, this graph does not determine a function.

The failure of the graph in Figure 10 to determine a function illustrates the following general rule:

> *A graph is the graph of a function if and only if every vertical line intersects the graph in at most one point.*

Functions Determined by Formulas. A formula such as

(1) $s = t^2 - 3t$

determines s as a function of t. To find the value of s when t has a certain value, such as $t = 5$, we let t take the value 5 and then compute s. Thus when $t = 5$,

$$s = 5^2 - 3 \cdot 5 = 25 - 15 = 10.$$

The reader should verify that

$s = 0$	when	$t = 0,$
$s = -2$	when	$t = 1,$
$s = -2$	when	$t = 2,$
$s = 0$	when	$t = 3.$

To draw the graph of this function, make a table of values and then proceed as in Section 1. This process is illustrated in Example 1 on page 16.

Mathematicians have an agreement that when a function is defined by a formula, the domain of that function is all values of the independent variable for which the formula makes sense, unless otherwise stated. Thus the domain of the function in Equation (1) is all numbers, but the domain of the function

$$s = t^2 - \frac{3}{t}$$

is all numbers except zero, because this formula does not make sense if $t = 0$ (division by zero is not defined). Formulas that arise in applied sciences (for example, *demand* as a function of *price* or *population* as a function of *time*) are usually valid only for limited values of the independent variable; these values would be the domain. In this book, we shall not explicitly specify the domain of each function we consider unless there is an important reason to do so.

Example 1. Draw the graph of the function $s = t^2 - 3t$.

SOLUTION.
Choose any convenient sequence of values for t. Then make a table showing the corresponding values of s. If we choose $-3, -2, -1, 0, 1,$ $2, 3, 4,$ and 5 as values for t, we obtain the following table:

t	-3	-2	-1	0	1	2	3	4	5
s	18	10	4	0	-2	-2	0	4	10

Draw a coordinate system. Since s is the dependent variable, it will be represented on the vertical axis. The horizontal axis always represents the independent variable, in this case t. Plot the points from the table and connect them by a smooth curve. The result is shown in Figure 11.

Example 2. A garden-supply store finds that the daily demand q in pounds for live earthworms depends on the price p in dollars per pound, according to the demand formula $q = 9 - p$ (for p between 0 and 9). Let R in dollars be the store's total daily revenue from earthworm sales.

a. Find a formula for R as a function of p.

b. Graph this function.

c. Find the approximate value of p that makes R a maximum.

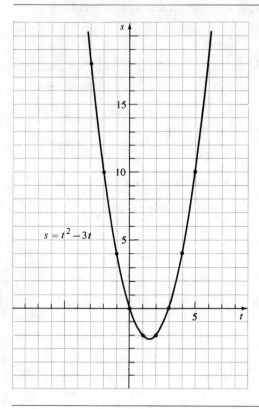

$$s = t^2 - 3t$$

Figure 11

SOLUTION.

a. First let us make sure we understand why R is a function of p. If the price p is set at a certain value, then the demand q will have a definite value determined by the demand formula. This means q pounds will be sold every day at the price of p dollars per pound. The daily total sales, or revenue, will therefore be $q \cdot p$ dollars. Thus R is uniquely determined by p; in fact, $R = pq$.

If we repeat this reasoning with formulas instead of words, we obtain

$$q = 9 - p$$

and

$$R = qp.$$

Now substitute the first equation into the second:

$$R = (9 - p)p \quad \text{or} \quad R = 9p - p^2$$

This formula expresses R in terms of p and is the required formula.

b. To graph the function $R = 9p - p^2$, make a table of values.

p	0	1	2	3	4	5	6	7	8	9
R	0	8	14	18	20	20	18	14	8	0

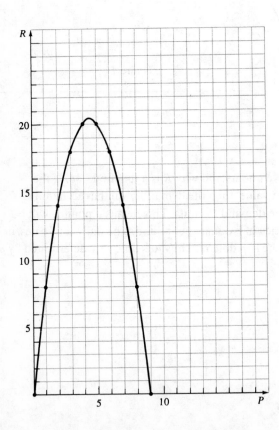

Figure 12

Since R is the dependent variable, it will be represented on the vertical axis. The independent variable is p; it will be represented on the horizontal axis. Draw a pR-coordinate system and plot the points from the table. The completed graph is shown in Figure 12.

c. From the graph we estimate that the maximum value of R is about $20\frac{1}{4}$ (dollars), and this occurs when p is about $4\frac{1}{2}$ (dollars).

Problems

1. Which of the graphs in Figures 13a, b, c, d are the graphs of functions?

2. In Figure 13a, what is the value of y when $x = 0$? When $x = 3$? When $x = 7$? When $x = -4$?

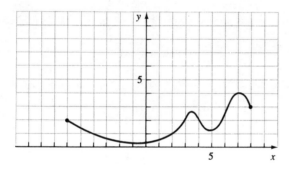

Figure 13a

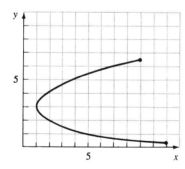

Figure 13b

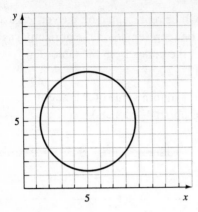

Figure 13c

3. In Figure 13d, is *s* the independent variable or the dependent variable? What is the value of *s* when $t = 6$? When $t = 3$? When $t = 7$?

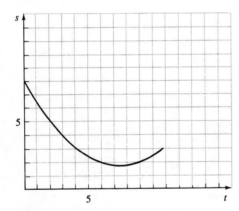

Figure 13d

4. Which of the following points lie on the graph in Figure 13b?

   A: (0,2) *B*: (5,1) *C*: (3,6) *D*: (2,4)

5. In Figure 13a, how much does *y* increase as *x* changes from 2 to 7? What is the maximum value of *y*? For what value of *x* does *y* attain its maximum value?

6. In Figure 13d, how much does *s* decrease as *t* changes from 3 to 6? What is the minimum value of *s*? For what value of *t* does *s* attain its minimum value?

In each of the following problems a function is given and an interval is specified as its domain. Sketch the graph in order to find, approximately, the maximum value of the dependent variable on this domain.

7. $y = x + 2$. Domain: $0 \leq x \leq 10$.

8. $s = t - 1$. Domain: $5 \leq t \leq 20$.

9. $R = 200 - 3p$. Domain: $10 \leq p \leq 50$.

10. $C = 60 - \dfrac{1}{2}q$. Domain: $20 \leq q \leq 100$.

11. $E = 5t - 4$. Domain: $3 \leq t \leq 10$.

12. $v = 10r - 5$. Domain: $0 \leq r \leq 5$.

13. $L = 70 - 2x$. Domain: $5 \leq x \leq 20$.

14. $y = 6x - x^2$. Domain: $0 \leq x \leq 5$.

15. $y = 10x - x^2$. Domain: $2 \leq x \leq 9$.

16. $y = 5x - x^2$. Domain: $2 \leq x \leq 5$.

17. $y = 8x - x^2$. Domain: $3 \leq x \leq 8$.

18. $y = 4x + 2 - x^2$. Domain: $0 \leq x \leq 3$.

19. $y = 14x - x^2 - 24$. Domain: $0 \leq x \leq 7$.

20. $y = 9x - x^2 - 2$. Domain: $1 \leq x \leq 6$.

21. $s = \sqrt{t + 4}$. Domain: $0 \leq t \leq 12$.

22. $V = \dfrac{20}{m}$. Domain: $10 \leq m \leq 40$.

23. $w = \dfrac{40}{u + 5}$. Domain: $0 \leq u \leq 35$.

24. $w = \dfrac{u + 101}{u + 1}$. Domain: $0 \leq u \leq 9$.

25. $y = 5x + 4x^2 - x^3$. Domain: $0 \leq x \leq 5$.

26. $y = 9x - x^2$. Domain: $0 \leq x \leq 3$.

In Section 3, we summarized the meaning of *function* as *a rule which assigns a unique value to each number in its domain.* If, for example, y is a function of x, then there is a rule which assigns a unique value, called the value of y, to each value of x that is in the domain. Let us choose a letter, such as f, to stand for the particular rule. We shall use this letter in the following way: To designate the value of y that is assigned to a certain value of x, say 3, write $f(3)$ (read "f of 3"). Thus $f(3)$ designates a particular number, namely, the value of y when x is 3.

Example 1. Let s be the following function of t:

$$s = t^2 + 3t$$

Designate this function by g. Find $g(2)$, $g(5)$, and $g(100)$.

SOLUTION.
The expression $g(2)$ stands for the value of s when $t = 2$. Since $s = 2^2 + 3 \cdot 2 = 10$ when $t = 2$, $g(2) = 10$.
    Similarly, $g(5) = 5^2 + 3 \cdot 5 = 40$, and $g(100) = 100^2 + 3 \cdot 100 = 10,300$.

The notation we have just introduced is called **functional notation.** It can be considered as an alternative to the notation using dependent variables. Table 8 shows how several statements would be expressed in each notation.

Table 8

Dependent Variable Notation	Functional Notation
y is a specified function of x	$y = f(x)$, where f is a specified function
the value of y when $x = 3$	$f(3)$
$y = 6$ when $x = 2$	$f(2) = 6$
subtract the value of y when $x = 1$ from the value of y when $x = 2$	$f(2) - f(1)$
graph the function $y = 2x^2 + 1$	graph the function $f(x) = 2x^2 + 1$

In modern mathematics, the functional notation is used much more frequently than the dependent variable notation. In the areas of science and business where mathematics is applied, the dependent variable notation is common.[8] The reader will find it essential to understand and use both forms of notation. The following example should help in this process. It reviews earlier techniques in terms of the new notation.

Example 2. Consider the function $F(x) = 9 - 2x$.

a. Make a table of values for $x = 0, 1, 4, 5, 7$.

b. Graph F.

c. Use the graph to find $F(6)$; then check the result by computation.

SOLUTION.

a. The required table of values is shown below.

x	0	1	4	5	7
$F(x)$	9	7	1	-1	-5

b. To draw the graph, construct a coordinate system. The independent variable, in this case x, is always represented on the horizontal axis. Plot the points obtained from the table of values with coordinates (0,9), (1,7), (4,1), (5,−1), and (7,−5); note that the first coordinate always refers to the independent variable. The plotted points are shown in Figure 14. The graph is a straight line.

c. To find $F(6)$ graphically, measure the ordinate (that is, the second coordinate) of the point on the graph which has 6 as its x-coordinate. Figure 14 shows that it is about -3. To check this result by computation, we see $F(6) = 9 - 2 \cdot 6 = 9 - 12 = -3$.

[8]This may be because the dependent variable notation helps keep track of the interpretation (price, weight, time, population, concentration, etc.) of the variables.

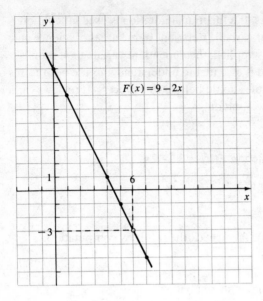

Figure 14

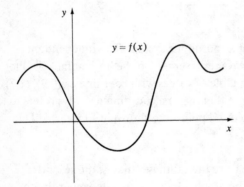

Figure 15a

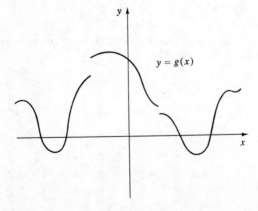

Figure 15b

Continuous Functions. Figures 15a and 15b show the graphs of two functions f and g. The function f in Figure 15a is called a **continuous function** because its graph has no breaks. The function g in Figure 15b is called a **discontinuous function**; its graph has breaks or *points of discontinuity*. In this course, we will be concerned almost exclusively with continuous functions.

In summary, we have discussed graphing and coordinate systems. The only technique we have at present for drawing graphs of continuous functions is plotting a number of points and then drawing a smooth curve through them. Functional notation has been introduced. It often provides a more convenient way to deal with functions. Note how it can be used to define the graph of a function: The graph of a function f is the set of all points (x, y) such that $y = f(x)$.

As you learn calculus, you will acquire improved techniques for constructing graphs.

Problems

1. If $h(t) = 3t + 5$, find $h(3)$, $h(100)$, and $h(-2)$.

2. If $g(x) = 2x^2 - 3x + 2$, find $g(0)$ and $g(4)$.

3. If $f(u) = \dfrac{u + 5}{u + 2}$, find $f(3)$, $f(1)$, and $f(0)$.

4. If $F(p) = p^2 - \frac{1}{2}p$, find $F(6)$.

5. If $f(x) = x^2 + 1$ and $g(x) = x^3 - 2$, find $f(3) - g(2)$.

6. If $Q(t) = (t - 1)(t + 1)$, show that $Q(a + 1) = a^2 + 2a$.

7. Let $G(x) = \sqrt{x(x + 6) + 9}$. Which of the following points lie on the graph of G: $(2,5)$, $(3,18)$, $(4,7)$?

8. Find four distinct points on the graph of the function G in Problem 7.

9. Find a formula for a function f so that $f(1) = 3$ and $f(2) = 5$. (Note: There are infinitely many correct answers.)

10. Let $f(x) = x^2 + x + 3$ and $g(x) = x^2 + 8$. Find the coordinates of a point that lies both on the graph of f and the graph of g. (Note: Use algebra; do not draw the graphs.)

11. Let $f(x) = x^2 - 2x + 2$. Sketch the graph. What value of x makes $f(x)$ a minimum? What is the value of this minimum?

12. Let $g(t) = 5t - t^2$. Sketch the graph of g. What value of t makes g a maximum? What is the value of this maximum?

13. Figure 16 shows the graph of a particular function f.
    a. Find $f(4)$.   b. Find $f(1)$.   c. Find $f(4) - f(1)$.
    d. Let $y = f(x)$; find the amount of increase for y as x changes from 1 to 4.
    e. True or false: The net increase in y as x changes from 2 to 5 is $f(5) - f(2)$.

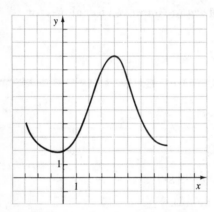

Figure 16

Section 5 □ Average Rate of Change

In daily life, we hear people speak of the exploding rate of increase of the world's population, the declining rate of growth of the economy, the phenomenal cure rate of a certain medical procedure, the stable rate of unemployment, etc. Such uses of the notion *rate of change* are sometimes not very precise. For example, when a television commentator announces that "inflation increased at a 3% rate this month," we cannot be sure if he means that prices increased by that amount, that the inflation of prices increased that amount, or, for that matter, whether 3% is a monthly rate or an annual rate.

In mathematics and its applications, a very precise definition of rate of change is needed. This subject is a main topic in calculus; when you have mastered it you will be able to speak knowingly about average rates of change, instantaneous rates of change, and even about rates of change of rates of change.

Average Rate of Change. Consider the following statement.

During June, sales increased at an average rate of $20 per day.

This statement does not mean that sales actually increased $20 each day in June. Rather, it means that the net increase in sales during the 30 days of June was as if the sales increased by $20 each day. The statement, therefore, really means only this: the net increase in sales during June was $600.

We now consider this concept in a general situation. Let y be a function of x. The Greek letter Δ (read "delta") is often used to denote a change in a quantity as follows:

Δx denotes a change in x;

Δy denotes the corresponding change in y.

(The symbol Δx stands for a number; it is not the product of Δ and x.)

The average rate of change of y during this change in x is $\dfrac{\Delta y}{\Delta x}$.

In the illustration above, let y (in dollars) be the total sales from January 1 to time x (in days) since January 1. During June, x changes by 30 days, so

$$\Delta x = 30.$$

The corresponding change in y is $600, so

$$\Delta y = 600.$$

The average rate of change of y is therefore

$$\frac{\Delta y}{\Delta x} = \frac{600}{30} = 20 \text{ (dollars per day)}.$$

Consider the general situation again: y is any function of x. Let x change from one value, say x_1, to a larger value, x_2. Then $\Delta x = x_2 - x_1$; it will be a positive number. (In this section, we need only consider the case $x_1 < x_2$. In the next section, we will use the Δx notation in more general situations where x might change from x_1 to a smaller value x_2; then Δx will be negative.) Let $y = y_1$ when $x = x_1$, and let $y = y_2$ when $x = x_2$. Then $\Delta y = y_2 - y_1$; it is a positive number if y increases, and it is negative if y decreases. The sign of the average rate of change $\dfrac{\Delta y}{\Delta x}$ therefore tells us whether y underwent a net decrease or net increase.

$$\frac{\Delta y}{\Delta x} > 0 \text{ implies } y \text{ underwent a net increase}$$

$$\frac{\Delta y}{\Delta x} < 0 \text{ implies } y \text{ underwent a net decrease}$$

$$\frac{\Delta y}{\Delta x} = 0 \text{ implies the net change of } y \text{ was zero}$$

Rate of increase and rate of decrease are simply special cases of *rate of change.*

Example 1. If $y = 3x^2 + 4$, find the average rate of change of y between $x = 3$ and $x = 5$.

SOLUTION.

Here x changes from 3 to 5.

$$x_1 = 3 \quad \text{and} \quad x_2 = 5, \quad \text{so} \quad \Delta x = x_2 - x_1 = 2$$

To find the change in y, we note the following.

When $x = 3, y = 31$

When $x = 5, y = 79$

Thus y changes from 31 to 79; hence

$$y_1 = 31 \quad \text{and} \quad y_2 = 79, \quad \text{so} \quad \Delta y = y_2 - y_1 = 48.$$

The desired rate of change is therefore

$$\frac{\Delta y}{\Delta x} = \frac{48}{2} = 24.$$

Example 2. If $q = 29 - 6p$, find the average rate of change of q between $p = 1$ and $p = 1.5$.

SOLUTION.

We have the following.

$$p_1 = 1 \quad \text{and} \quad p_2 = 1.5, \quad \text{so} \quad \Delta p = p_2 - p_1 = 0.5$$

When $p = 1, q = 23$; hence $q_1 = 23$. When $p = 1.5, q = 20$; hence $q_2 = 20$.

$$\Delta q = q_2 - q_1 = 20 - 23 = -3$$

The desired rate of change of q is therefore $\dfrac{\Delta q}{\Delta p} = \dfrac{-3}{0.5} = -6$. (*Note.* This rate of change of q is negative. This signifies that q underwent a net decrease as p changed from 1 to 1.5.)

Example 3. If $s = 5t - 7$, find the average rate of change of s between $t = 0$ and $t = 2$.

SOLUTION.
We have the following.

$$t_1 = 0 \quad \text{and} \quad t_2 = 2, \quad \text{so} \quad \Delta t = 2$$

When $t = 0, s = -7$; hence $s_1 = -7$. When $t = 2, s = 3$; hence $s_2 = 3$.

$$\Delta s = s_2 - s_1 = 3 - (-7) = 3 + 7 = 10$$

The desired rate of change of s is therefore $\dfrac{\Delta s}{\Delta t} = \dfrac{10}{2} = 5.$

The functions dealt with in mathematical applications usually represent quantities in certain specified units. In such cases, the average rate of change is also expressed in units:

$$\frac{\Delta y}{\Delta x} \text{ units of } y \text{ per unit of } x.$$

For example, if y represents the cost in *dollars* to produce x *tons* of steel then the average rate of change of y is expressed in *dollars per ton.*

Example 4. A poor sugar harvest caused the price of sugar to rise steeply for a six-month period following July 1, 1974. The price P in dollars for a 5-pound bag t months after that date was approximately

$$P = 0.8 + 0.5t.$$

Find the average rate of change of P for the period August 1, 1974 to October 1, 1974.

SOLUTION.

August 1, 1974 corresponds to $t = 1$. October 1, 1974 corresponds to $t = 3$. The average rate of change of P between $t = 1$ and $t = 3$ is computed as follows.

$$t_1 = 1 \quad \text{and} \quad t_2 = 3, \quad \text{so} \quad \Delta t = 2$$

When $t = 1$, $P = 1.3$; hence $P_1 = 1.3$. When $t = 3$, $P = 2.3$; hence $P_2 = 2.3$.

$$\Delta P = P_2 - P_1 = 2.3 - 1.3 = 1$$

The desired rate of change of P is therefore

$$\frac{\Delta P}{\Delta t} = \frac{1}{2} = 0.5 \text{ dollars per month.}$$

Functional Notation. Let us write the average rate of change $\dfrac{\Delta y}{\Delta x}$ in functional notation.

Let $y = f(x)$ be the given function. The quantities used to define rate of change and their equivalent expressions in functional notation are given in Table 9.

Table 9

Dependent Variable Notation		Functional Notation
$y = y_1$	when $x = x_1$	$y_1 = f(x_1)$
$y = y_2$	when $x = x_2$	$y_2 = f(x_2)$
	Δy	$f(x_2) - f(x_1)$
	Δx	$x_2 - x_1$
	$\dfrac{\Delta y}{\Delta x}$	$\dfrac{f(x_2) - f(x_1)}{x_2 - x_1}$

Thus the average rate of change of $y = f(x)$ from x_1 to x_2 can be written as

$$\frac{f(x_2) - f(x_1)}{x_2 - x_1}.$$

Problems

In Problems 1–11, find the average rate of change of each function.

1. $y = 5x^2 - 4$ between $x = 2$ and $x = 4$.

2. $v = 60 - 3u^2$ between $u = 1$ and $u = 3$.

3. $Q = 3 - 6p^2$ between $p = 1$ and $p = 4$.

4. $C = 9q + 5$ between $q = 2$ and $q = 7$.

5. $f(x) = 6x - 1$ between $x = 4$ and $x = 5$.

6. $g(t) = 8 - 3t$ between $t = 6$ and $t = 8$.

7. $s = 16t^2$ between $t = 1.5$ and $t = 2$.

8. $F(p) = \dfrac{12}{p + 2}$ between $p = 1$ and $p = 4$.

9. $y = \dfrac{2}{3}x - 1$ between $x = 12$ and $x = 18$.

10. $L(q) = \sqrt{q + 3}$ between $q = 1$ and $q = 6$.

11. $D = 2x^3 + 1$ between $x = 2$ and $x = 4$.

12. Economist Henry Schultz found that during the period 1915–1929, the per capita consumption C (in pounds) of sugar and the price p (in dollars) for one pound of sugar were related, approximately, by the formula $C = 135 - 8p$.[9] Find the average rate of change of C between $p = 2$ and $p = 4$. Give your answer in the proper units.

[9]H. Schultz, *The Theory and Measurement of Demand*, (Chicago: University of Chicago Press, 1938), p. 815.

13. The energy consumption c of a nation has been estimated as a function of the gross national product P. Suppose $c = 18P^2 + 2P$, where c is measured in millions of barrels per day of oil equivalent and P is measured in trillions of dollars. Economists call the rate of change of c the *propensity to consume*. Find the average propensity to consume between $P = 1$ and $P = 1.5$.

14. A rock is dropped from a high cliff. In t seconds, it will fall s feet, where $s = 16t^2$. The average rate of change of s is called *average velocity*. Find the average velocity between $t = 4$ and $t = 4.5$ and between $t = 0$ and $t = 0.5$.

15. Suppose that the height h in inches of a corn stalk t days after the seed germinates is $h = 8\sqrt{t} - 1$. Find the average growth rate between days 4 and 9 and between days 81 and 100.

16. If a farmer plants x acres of soybeans, his profit will be $f(x)$ dollars, where $f(x) = 1800x - 9x^2$. Find the average rate of change of profit between $x = 10$ and $x = 20$; between $x = 90$ and $x = 100$; and between $x = 150$ and $x = 160$.

17. A hospital finds that if it invests c dollars in ultraviolet lighting, it can reduce the bacteria count on a sample surface to N, where $N = \dfrac{50{,}000}{0.2c + 1000}$. Find the average rate of change of N between $c = 10{,}000$ and $c = 20{,}000$.

18. The rate of change of velocity is called *acceleration*. If the velocity v in feet per second of a particle t seconds after the motion starts is given by the formula $v = 2t^2 + t$, find the average acceleration between $t = 1$ and $t = 4$.

19. The rate of change of price is called *inflation*. If the price p in dollars after t years is $p = 3t^2 + t + 1$, find the average inflation from $t = 2$ to $t = 4$.

Section 6 □ Slope

Figure 17 shows four lines. If we wish to describe the relative steepness of these lines, we would say that in moving along each line from left to right, A_1A_2 rises more steeply than B_1B_2 and D_1D_2 falls more steeply than C_1C_2.

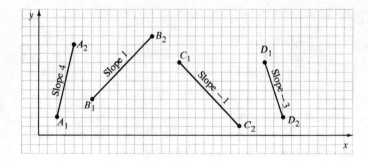

Figure 17

In mathematics, there is an exact numerical measurement of steepness called **slope**. If the slope is a positive number, the line is rising going along it from left to right—the larger the number, the greater the steepness. If the slope is a negative number, the line is falling from left to right. Note how the steepness is reflected in the values of the slopes given in Figure 17.

Vertical lines are an exception; *no value of slope is assigned to a vertical line.* This is not surprising; for instance, it makes no sense to say that a vertical line is rising or falling when moving along the line from left to right, as there is no left-to-right direction on such a line.

The mathematical definition of slope is as follows:

To find the slope of a nonvertical line, pick any two distinct points on the line and call these points P_1: (x_1, y_1) and P_2: (x_2, y_2); then the slope of the line is the number

$$\frac{y_2 - y_1}{x_2 - x_1}.$$

This definition of slope can be expressed nicely using $\Delta x, \Delta y$-notation. For any value of x, the given line determines a unique value of y. As x changes from x_1 to x_2, y will change from y_1 to y_2. Thus $\Delta x = x_2 - x_1$ and $\Delta y = y_2 - y_1$ and the following formula is obtained.

Slope of a Nonvertical Line

$$(1) \quad \text{Slope} = \frac{y_2 - y_1}{x_2 - x_1} = \frac{\Delta y}{\Delta x}$$

Figure 18 illustrates the interpretation of slope as "rise over run" in the case where Δx and Δy are positive. Note that Formula 1 is valid for any distinct values of x_1 and x_2; in particular, Δx is allowed to be negative.

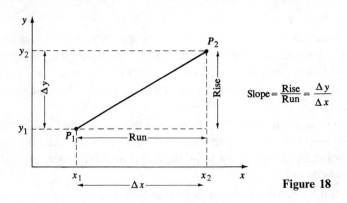

$$\text{Slope} = \frac{\text{Rise}}{\text{Run}} = \frac{\Delta y}{\Delta x}$$

Figure 18

Example 1. Check that the slopes in Figure 17 are correctly labeled.

SOLUTION.
Compare Figures 17 and 19. For A_1A_2, the change in x is $\Delta x = 2$ and the change in y is $\Delta y = 8$. Therefore the slope of A_1A_2 is $\dfrac{\Delta y}{\Delta x} = \dfrac{8}{2} = 4$.

For B_1B_2, we have $\Delta x = 7$ and $\Delta y = 7$. Therefore the slope of B_1B_2 is $\dfrac{\Delta y}{\Delta x} = \dfrac{7}{7} = 1$.

For C_1C_2, the change in x is $\Delta x = 7$. The change in y is $\Delta y = -7$. This change is negative because y has decreased; explicitly, $\Delta y = y_2 - y_1 = 1 - 8 = -7$. Therefore the slope of C_1C_2 is $\dfrac{\Delta y}{\Delta x} = \dfrac{-7}{7} = -1$.

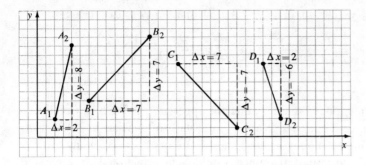

Figure 19

For D_1D_2, we have $\Delta x = 2$ and $\Delta y = -6$. Therefore the slope of D_1D_2 is $\dfrac{\Delta y}{\Delta x} = \dfrac{-6}{2} = -3$.

Example 2. Find the slope of the line between the points with coordinates (8,12) and (11,27).

SOLUTION.
We apply Formula 1 with $x_1 = 8$, $y_1 = 12$, $x_2 = 11$, and $y_2 = 27$.

$$\text{slope} = \frac{\Delta y}{\Delta x} = \frac{y_2 - y_1}{x_2 - x_1} = \frac{27 - 12}{11 - 8} = \frac{15}{3} = 5$$

Note that the same result occurs if we take (x_1, y_1) to be (11,27) and (x_2, y_2) to be (8,12).

$$\text{slope} = \frac{\Delta y}{\Delta x} = \frac{y_2 - y_1}{x_2 - x_1} = \frac{12 - 27}{8 - 11} = \frac{-15}{-3} = 5$$

Example 3. Find the slope of the line that passes through the points $(29, -3)$ and (1,4).

SOLUTION.

If we take $(x_1, y_1) = (29, -3)$ and $(x_2, y_2) = (1, 4)$, then Formula (1) gives

$$\text{slope} = \frac{\Delta y}{\Delta x} = \frac{4 - (-3)}{1 - 29} = \frac{4 + 3}{1 - 29} = \frac{7}{-28} = -\frac{1}{4}.$$

We have mentioned that no value of slope is assigned to a vertical line. What would happen if one tried to apply Formula 1 to a vertical line? On such a line, all points have the same x-coordinate. Therefore $\Delta x = 0$ for any two points on the line; Formula 1, if applied, would yield a fraction with zero in the denominator. Since division by zero is excluded in arithmetic, we say that *slope is not defined for vertical lines*. Some mathematicians prefer to say that vertical lines have "infinite slope," since lines that are nearly vertical have a very large numerical slope.

Problems

1. Find the slopes of lines L_1, L_2, L_3, L_4, L_5, and L_6 in Figure 20.

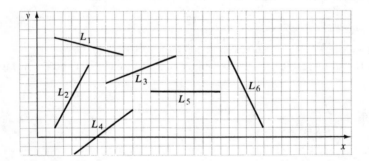

Figure 20

2. Construct a coordinate system on graph paper and draw two different lines, each with slope 2, two lines with slope $-\frac{1}{3}$, and two lines with slope zero. If two lines are parallel do they have the same slope? If two lines have the same slope, are they parallel?

3. Calculate the slope of the line which passes through the given points.

   a. (3,6) and (7,18).   b. (5,7) and (2,1).   c. (4,10) and (6,13).

   d. (8,8) and (1,3).   e. (2,37) and (20,1).   f. (7,9) and (2,12).

   g. (5,7) and (3,7).   h. (7,5) and (7,3).   i. $(6,-7)$ and $(8,-5)$.

   j. $(5,-5)$ and (2,1).

4. Construct a coordinate system. Draw the line L with slope $\frac{1}{3}$ which passes through the point (0,3). From your sketch, estimate the y-coordinate of the point on L which has x-coordinate 4. Can you figure out an arithmetic way to calculate this value exactly?

5. Line PQ has slope 5. P has coordinates (2,7). The x-coordinate of Q is 3. What is the y-coordinate of Q?

6. If a line is horizontal, what is its slope?

In Problems 7–15, the points P_1: (x_1,y_1) and P_2: (x_2,y_2) lie on the graph of the given function. Use the information supplied to find the slope of line P_1P_2.

7. $g(x) = \sqrt{x}$; $x_1 = 4$, $x_2 = 25$.

8. $h(x) = x^3$; $x_1 = 0$, $x_2 = 2$.

9. $f(x) = x^2$; $x_1 = 1$, $x_2 = 3$.

10. $f(x) = x^2$; $x_1 = 1$, $x_2 = 2$.

11. $f(x) = x^2$; $x_1 = 1$, $x_2 = 1.1$.

12. $f(x) = x^2$; $x_1 = 1$, $x_2 = 1.01$.

13. $f(x) = x^2$; $x_1 = 1$, $x_2 = 1.001$.

14. $f(x) = x^2$; $x_1 = 1$, $x_2 = 1.0001$.

15. $f(x) = x^2$; $x_1 = 1$, $x_2 = 1.00001$.

Section 7 □ Average Rate of Change—Graphical Interpretation

As you learn calculus, you will observe that a certain pattern occurs again and again. Almost every new concept will be discussed from two viewpoints—a graphical discussion (the "geometric interpretation") and an algebraic discussion (the "analytic interpretation"). Let us see how this works for the concept of average rate of change.

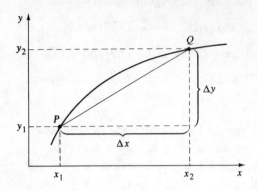

Figure 21

Consider some function; suppose y is a function of x. Then suppose y changes from y_1 to y_2 when x changes from x_1 to x_2. Recall that

(1) $$\frac{\Delta y}{\Delta x} = \frac{y_2 - y_1}{x_2 - x_1}$$

is the average rate of change of y during this change in x. This is the *analytic treatment* of average rate of change.

Now consider the graph of this function y. Suppose it is the curve shown in Figure 21. The points P: (x_1, y_1) and Q: (x_2, y_2) lie on the graph. The slope of line PQ is

$$\frac{\Delta y}{\Delta x} = \frac{y_2 - y_1}{x_2 - x_1},$$

which is the same value as the rate of change in Equation 1. Thus rates of change can be found by a *graphical method*.

> *To find the average rate of change of y as x changes from x_1 to x_2, mark the points P and Q on the graph where $x = x_1$ and $x = x_2$. Estimate the slope of the line PQ. This slope is the desired rate of change.*

This is the *geometric interpretation* of average rate of change.

Example 1.   Figure 22 shows the daily profit p in dollars of a company at various months t during the year.

a.   Use the graphical method to find the average rate of change of profit between $t = 1$ and $t = 6$.

b.   Repeat Part a between $t = 7$ and $t = 11$.

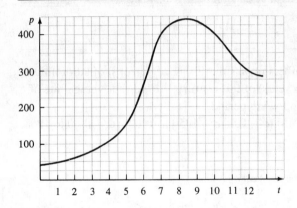

Figure 22

SOLUTION.

a. Mark the points P and Q on the graph where $t = 1$ and $t = 6$ (see Figure 23). Estimate the slope of line PQ. Figure 23 shows that the rise of PQ is approximately $\Delta p = 200$, and the run is approximately $\Delta t = 5$. Thus the slope is $\dfrac{\Delta p}{\Delta t} = \dfrac{200}{5} = 40$. Therefore the average rate of change of daily profit during the period from $t = 1$ to $t = 6$ was about 40 per month.

b. Mark the points P' and Q' on the graph where $t = 7$ and $t = 11$ (see Figure 23). We see immediately that the slope is negative; it is $\dfrac{\Delta p}{\Delta t} = \dfrac{-50}{4}$ $= -12.5$. Therefore the average rate of change of p between $t = 7$ and $t = 11$ is $-\$12.5$ per month. In other words, during the period $t = 7$ to $t = 11$ the profits decreased at an average rate of $\$12.50$ per month.

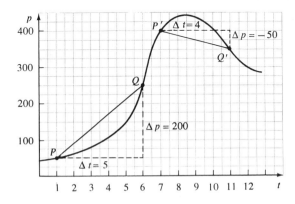

Figure 23

Problems

1. Figure 24 shows the membership *m* (in millions of persons) of labor unions during various times *t* (in years) of this century.[10] Find the average rate of change of *m* during the following periods.
   a. 1910 to 1925 **b.** 1940 to 1950 **c.** 1950 to 1960 **d.** 1960 to 1965.

Figure 24

2. A 30 square inch skin wound was treated with a vitamin ointment. The area *A* (in square inches) of the healed skin after *t* days is shown in Figure 25. Find the average rate of healing in each of the following time periods:
   a. between days 2 and 5
   b. between days 5 and 20
   c. between days 20 and 25

3. Figure 26 shows the total national income *u* and the total income *v* of the poorer 50% of the population, where *u* and *v* are in billions of dollars. Find the average rate of change of *u* and of *v* between 1960 and 1970.

[10]Data taken from *The World Almanac*, 1976.

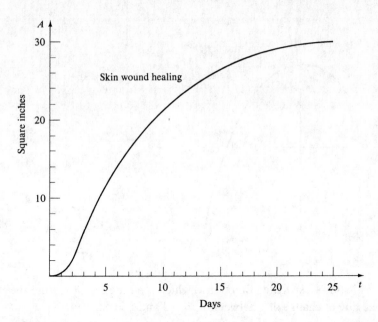

Figure 25

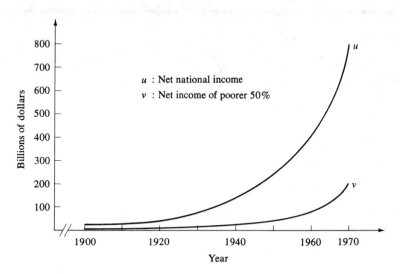

Figure 26

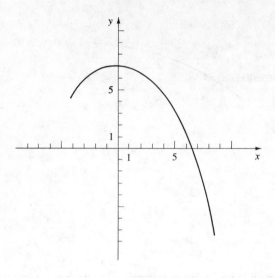

Figure 27

4. Figure 27 shows a specified function y, which is a function of x. Find the average rate of change of y between $x = -3$ and $x = 8$.

5. Suppose c is the function of q such that $c = 2q^2 + q + 1$. In order to find the average rate of change of c between $q = 1$ and $q = 3$, would it be more efficient to use the graphical method or the analytic method? Find that rate of change by whichever method you chose.

Section 8 □ Linear Functions

Linear functions represent a very basic class of functions.

Linear Functions
A linear function is a function of the form

$$f(x) = mx + b$$

where m and b are constants.

For example, $f(x) = 2x + 3$, $y = \frac{1}{2}x - 5$, $c = -2p + 4$, $v = 32t$, and $g(t) = 2$ are all linear functions.

Linear functions are easy to graph because the graph of a linear function is a straight line. They are used in many mathematical applications, sometimes to describe an exact relationship (see Example 1), and sometimes to approximate such a relationship (see Example 5).

Example 1. Two friends decide to manufacture ceramic urns. They will need a kiln which costs $150 and a potter's wheel which costs $250. A dozen urns require $125 for clay plus $75 for glaze. Find the function which gives the total cost c in dollars for making x dozen urns. Graph the function.

SOLUTION.

To make x dozen urns requires the following.

> $125x$ dollars for clay
>
> $75x$ dollars for glaze
>
> 150 dollars for kiln
>
> 250 dollars for potter's wheel.

The total cost for x dozen urns is therefore

$$c = 125x + 75x + 150 + 250,$$

or

$$c = 200x + 400.$$

Note that c is a linear function of x. Therefore, the graph will be a straight line. To sketch the graph, plot two points and draw a straight line through them. For example, if $x = 0$, then $c = 400$; so $(0,400)$ is on the graph. If $x = 5$, then $c = 1400$; so $(5,1400)$ is on the graph. The graph is shown in Figure 28.

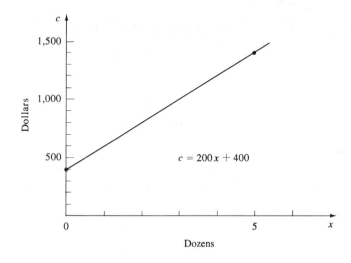

Figure 28

The point where a nonvertical line crosses the y-axis is called the *y-intercept* of the line. For example, the line in Figure 28 has (0,400) as its *y*-intercept.

Slope-intercept Formula.

The graph of $y = mx + b$ is a straight line having slope m and y-intercept $(0,b)$.

To verify the formula, check that $y = b$ when $x = 0$; thus $(0,b)$ lies on the line. The point $(0,b)$ also lies on the y-axis since its x-coordinate is zero. Therefore $(0,b)$ is the y-intercept. Now pick any other point on the line. For instance, when $x = 1$, $y = m + b$, so $(1, m + b)$ lies on the line. The slope between the points $(0,b)$ and $(1, m + b)$ is $\dfrac{\Delta y}{\Delta x} = \dfrac{m + b - b}{1 - 0} = m$. Thus the slope of the line is m.

Let us note the special case $m = 0$ of the slope-intercept formula: the graph of $y = b$ is a horizontal line with y-intercept $(0,b)$.

Example 2.

a. Describe the graph of $y = -2x + \frac{1}{2}$.

b. Describe the graph of $u = 8t - 7$.

c. If $q = 3p + 5$, find the average rate of change of q between any two values of p.

SOLUTION.

a. From the slope-intercept formula with $m = -2$ and $b = \frac{1}{2}$, we see that the graph of $y = -2x + \frac{1}{2}$ is a straight line of slope -2; its y-intercept is $(0,\frac{1}{2})$.

b. We can use the slope-intercept formula even though the dependent variable is u rather than g and the independent variable is t rather than x. The graph is a straight line of slope 8; it intercepts the vertical u-axis at the point $(0, -7)$.

c. The graph of $q = 3p + 5$ is a straight line of slope 3. Therefore the average rate of change of q (during any change in p) must also be 3.

We have seen that it is easy to draw the graph of a given linear function. Now we consider the reverse problem: given a straight line graph, find the corresponding linear function or equation.

Point-slope Formula.

The equation for the straight line which has slope m and passes through the point (x_1, y_1) is

$$y - y_1 = m(x - x_1).$$

To derive this formula, consider a general point (x, y) on the given line. The slope between (x, y) and (x_1, y_1) is $\dfrac{y - y_1}{x - x_1}$, which must be equal to the given slope m. Thus $\dfrac{y - y_1}{x - x_1} = m$; multiply both sides of this equation by $x - x_1$ and we obtain the point-slope formula.

Example 3.   A straight line passes through the point $(2,15)$. The slope of the line is 4. Find the equation of the line.

SOLUTION.
Applying the point-slope formula with $m = 4$, $x_1 = 2$, $y_1 = 15$, we obtain $y - 15 = 4(x - 2)$ as an equation of the line. This equation can be simplified somewhat.

$$y - 15 = 4(x - 2), \qquad y - 15 = 4x - 8, \qquad y = 4x - 8 + 15$$

or, finally, $y = 4x + 7$.

Example 4.   Suppose the demand y for a commodity is a linear function of the price x. Find the formula for y in terms of x if

$$y = 17 \qquad \text{when} \qquad x = 2$$

and

$$y = 5 \qquad \text{when} \qquad x = 6.$$

SOLUTION.
The graph will be a straight line since the function is linear. We know that the two points $(2,17)$ and $(6,5)$ lie on the line, and we want to find an equation of the line.

To use the point-slope formula, we will first need to find the slope m of the line. We can calculate m from the two given points.

$$m = \frac{\Delta y}{\Delta x} = \frac{5 - 17}{6 - 2} = \frac{-12}{4} = -3$$

We can now apply the point-slope formula with $m = -3$, $x_1 = 2$, and $y_1 = 17$. (We could choose $x_1 = 6$ and $y_1 = 5$; our answer would be the same.)

$$y - 17 = -3(x - 2),$$
$$y - 17 = -3x + 6,$$
$$y = -3x + 6 + 17,$$

or,

$$y = -3x + 23$$

Linear Regression.   A psychologist is interested in predicting a student's college grade point average (GPA) from that student's college entrance examination (ACT) score. He examines the records of seven college seniors and tabulates their ACT scores (denoted by x) and GPA (denoted by y).

x	20	21	22	23	24	25	26
y	2.4	2.2	2.6	2.9	3.2	3.0	3.2

These values of x and y are plotted on a graph and the best fitting straight line is drawn (see Figure 29). The equation $y = f(x)$ of the straight line is found; it provides a formula for predicting y (GPA) whenever x (ACT score) is known.

This process of finding a linear function to approximate an empirical relationship is called **linear regression.** As presented above, it is rather inexact since "best fitting straight line" is a subjective term. In Section 7 of Chapter 7, we shall show how calculus is used to overcome that difficulty.

Example 5.   Find the formula $y = f(x)$ referred to above. Use it to predict what GPA a student will earn if he enters college with an ACT score of 30.

SOLUTION.
To find the equation of the line in Figure 29, let us find its slope and a point on it and then use the point-slope formula. The slope between $x = 20$ and $x = 28$ is

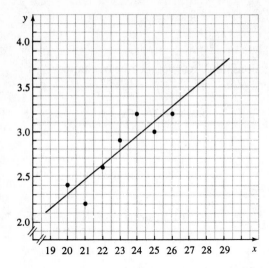

Figure 29

$$m = \frac{\Delta y}{\Delta x} = \frac{3.6 - 2.3}{28 - 20} = \frac{1.3}{8} \approx .16.$$

The point (20,2.3) lies on the line, so by the point-slope formula,

$$y - 2.3 = 0.16(x - 20).$$

This simplifies to

$$y = 0.16x - 0.9.$$

If $x = 30$, this formula predicts that the GPA will be

$$y = (0.16)(30) - 0.9 = 3.9.$$

Problems

In Problems 1–11, find the slope and the y-intercept; then sketch the graph.

1. $y = 9x - 2.$

2. $y = 12.$

3. $y = 4 + 3x.$

4. $y = 5 - \dfrac{x}{2}.$

5. $y + 2x = 7.$

6. $y = \dfrac{3x}{5} + 1.$

7. $x = y.$

8. $y = 5x.$

9. $y = 2$. **10.** $x + 3y = 1$.

11. $x + 3y = 0$.

12. For each of the following functions, find the average rate of change of y during any change in x.

a. $y = 10x - 25$ **b.** $y = 2 - 5x$

c. $y = \frac{1}{2}x + 3$ **d.** $y = 4 - \dfrac{x}{3}$.

13. Find an equation for the line that passes through the given point or points and, where specified, has the given slope.

a. $(0,3)$, slope 6   **b.** $(0,2)$, slope $-\frac{1}{2}$

c. $(2,7)$, slope 1   **d.** $(2,7)$, slope -1

e. $(2,7)$, slope 3   **f.** $(2,7)$, slope 0

g. $(2,7)$ and $(10,3)$ **h.** $(2,7)$ and $(-3,-3)$

14. For tax purposes, a construction company estimates that the useful life of a \$35,000 bulldozer is 7 years. Find a formula for the value y in dollars of the bulldozer x years after purchase if linear depreciation is used and if the salvage value is zero (that is, y is a linear function of x; the value of y when $x = 7$ is called the *salvage value*).

15. Solve Problem 14 if the salvage value is \$7000.

16. If the price of a certain commodity is x, then the quantity that producers will supply is $S(x) = 2x$ and the quantity consumers will demand is $D(x) = -4x + 300$. Graph the supply curve and demand curve on a single graph. Find the *equilibrium price* (the x-coordinate of the intersection point) and the *equilibrium quantity* (the y-coordinate of the intersection point).

17. Let x be the average grade that a class of calculus students earned on standardized examinations. The class evaluated their instructor's teaching; let y be the average grade the instructor received. Actual data was collected;[11] the following values of x and y occurred in twelve different classes, where x and y are on the scale $0.0 = $ F to $4.0 = $ A.

x	2.17	2.89	3.20	3.34	3.35	3.42	3.45	3.50	3.62	3.77	3.80	3.98
y	3.54	3.52	3.50	3.43	3.30	3.32	3.40	3.36	3.21	3.35	3.35	3.02

[11] *Science*, Vol. 177, No. 4055, pp. 1164–1166.

Graph the data, draw the best fitting straight line, and obtain an approximate formula for predicting y in terms of x. (The negative slope was a surprise—it stimulated further research about the validity of student evaluations of teachers.)

Summary

In this chapter, we have studied functions. Their properties can often be studied in two ways: from their graphs or from their equations. A very important property of functions is the **average rate of change.** You should be able to calculate rate of change both graphically and algebraically. The average rate of change can be expressed in the two notational systems as

$$\frac{\Delta y}{\Delta x} \quad \text{or} \quad \frac{f(x + \Delta x) - f(x)}{\Delta x};$$

You should become thoroughly familiar with each, since they are used equally often.

     Linear functions are among the simplest functions. Their graphs are straight lines and have a **slope** that can be defined in an elementary manner. You should be able to find the equation of a line when you know two points on the line, or one point and the slope. In the next chapter we shall see how calculus can be used to find the slope of a graph that is not a straight line.

Miscellaneous Problems

1. Let $y = f(x)$ where $f(x) = x - x^2/8 + 31/2$.
   a. Find the value of y when $x = 2$.
   b. Find $f(6)$ and $f(-10)$.
   c. Make a table of values for $x = -10, -6, -2, 0, 2, 6, 10, 14, 18$.
   d. Plot the points from your table (Part c) and sketch the graph of f.
   e. Estimate $f(12)$ from your graph (Part d); then calculate $f(12)$ exactly.
   f. Use the graph (Part d) to find, approximately, the values of x for which $f(x) = 0$. (That is, solve the equation $x - x^2/8 + 31/2 = 0$ graphically.)
   g. What value of x makes $f(x)$ a maximum?
   h. What is this maximum value of f?
   i. Estimate Δy as x changes from 0 to 4.
   j. Estimate Δy as x changes from 7 to 13.

2. Figure 30 is the graph of a function $q = F(p)$.
   a. Is p the dependent or independent variable?
   b. Estimate $F(0)$, $F(1)$, $F(2)$, $F(3)$, $F(5)$, and $F(9)$.
   c. Solve the equation $F(p) = 0$ for p.
   d. Is $(6,5)$ on the graph of F?
   e. Is $(5,6)$ on the graph of F?
   f. Estimate the value of q when $p = 7$.
   g. Solve the equation $F(p) = 1$ for p.
   h. Solve the equation $F(p) = 2$ for p.
   i. Find Δq, the change in q, as p changes from 2 to 5.
   j. Find Δq when p changes from 5 to 9.

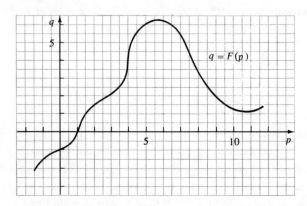

Figure 30

3. In psychology, a mathematical model of memory predicts that if a subject spends t hours memorizing a certain long list of words, then the percentage s of words remembered is $s = \dfrac{90t + 84}{t + 2}$. Find the average rate of change of s in each case.
   a. As t changes from $t = 2$ to $t = 6$
   b. As t changes from $t = 2$ to $t = 14$
   c. As t changes from $t = 6$ to $t = 14$

4. The graph of $q = F(p)$ is shown in Figure 30. Find the average rate of change of q for each case.
   a. Between $p = 3$ and $p = 5$ b. Between $p = 3$ and $p = 9$
   c. Between $p = 5$ and $p = 9$

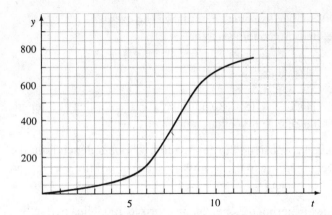

Figure 31

5. A rock is thrown vertically upward. Its height after t seconds is
$h(t) = 128t - 16t^2$ feet. The rate of change of height is called *speed.* Find
the average speed in each case.
   a. Between $t = 0$ and $t = 1$ **b.** Between $t = 1$ and $t = 2$
   c. Between $t = 0$ and $t = 2$ **d.** Between $t = 3$ and $t = 4$

6. Graph the function h in Problem 5, and find the value of t at which the
rock reached its maximum height.

7. The velocity v in feet per second of a rocket t seconds after launching is
$v = 4t^3 + 2t^2$. The rate of change of velocity is called *acceleration.* Find
the average acceleration in each case.
   a. Between $t = 0$ and $t = 2$ **b.** Between $t = 2$ and $t = 4$.

8. Beginning at midnight, a record was kept of cars crossing a certain
intersection. The total number y of cars after t hours is shown by the
graph in Figure 31. The rate of change of y is called *traffic flow.* Find the
average traffic flow in each case.
   a. Between midnight and 6 A.M.
   b. Between 6 A.M. and 9 A.M.
   c. Between 9 A.M. and noon

9. Suppose that the number B of bacteria in a Petri dish after t hours is
given by the formula $B = 1 + t + t^2/2 + t^3/6$. The rate of change of B
is called the *growth rate* of this culture. Find the average growth rate in
each case.
   a. Between $t = 3$ and $t = 6$ **b.** $t = 6$ and $t = 12$

10. Let q be the quantity of a commodity that will be demanded by consumers when the price of the commodity is p. If p changes by an amount

Δp, then q will change by some amount Δq. The *average elasticity of demand* is defined by economists to be

$$\frac{p}{q} \cdot \frac{\Delta q}{\Delta p}.$$

If the demand law is $q = 15 - 3p$, find the average elasticity of demand when $p = 3$ for a change in p of amount $\Delta p = 2$.

11. The construction cost c in dollars for a sailing schooner depends on the length l in feet according to the formula $c = 50l^2$. Find the average rate of change of c between $l = 30$ and $l = 40$.

12. In a psychology experiment, a person memorized 100 items perfectly. After 12 days, the person could remember only 40 of the items. Let r be the number of items remembered after t days. Find a formula for r as a function of t if it is assumed that the function is linear. How many items will be remembered after 17 days?

13. Suppose that the duration l of immunity to a certain disease is a linear function of the amount x of vaccine administered. If 2 cubic centimeters of vaccine gives 3 years of immunity, find a formula for l in years as a function of x in cubic centimeters. (*Note:* $l = 0$ when $x = 0$.)

14. The demand q for a certain commodity is a linear function of the price p. The average rate of change of q is always $\dfrac{\Delta q}{\Delta p} = -\dfrac{1}{2}$. If $q = 68$ when $p = 24$, find a formula for q as a function of p.

15. The amount of participation of the third-best student in a seminar was measured and denoted by r. The number of students in the seminar was denoted by n. After measurements in many seminars, sociologists found that r was a linear function of n.[12] If $r = .5908$ when $n = 4$ and $r = .7284$ when $n = 12$, find $\dfrac{\Delta r}{\Delta n}$. Also find the formula for r in terms of n (called the *Stephen-Mischler formula*).

---

[12]J. S. Coleman, *Introduction to Mathematical Sociology* (London: The Free Press of Glencoe Collier-Macmillan Ltd., 1964), p. 29.

two

Differentiation

The notion of *average rate of change* introduced in Chapter 1 will be refined to that of **instantaneous rate of change,** also called the **derivative.** This chapter deals mostly with techniques; additional applications will be treated in following chapters.

Section 1 □ Tangent Lines

Consider the following statement carefully.

> *Figure 1 shows a curve C, points P and Q on C, and two lines L_1 and L_2 which pass through P. Line L_1 is tangent to C at P; line L_2 is not tangent to C at P.*

Different students react in varied ways to the above statement. Some students seem to have an intuitive sense of why L_1 is called a tangent line whereas L_2 is not. Other students are completely bewildered by the statement; they do not see any essential difference between the lines L_1 and L_2. To which group do you belong? To check, draw a line through point Q in Figure 1 so that the line you draw is tangent to the curve C at Q. Then compare your line with the actual tangent line shown in Figure 2.

These different reactions do not reflect mathematical ability but rather seem to depend on previous experiences. For example, if you experienced the first reaction, you may be familiar with navigation, graphic arts, woodworking, mechanics, or other fields in which tangent lines play a direct or indirect role.

The purpose of this section is to help you develop an intuitive notion of *tangent line.* After studying this section, you should be able to draw an accurate tangent line to a smooth curve at any given point on the curve.

Even if you already have an intuitive understanding of tangent lines, you will benefit from a careful consideration of the *approximation method* discussed after Example 1. This method contains some fundamental ideas of calculus.

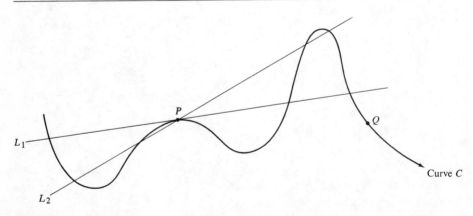

Figure 1

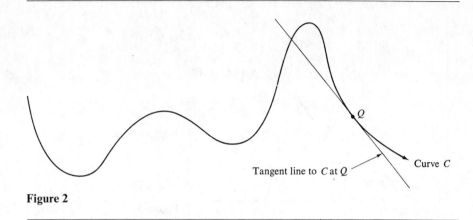

Figure 2

Example 1. A race-car driver might explain tangent lines in the following way. Imagine that Figure 1 is a bird's-eye view of a curvy race track C. Imagine a car going full speed (in racing jargon: "at the ragged edge") along this road; suppose it is traveling in a general left-to-right direction. At P there is an oil slick on the road. As the car approaches P, it is in a right turn. When it hits the oil slick at P, the tires will lose their traction and the car will "spin out" in a straight line in the exact direction that the car was moving when it hit P.

An experienced racing driver intuitively feels that the spin-out will be along L_1, not L_2. He might define tangent line as the line of spin-out, or the line which has the same direction as the curve at the given point. Move your pencil along C to imitate the path of the car as it approaches P and then spins out along L_1. Perhaps you can acquire the knack of drawing tangent lines this way. Try it at point Q in Figure 1; check your answer by looking at Figure 2.

Approximation Method. There is an approximation method for drawing tangent lines. It will not give the exact tangent line, but will produce a result that can be made as close to exact as you wish.

Consider the problem of drawing a tangent line to the graph G (see Figure 3) at the point A. To solve the problem, pick a point B on G, fairly close to A. Draw the line AB, it is called a **secant line.** A *secant line* is a line which crosses a curve in at least two points. Secant AB serves as an approximation to the desired tangent line. A better approximation can be obtained by choosing B closer to A. Figure 4 shows how the exact tangent line is approximated more and more closely by the secant lines.

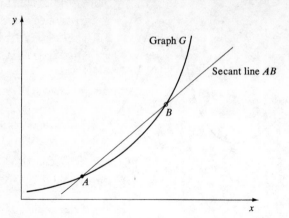

Graph G

Secant line AB

B

A

x

Figure 3

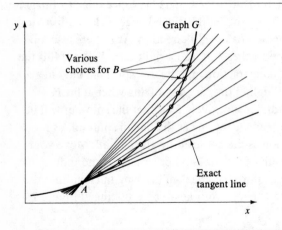

Graph G

Various
choices for B

Exact
tangent line

A

x

Figure 4

Example 2. Figure 5 shows the graph of the function $y = x^2$. Find the slope of the line that is tangent to this graph at the point (1,1).

SOLUTION.
The analytical solution to this problem will be given in Section 2. To solve the problem geometrically, draw the tangent line as accurately as you can. Then measure its slope. From Figure 6, we obtain the answer.

$$\text{slope of tangent line} = \frac{\text{rise}}{\text{run}} = \frac{1.0}{.5} = 2$$

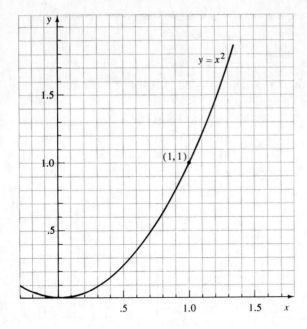

Figure 5

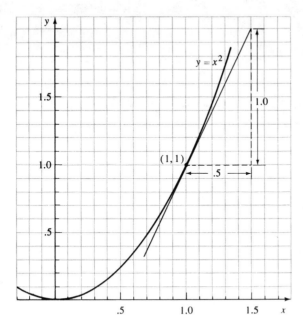

Figure 6

Problems

1. Figure 7 shows the graph of a specific function. At each of the following points, draw the tangent line and estimate its slope.
   a. At point A.   b. At point B.
   c. At point C.   d. At point P.
   e. At point Q.   f. At point R.
   (*Note:* Problems 2–6 will be treated again in the next section. Save your solutions for use at that time.)

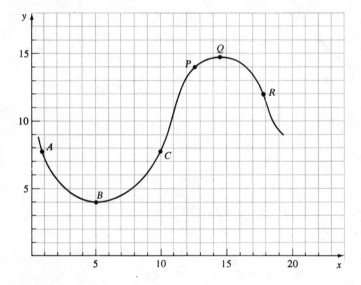

Figure 7

2. Figure 5 shows the graph of the function $y = x^2$. Draw the tangent line at the point (.5,2.5) and estimate its slope.

3. Repeat Problem 2 for the point (.2,.4).

4. Repeat Problem 2 for the point (.7,.49).

5. Estimate the slope of the line tangent to the graph of $y = x^2/2 + x$ at the point $(1, \frac{3}{2})$.

6. Repeat Problem 5 for the point $(\frac{1}{2}, \frac{5}{8})$.

Section 2 □ Instantaneous Rates of Change

Statements such as "during June, oil reserves increased at an average rate of 2 million barrels per month" have been given a clear meaning by our study of average rates of change. But what exactly is meant by the statement "at noon on June 1, oil reserves were increasing at the rate of 1 million barrels per month?"

The second statement involves a new concept of rate of change at a particular instant rather than the average rate of change during an interval. This new concept is called *instantaneous rate of change.*

Figure 8 shows the oil reserves r in millions of barrels at time t in months. The average rate of change of r between $t = 5$ (June 1) and $t = 6$ (July 1) is the slope of the secant line PQ. This slope indicates the average rate at which the graph was rising during the interval $t = 5$ to $t = 6$.

Figure 9 shows the tangent line to the graph at the point where $t = 5$ (June 1). The slope of this line indicates how fast the graph was rising at the instant when $t = 5$. This slope is called the instantaneous rate of change of r when $t = 5$.

In general, let $y = f(x)$ be any function and let x_0 be any value of x.

To find the instantaneous rate of change of y when $x = x_0$, mark the point P on the graph where x has the value x_0 (see Figure 10.) Draw the line L that is tangent to the graph at P. The slope of L is the desired instantaneous rate of change.

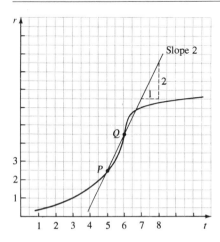

Figure 8

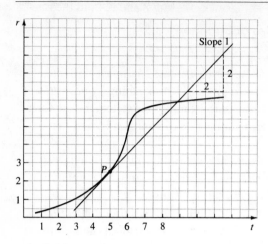

Figure 9

The above description will serve as a geometric definition of *instantaneous rate of change*. This notion is used so frequently that the adjective instantaneous is often omitted. When units are given for x and y, the rate of change of y is expressed in units of y per unit of x, just as in the case of average rates of change.

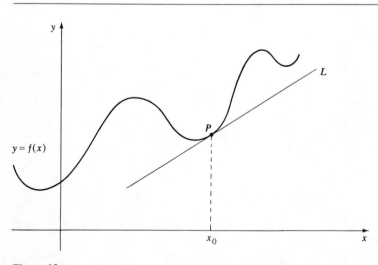

Figure 10

Example 1. Let r be the function of t given in Figures 8 and 9. Find the rate of change of r when $t = 3$.

SOLUTION.
Mark the point on the graph where $t = 3$ and draw the tangent line at this point. Figure 11 shows that this line has slope $\frac{1}{2}$. Therefore the desired rate of change is $\frac{1}{2}$ million barrels per month.

Example 2. Consider the function $y = x^2$. Find the rate of change of y when $x = 1$.

SOLUTION.
At present we must solve the problem geometrically – an analytical method will be given later in this chapter. We construct a table of values, as shown below, and draw the graph (see Figure 12). Mark the point P on the graph where $x = 1$. Draw the tangent line L at P. We see from the figure that L has slope 2. The desired rate of change is therefore 2.

x	0	0.1	0.2	0.3	0.4	0.5	0.6	0.7	0.8	0.9	1.0	1.1	1.2
y	0	0.01	0.04	0.09	0.16	0.25	0.36	0.49	0.64	0.81	1.0	1.21	1.44

The Freely-falling Body. Historically, the most important example of instantaneous rate of change occurs in the problem considered by Galileo (1564–1642) of finding the speed of a freely-falling body at any given instant of time.

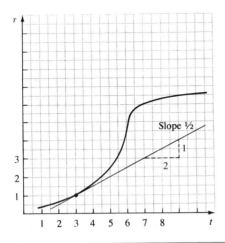

Figure 11

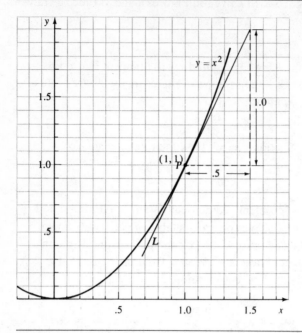

Figure 12

Suppose a stone is dropped from a tall tower. We observe that the stone falls faster and faster as it nears the ground. Indeed, it hardly hurts at all to be hit by a stone that has fallen 3 inches; to be hit by a stone that has fallen 100 feet is quite a different matter.

How fast is the stone falling 1 second after being released? The question seems to make good sense at first; on second thought, we realize that it is difficult to define *speed* precisely in this case, since it seems to be constantly changing. Constant speed is calculated by the formula

$$\text{speed} = \frac{\text{distance traveled}}{\text{time of travel}}.$$

Our problem is to calculate (or even define) speed when it is not constant.

Let s be the number of feet the stone has fallen after t seconds. By careful measurement, one might obtain the following table of values for s and t.

t	0	0.5	1.0	1.5	2.0	2.5	3.0	3.5	4.0
s	0	4	16	36	64	100	144	196	256

When these points are used to sketch a graph of s, the curve in Figure 13 is obtained.

The average speed (compare with Problem 5 on page 51) during the time interval $t = 1$ to $t = 4$ is

$$\frac{\Delta s}{\Delta t} = \frac{256 - 16}{4 - 1} = \frac{240}{3} = 80 \text{ feet per second}$$

This average speed is represented in Figure 13 by the slope of the secant line PQ. Other average speeds are:

Between $t = 1$ and $t = 3$: $\frac{\Delta s}{\Delta t} = \frac{144 - 16}{3 - 1} = 64$ feet per second

Between $t = 1$ and $t = 2$: $\frac{\Delta s}{\Delta t} = \frac{64 - 16}{2 - 1} = 48$ feet per second

Between $t = 1$ and $t = 1.5$: $\frac{\Delta s}{\Delta t} = \frac{36 - 16}{1.5 - 1} = 40$ feet per second

These average speeds are represented in Figure 13 by the slopes of secant lines PR, PS, and PT.

The key to solving the problem is this idea: these average speeds, as Δt becomes smaller and smaller, will become better and better approximations to

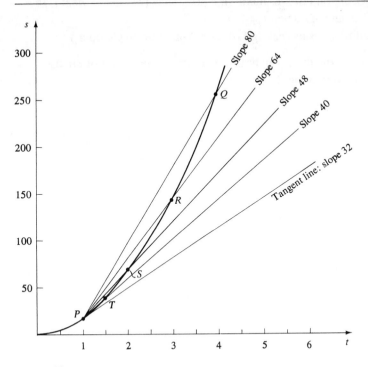

Figure 13

the exact (or instantaneous) speed when $t = 1$. If we accept this idea, we must conclude that the slope of the tangent line at P represents the instantaneous speed of the stone when $t = 1$. Figure 13 shows that this tangent line has slope 32. Hence the stone was traveling at 32 feet per second at the instant $t = 1$.

Problems

1. Figure 12 shows the graph of the function $y = x^2$.
    a. What is the rate of change of y when $x = \frac{1}{2}$?
    b. How fast is y increasing when $x = .8$ (that is, what is the rate of change of y when $x = .8$)?
    c. What is the rate of change of y when $x = 0$?
    d. How fast is y increasing when $x = \frac{1}{4}$?

2. Figures 8, 9, and 11 show the oil reserves r in millions of barrels at time t in months.
    a. Find the rate of change of r when $t = 8$.
    b. How fast are oil reserves increasing on March 1? (That is, what is the rate of change of r when $t = 2$?)
    c. Were oil reserves increasing faster on April 1 or on October 1?

3. Figure 14 shows the mass w in grams of a tumor after t days of intense gamma radiation treatment.

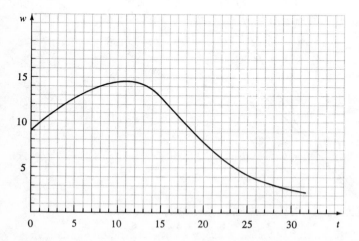

Figure 14

a. How fast is the tumor growing when $t = 2$?
b. How fast is the tumor growing when $t = 11$?
c. What is the rate of change of w when $t = 15$?
d. What is the rate of change of w when $t = 20$?
e. How fast is the tumor shrinking when $t = 25$?

4. How fast is a stone traveling 2 seconds after it is dropped from a high tower? After 3 seconds? (Use Figure 13.)

5. Figure 15 shows the number of unemployed workers in the United States between 1970 and 1973.
   a. How fast was unemployment increasing on January 1, 1970?
   b. How fast was unemployment decreasing on January 1, 1973?

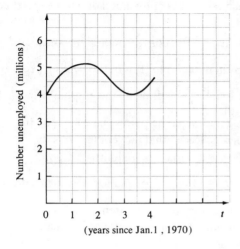

(years since Jan.1 , 1970)   **Figure 15**

6. Figure 24 of Chapter 1 (see page 40) shows labor union membership during this century.
   a. How fast was membership increasing when $t = 1940$?
   b. Was membership increasing faster when $t = 1940$ or when $t = 1970$?
   c. Was union membership larger when $t = 1940$ or when $t = 1970$?
   d. What was the rate of change of union membership when $t = 1965$?

7. Figure 25 of Chapter 1 (see page 41) shows the area A of a skin wound t days after treatment.
   a. How fast was the wound healing when $t = 4$?
   b. How fast was the wound healing when $t = 20$?

8. Figure 26 of Chapter 1 (see page 41) shows the net national income u and the net income v of the poorer 50% of the population. In 1960, which was increasing faster, u or v?

Section 3 □ Limits. The Δ-Process

Consider a function $y = f(x)$ and a particular value of x, say $x = x_0$. The **instantaneous rate of change** of y when $x = x_0$ (or, equivalently, the slope of the line tangent to the graph of $y = f(x)$ at the point where $x = x_0$) is more frequently called the **derivative of y when $x = x_0$**. Symbolically, it is denoted by

$$\frac{dy}{dx}\bigg|_{x=x_0}$$

or simply by $\dfrac{dy}{dx}$ when the particular value x_0 is understood or will be specified

later. (The notation $\dfrac{dy}{dx}$ does not indicate a fraction; $\dfrac{dy}{dx}$ should be considered

as a single symbol.) Another common notation for this derivative is $f'(x_0)$.

In the previous section, we found $\dfrac{dy}{dx}$ by drawing the graph of $y = f(x)$

and measuring the slope of the tangent line at the point where $x = x_0$. In Figure 16 this tangent line is labeled L. Recall that tangent line L is the limiting position of secant line AB as B approaches A along the graph.

Therefore $\dfrac{\Delta y}{\Delta x}$ is an approximation to $\dfrac{dy}{dx}$; this approximation becomes better

and better as Δx is chosen closer and closer to, but different from, zero.

The preceding sentence is of fundamental importance in calculus. It is usually stated using the word **limit**[1] as follows:

$\dfrac{dy}{dx}$ is the *limit* of $\dfrac{\Delta y}{\Delta x}$ as Δx approaches zero.

In symbols, it is written

(1) $\qquad \dfrac{dy}{dx} = \lim_{\Delta x \to 0} \dfrac{\Delta y}{\Delta x}$

[1] The use of limits in more general situations will be discussed in Section 5 of this chapter.

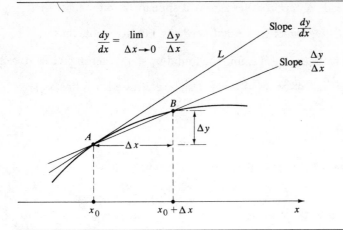

$$\frac{dy}{dx} = \lim_{\Delta x \to 0} \frac{\Delta y}{\Delta x}$$

Slope $\dfrac{dy}{dx}$

Slope $\dfrac{\Delta y}{\Delta x}$

Figure 16

or

$$\frac{\Delta y}{\Delta x} \to \frac{dy}{dx} \text{ as } \Delta x \to 0.$$

Let us work out a numerical illustration of Equation 1. We shall use the function $y = x^3$ at the value $x_0 = 4$. That is, let $x = 4$ and then calculate $\dfrac{\Delta y}{\Delta x}$ for various changes Δx in this value of x. For example, if $\Delta x = \frac{1}{2}$, then we can use a hand calculator to obtain

$$\Delta y = (4\tfrac{1}{2})^3 - 4^3 = 91.125 - 64 = 27.125,$$

and so

$$\frac{\Delta y}{\Delta x} = \frac{27.125}{0.5} = 54.25.$$

The entries in Table 1 were obtained in a similar manner.

Table 1

Δx	$\dfrac{\Delta y}{\Delta x}$	Δx	$\dfrac{\Delta y}{\Delta x}$
0.5	54.25	−0.5	42.25
0.02	48.2404	−0.02	47.7604
0.001	48.012	−0.001	47.988
0.0002	48.0024	−0.0002	47.9976
0.00003	48.0003	−0.00003	47.9997

As $\Delta x \to 0$ (read "Δx approaches zero"), it appears from the table that $\dfrac{\Delta y}{\Delta x} \to 48$. This is actually correct, as we shall see later. For now, note that we discovered this limit by guessing; it cannot be found by simply setting $\Delta x = 0$.

The fact that $\dfrac{\Delta y}{\Delta x} \to 48$ as $\Delta x \to 0$ may also be expressed as follows:

$$\lim_{\Delta x \to 0} \frac{\Delta y}{\Delta x} = 48.$$

By Equation 1, the derivative of the function $y = x^3$, when x is 4, is 48; in symbols,

$$\frac{dy}{dx} = 48 \qquad \text{when } x = 4.$$

The search by mathematicians for a precise definition of the term *limit* has a long and interesting history. It was easy for seventeenth-century scientists to explain the meaning of Equation 1 by using intuitive notions or physical terms such as motion. For example, in Figure 16, one can imagine the distance Δx shrinking to zero, the point B thereby being forced to move along the curve to A, and consequently the slope $\dfrac{\Delta y}{\Delta x}$ of AB becoming the slope $\dfrac{dy}{dx}$ of L. But how can one explain this using only mathematical terms? The German mathematician Gottfried Leibnitz (1646–1716) used a concept of "infinitesimals" which caused much controversy. Until the time when Augustin Cauchy (1789–1857) and Karl Weierstrass (1815–1897) found a rigorous way to treat the subject of limits, the calculus that Isaac Newton (1642–1727) and Leibnitz invented possessed a mysterious and supernatural flavor. We shall not present the rigorous definition of limit. The intuitive description above and the illustrative examples below will provide you with the basic understanding you need.

Example 1. Find the values of each limit.

a. $\lim\limits_{\Delta x \to 0} [3 + \Delta x]$.

b. $\lim\limits_{\Delta u \to 0} [8 + 6\,\Delta u]$.

c. $\lim\limits_{\Delta x \to 0} \left[\dfrac{3\Delta x + (\Delta x)^2}{\Delta x} \right]$.

SOLUTION.

a. As Δx shrinks to zero, $3 + \Delta x$ approaches 3. Therefore $\lim\limits_{\Delta x \to 0} [3 + \Delta x] = 3$.

b. As Δu shrinks to zero, $6\Delta u$ also shrinks to zero. Therefore $\lim\limits_{\Delta u \to 0} [8 + 6\Delta u]$ $= 8$.

c. We cannot find this limit by letting Δx become zero because we would obtain the meaningless expression $\frac{0}{0}$. Thus we will simplify the quantity in brackets by using algebra.

$$\frac{3\Delta x + (\Delta x)^2}{\Delta x} = \frac{3\Delta x}{\Delta x} + \frac{(\Delta x)^2}{\Delta x} = 3 + \Delta x$$

Thus

$$\lim_{\Delta x \to 0} \left[\frac{3\Delta x + (\Delta x)^2}{\Delta x} \right] = \lim_{\Delta x \to 0} [3 + \Delta x] = 3.$$

The Δ-process. We shall now use Equation 1 to calculate derivatives exactly. This method is called the **Δ-process** and is the analytic counterpart to the geometric method of Section 2 where derivatives (that is, rates of change) were calculated by sketching tangent lines and measuring their slopes.

To illustrate the Δ-process, consider this problem. If $y = 2x^2 - 1$, find $\frac{dy}{dx}$ when $x = 3$.

The method of solution is to compute $\frac{\Delta y}{\Delta x}$, and then simplify the result

so that

$$(2) \quad \lim_{\Delta x \to 0} \frac{\Delta y}{\Delta x}$$

can be found. The value of this limit will be the desired answer. Now, if $x = 3$ changes by an amount Δx, then Δy is found as follows.

When $x = 3$, $y = 2(3)^2 - 1$; this is y_1.

When $x = 3 + \Delta x$, $y = 2(3 + \Delta x)^2 - 1$; this is y_2.

Therefore $\Delta y = y_2 - y_1 = [2(3 + \Delta x)^2 - 1] - [2(3)^2 - 1]$

$$\frac{\Delta y}{\Delta x} = \frac{[2(3 + \Delta x)^2 - 1] - [2(3)^2 - 1]}{\Delta x}$$

$$= \frac{[2(9 + 6\Delta x + (\Delta x)^2) - 1] - [17]}{\Delta x}$$

$$= \frac{12\Delta x + 2(\Delta x)^2}{\Delta x}$$

$$= 12 + 2(\Delta x)$$

Therefore,

$$\lim_{\Delta x \to 0} \frac{\Delta y}{\Delta x} = \lim_{\Delta x \to 0} [12 + 2(\Delta x)] = 12.$$

Hence the answer is

$$\frac{dy}{dx} = 12 \quad \text{when} \quad x = 3.$$

Example 2. Let $y = x^2$. Use the Δ-process to calculate each of the following.

a. $\dfrac{dy}{dx}$ when $x = 4$.

b. $\dfrac{dy}{dx}$ when $x = x_0$.

c. $\dfrac{dy}{dx}$ when $x = 5$.

SOLUTION.

a. Suppose $x = 4$ changes by an amount Δx.

When $x = 4$, $y = 4^2$; this is y_1.
When $x = 4 + \Delta x$, $y = (4 + \Delta x)^2$; this is y_2.
Therefore $\Delta y = y_2 - y_1 = (4 + \Delta x)^2 - 4^2$.

$$\frac{\Delta y}{\Delta x} = \frac{(4 + \Delta x)^2 - 4^2}{\Delta x} = \frac{16 + 8\Delta x + (\Delta x)^2 - 16}{\Delta x} = 8 + \Delta x$$

$$\lim_{\Delta x \to 0} \frac{\Delta y}{\Delta x} = \lim_{\Delta x \to 0} [8 + \Delta x] = 8.$$

The answer is: $\dfrac{dy}{dx} = 8$ when $x = 4$.

b. We repeat the reasoning in Part a using $x = x_0$ instead of $x = 4$.

When $x = x_0$, $y = x_0^2$; this is y_1.
When $x = x_0 + \Delta x$, $y = (x_0 + \Delta x)^2$; this is y_2.

Therefore $\Delta y = (x_0 + \Delta x)^2 - x_0^2$.

$$\frac{\Delta y}{\Delta x} = \frac{(x_0 + \Delta x)^2 - x_0^2}{\Delta x} = \frac{x_0^2 + 2x_0\Delta x + (\Delta x)^2 - x_0^2}{\Delta x}$$

$$= 2x_0 + \Delta x$$

$$\lim_{\Delta x \to 0} \frac{\Delta y}{\Delta x} = \lim_{\Delta x \to 0} [2x_0 + \Delta x] = 2x_0$$

The answer is:

(3) $\quad \dfrac{dy}{dx} = 2x_0 \quad$ when $\quad x = x_0$.

c.   To find $\dfrac{dy}{dx}$ when $x = 5$ we merely substitute into (3). We obtain the

answer: $\dfrac{dy}{dx} = 10$ when $x = 5$.

The notation $\dfrac{dy}{dx}$ for derivatives changes in the obvious way when letters other than x and y are used for the independent and dependent variables. For example, suppose s is a function t. Then $\dfrac{\Delta s}{\Delta t}$ would denote the average rate of change of s when some value of t changes by an amount Δt (see Chapter 1, Section 5), $\dfrac{ds}{dt}$ would denote the derivative (instantaneous rate of change of s), and the analogue of Equation 1 would be

$$\frac{ds}{dt} = \lim_{\Delta t \to 0} \frac{\Delta s}{\Delta t}.$$

Example 3.   Reserves of a certain chemical vary according to the law $A = 18/t$, where A in tons is the amount on reserve at time t in years. Find the rate of change of A when $t = 3$.

SOLUTION.
Suppose $t = 3$ changes by an amount Δt.

When $t = 3, A = \dfrac{18}{3} = 6$; this is A_1.

When $t = 3 + \Delta t$, $A = \dfrac{18}{3 + \Delta t}$; this is A_2.

Therefore $\Delta A = A_2 - A_1 = \dfrac{18}{3 + \Delta t} - 6$.

$$\dfrac{\Delta A}{\Delta t} = \dfrac{\dfrac{18}{3 + \Delta t} - 6}{\Delta t} = \dfrac{\dfrac{18 - 6(3 + \Delta t)}{3 + \Delta t}}{\Delta t} = \dfrac{-6}{3 + \Delta t}$$

$$\dfrac{dA}{dt} = \lim_{\Delta t \to 0} \left[\dfrac{-6}{3 + \Delta t} \right] = \dfrac{-6}{3} = -2.$$

The rate of change of A when $t = 3$ is -2 tons per year. The negative sign indicates that A is decreasing. We could also say that when $t = 3$, A is decreasing at the rate of 2 tons per year.

Problems

In Problems 1–22, use the Δ-process to find $\dfrac{dy}{dx}$ when $x = x_0$ and when $x = 2$.

1. $y = x^2 + 3$.

2. $y = 2 + x^2$.

3. $y = 5x^2$.

4. $y = 2 + x$.

5. $y = 5x^2 - 1$.

6. $y = 2$.

7. $y = 1 - x^2$.

8. $y = x^3$.

9. $y = 1 - 5x^2$.

10. $y = 2 - x^2$.

11. $y = 3x$.

12. $y = 2 - x$.

13. $y = 3x + 5$.

14. $y = 7x^2$.

15. $y = 5 - \dfrac{16}{x}$.

16. $y = 5$.

17. $y = -3x$.

18. $y = x$.

19. $y = 2$.

20. $y = 2x$.

21. $y = -3$.

22. $y = -2x$.

23. Use the Δ-process to find the exact slope of the line tangent to the graph of $y = x^3$ at the point $(2,8)$.

24. Use the Δ-process to obtain exact answers to Problems 2, 3, and 4 in Section 1 of this chapter.

25. The demand law for a certain commodity is $q = 100 - p^2$, where q is measured in gallons, and p in dollars. Find the rate of change of q in each case.

   a. When $p = p_0$.   b. When $p = 5$.

26. Use the Δ-process to obtain exact answers to Problems 5 and 6 in Section 1 of this chapter.

Section 4 □ Rules for Differentiation

If we tried to use the Δ-process to calculate the derivative of a complicated function such as $y = (3x^8 + 2x^2 - 1)^3$, the algebraic difficulties would be almost insurmountable. In this and the next section, we develop rules for calculating such derivatives very easily.

Rule for Differentiating x^n. In Part b of Example 2 in the preceding section, we found that if $y = x^2$, then $\dfrac{dy}{dx} = 2x_0$ when $x = x_0$. We used x_0 here to denote a particular value of x, but we could just as well have used x itself. Then we would state this result:

If $y = x^2$, then $\dfrac{dy}{dx} = 2x$.

Part d of Problem 2 in the preceding section showed:

If $y = x^3$, then $\dfrac{dy}{dx} = 3x^2$.

These two results are special cases of the following general rule for any positive integer n.

Rule for Differentiating x^n

(1)   If $y = x^n$ then $\dfrac{dy}{dx} = nx^{n-1}$.

We can prove Rule 1 using the Δ-process.

We have $y_1 = x_0^n$, $y_2 = (x_0 + \Delta x)^n$, and $\Delta y = (x_0 + \Delta x)^n - x_0^n$. Now rewrite Δy by using the *binomial theorem*[2].

$$(x_0 + \Delta x)^n = x_0^n + nx_0^{n-1}\Delta x + \frac{n(n-1)}{1 \cdot 2}x_0^{n-2}(\Delta x)^2 + \cdots + (\Delta x)^n$$

Thus

$$\frac{\Delta y}{\Delta x} = nx_0^{n-1} + \frac{n(n-1)}{1 \cdot 2}x_0^{n-2}\Delta x + \cdots + (\Delta x)^{n-1}.$$

In the sum on the right, all terms except the first approach zero as Δx approaches zero. Therefore

$$\frac{dy}{dx} = \lim_{\Delta x \to 0} \frac{\Delta y}{\Delta x} = nx_0^{n-1}$$

Example 1. If $y = x^4$ find each of the following.

a. $\dfrac{dy}{dx}$.

b. $\dfrac{dy}{dx}$ when $x = 2$.

SOLUTION.

a. Apply Rule 1 when $n = 4$.

If $y = x^4$, then $\dfrac{dy}{dx} = 4x^3$.

b. We let $x = 2$ in the above expression for $\dfrac{dy}{dx}$.

When $x = 2$, $\dfrac{dy}{dx} = 4(2)^3 = 32$.

---

[2]The *Binomial Theorem* states that, for a fixed positive integer n,

$$(a + b)^n = a^n + C_1 a^{n-1}b + C_2 a^{n-2}b^2 + \cdots + C_{n-1}ab^{n-1} + C_n b^n$$

where the coefficients $C_1, C_2, \ldots, C_n$ are given by

$$C_1 = \frac{n}{1}, C_2 = \frac{n(n-1)}{1 \cdot 2}, \ldots, C_{n-1} = \frac{n(n-1)\cdots 2}{1 \cdot 2 \cdots (n-1)} = n,$$

$$C_n = \frac{n(n-1)\cdots 2 \cdot 1}{1 \cdot 2 \cdots (n-1)n} = 1.$$

Example 2. Suppose that the M_2 *money supply*[3] S at time t is $S = t^3$. Find the rate of change of S when $t = 4$.

SOLUTION.

We use Rule 1 with y, x, and n replaced by S, t, and 3. After these substitutions, Rule 1 becomes:

If $S = t^3$, then $\dfrac{dS}{dt} = 3t^2$.

Hence, when $t = 4$, $\dfrac{dS}{dt} = 3(4)^2 = 48$.

(*Note:* We first find $\dfrac{dS}{dt}$ for a general t, and then we let $t = 4$. It would not work

to first set $t = 4$ in the equation of the function $S = t^3$ and then try to find $\dfrac{dS}{dt}$.)

Rule 1 applies when $n = 1$ if we recall from algebra that $x^1 = x$ and $x^0 = 1$. Thus if $n = 1$, Rule 1 yields the following.

(2)   If $y = x$, then $\dfrac{dy}{dx} = 1$.

Note the graphical meaning of (2): the tangent line to the graph of $y = x$ at any point has slope 1. This fact can be verified directly, since the graph of $y = x$ is a straight line of slope 1.

Rule for a Constant Multiplier. This rule says that multiplication of a function by a constant k has the effect of multiplying the derivative by k. To put this into symbols, let u be a function of x.

Rule for a Constant Multiplier

(3)   If $y = ku$, then $\dfrac{dy}{dx} = k\dfrac{du}{dx}$, k a constant.

---

[3] M_2 *money supply* is the sum of all coin and currency in circulation outside the banks plus the balances of all privately held deposit accounts.

Rule 3 can be proved as follows. Let x change from x_1 to x_2. Suppose this causes u to change from u_1 to u_2 and y to change from y_1 to y_2. Then $\Delta y = y_2 - y_1 = ku_2 - ku_1$ $= k(u_2 - u_1) = k\Delta u$, and so $\dfrac{\Delta y}{\Delta x} = \dfrac{k\Delta u}{\Delta x}$. If we let $\Delta x \to 0$, this last equation yields

$$\frac{dy}{dx} = k\frac{du}{dx}.$$

Example 3.

If $\quad y = 6x^3$, $\qquad$ then $\qquad \dfrac{dy}{dx} = 6(3x^2) = 18x^2$.

If $\quad y = 13x^5$, $\qquad$ then $\qquad \dfrac{dy}{dx} = 13(5x^4) = 65x^4$.

If $\quad y = \dfrac{7}{5}x^{100}$, $\qquad$ then $\qquad \dfrac{dy}{dx} = \dfrac{7}{5}(100x^{99}) = 140x^{99}$.

If $\quad y = -\sqrt{2}x^6$, $\qquad$ then $\qquad \dfrac{dy}{dx} = -\sqrt{2}(6x^5) = -6\sqrt{2}x^5$.

The Derivative of a Constant Function. Let c be a fixed number. The graph of $y = c$ is a horizontal line which intersects the y-axis at $(0,c)$. Since $\dfrac{dy}{dx}$ is the slope of this graph, we must have $\dfrac{dy}{dx} = 0$. This is our next rule.

Rule for Differentiating a Constant Function

(4) $\quad$ If $\quad y = c,$ $\qquad$ then $\qquad \dfrac{dy}{dx} = 0.$

Rule for Differentiating a Sum.

> *The derivative of a sum (or difference) of functions is the sum (or difference) of their separate derivatives.*

In symbols, this rule would be expressed as follows.

Rule for Differentiating a Sum

Let u and v be functions of x.

(5)   If $y = u \pm v$ then $\dfrac{dy}{dx} = \dfrac{du}{dx} \pm \dfrac{dv}{dx}$.

To prove Rule 5, suppose x changes from x_1 to x_2. This causes u and v to change; suppose u changes from u_1 to u_2 and v changes from v_1 to v_2. Then $\Delta y = y_2 - y_1$ $= (u_2 + v_2) - (u_1 + v_1) = (u_2 - u_1) + (v_2 - v_1) = \Delta u + \Delta v$. Hence

$$\frac{\Delta y}{\Delta x} = \frac{\Delta u}{\Delta x} + \frac{\Delta v}{\Delta x}.$$

Let $\Delta x \to 0$ in the equation above, to obtain

$$\frac{dy}{dx} = \frac{du}{dx} + \frac{dv}{dx}.$$

Example 4. To differentiate $y = x^3 + x^8$, we use Rule 5 with $u = x^3$ and $v = x^8$. Then $\dfrac{du}{dx} = 3x^2$ and $\dfrac{dv}{dx} = 8x^7$, and so Rule 5 yields:

     If $y = x^3 + x^8$,   then $\dfrac{dy}{dx} = 3x^2 + 8x^7$.

Differentiation of Polynomials.   Repeated use of these rules make it possible to find the derivatives of functions such as

(1) $y = 5x^{10} - 3x^2 + x + 5$,

(2) $P = 3u^2 + 2u - 1$,

(3) $s = \frac{1}{2}t^2 - 3t + 2$.

     These functions (sums and differences of constants times nonnegative integral powers of the independent variable) are called **polynomial functions.**

Example 5. For the polynomial functions 1, 2, and 3 above, find the following derivatives.

a. $\dfrac{dy}{dx}$.   **b.** $\dfrac{dy}{dx}$ when $x = 1$.

c. $\dfrac{dP}{du}$ when $u = 4$.   **d.** $\dfrac{ds}{dt}$ when $t = 2$.

SOLUTION.

a. $\dfrac{dy}{dx} = 5(10x^9) - 3(2x^1) + 1 + 0$

$= 50x^9 - 6x + 1.$

b. When $x = 1$, $\dfrac{dy}{dx} = 50(1)^9 - 6(1) + 1 = 45.$

c. We *cannot* begin by setting $u = 4$. Instead, we must calculate $\dfrac{dP}{du}$ and then set $u = 4$ in that result. Thus

$\dfrac{dP}{du} = 3(2u^1) + 2(1) - 0 = 6u + 2,$

and so, when $u = 4$,

$\dfrac{dP}{du} = 6(4) + 2 = 26.$

d. We must first find $\dfrac{ds}{dt}$.

$\dfrac{ds}{dt} = \tfrac{1}{2}(2t^1) - 3(1) + 0 = t - 3$

Therefore, when $t = 2$,

$\dfrac{ds}{dt} = 2 - 3 = -1.$

Example 6. The purpose of this example is to show the use of the functional notation for derivatives.

a. If $f(x) = \tfrac{1}{3}x^6 - 3x^2 + 1$, find $f'(2)$.

b. If $g(t) = 7t^4 - 2t + 5$, find $g'(t)$ and $g'(3)$.

SOLUTION.

a. As usual, we find $f'(x)$ first, and then let $x = 2$.

$f'(x) = \tfrac{1}{3}(6x^5) - 3(2x^1) + 0 = 2x^5 - 6x,$

so, when $x = 2$, we have

$f'(2) = 2(2)^5 - 6(2) = 52.$

b. The derivative of g is

$$g'(t) = 7(4t^3) - 2(1) + 0 = 28t^3 - 2,$$

so, when $t = 3$, we have

$$g'(3) = 28(3)^3 - 2 = 754.$$

Problems

1. Differentiate the following functions.

a. $y = x^{35}$.

b. $y = 5x^9$.

c. $y = 5x^6 - 3x^4 + 2x - 1$.

d. $P = (1.32)t^2 + (4.06)t - 19.21$.

e. $C = \frac{1}{2}u^4 - \frac{1}{3}u^3 + \frac{1}{8}$.

f. $C = 2\pi r$.

g. $A = \pi r^2$.

h. $y = \dfrac{2x^3 + x + 1}{3}$.

i. $y = \dfrac{5t^2}{3} + \dfrac{2t}{5}$.

j. $s = \frac{1}{3}(u^{27} - \sqrt{2})$.

2. If $L = 3p^{25} - 127$, find $\dfrac{dL}{dp}$.

3. If $y = 7x^3 + 2x + 1$, find $\dfrac{dy}{dx}$ when $x = 5$.

4. If $y = 7x^4 - 3x^3 + 5x^2 - 2x + 1$, find $\dfrac{dy}{dx}$ when $x = 2$.

5. If $E = 8x^5 - 9x^4 + 25x$, find $\dfrac{dE}{dx}$ when $x = 0$.

6. If $P = 7x^3 + x^2 - 12x - 10$, find $\dfrac{dP}{dx}$ when $x = 5$.

7. If $w = 0.6v^3 - 0.8v$, find $\dfrac{dw}{dv}$ when $v = 4$.

8. If $R = \frac{1}{2}v^6 - 32v^2 + v$, find each of the following.

a. $\dfrac{dR}{dv}$ when $v = 0$.

b. $\dfrac{dR}{dv}$ when $v = 1$.

c. $\dfrac{dR}{dv}$ when $v = -1$.

9. If $f(x) = 9x^3 - 12x + 5$, find $f'(x)$ and $f'(6)$.

10. If $f(x) = \frac{1}{3}x^2 - \frac{2}{5}x + \frac{1}{4}$, find $f'(\frac{2}{3})$.

11. If $h(t) = 12t^3 - t + 5$, find $h'(\frac{1}{4})$.

12. If $F(x) = x^4 - 2x^2$, find $F'(5)$.

13. If $g(u) = \frac{1}{3}u^9 - 2u + 12$, find $g'(1)$.

14. Highway engineers normally assume that the distance F (in feet) required for stopping a car is related to the speed V (in miles per hour) of the car according to a law of the form $F = kv^2$, where k is a constant that depends on the road surface material. A test on a certain road surface yields $F = 36$ when $V = 30$. Evaluate k and then find the rate of change of F with respect to V in each case.
    a. $V = 20$.   b. $V = 50$.   c. $V = 100$.

15. An antibiotic is introduced into a Petri dish. The number of bacteria in the dish t hours later can be determined by the function

    $$B(t) = 37500 + 2500t - 125t^2$$

    for values of t with $0 \le t \le 30$. How fast is the bacteria count changing when $t = 8$? When $t = 11$?

16. Suppose a management consulting firm advises a company that the company's monthly profit P will depend on its monthly advertising expenditures x according to a formula

    $$P = 500 + 40x - x^2,$$

    where x and P are in units of thousands of dollars. If the current expenditure is $21,000 per month, will profits increase or decrease if advertising expenditures are increased? (*Hint:* Restate the problem. Is $\dfrac{dP}{dx}$ positive or negative when $x = 21$?)

17. Find the equation of the line that is tangent to the graph of $y = 3x^2 - 5x + 4$ at the point on the graph where $x = 2$.

In this section, we examine the concept of limit more closely. Our informal definition is the following. Let $y = f(x)$ be a function. If the values of y get closer and closer to a number A as the values of x get closer and closer to x_0, then we say the y, or $f(x)$, *approaches the limit A as x approaches* x_0. This is expressed symbolically in any of the following four ways.

$$y \to A \quad \text{as} \quad x \to x_0$$

$$f(x) \to A \quad \text{as} \quad x \to x_0$$

$$\lim_{x \to x_0} y = A$$

$$\lim_{x \to x_0} f(x) = A$$

Example 1 (An obvious limit). Find $\lim_{x \to 5} \dfrac{2x + 8}{3x - 12}$.

SOLUTION.

Let $y = \dfrac{2x + 8}{3x - 12}$, and consider the behavior of y as the values of x get closer and closer to 5. As x approaches 5, the numerator $2x + 8$ gets closer and closer to $2 \cdot 5 + 8$, or 18. The denominator approaches $3 \cdot 5 - 12$, or 3. Thus the quotient $(2x + 8)/(3x - 12)$ gets closer and closer to $18/3$, or 6. We conclude that

$$\lim_{x \to 5} \frac{2x + 8}{3x - 12} = 6.$$

The preceding example illustrates the reasoning that justifies the following theorem.

[4]This section may be covered at a later time if the instructor wishes. The material on infinite limits will be used briefly in Chapters 5 and 8. The remaining material is optional.

Limit Theorem. Let f and g be functions such that $\lim_{x \to x_0} f(x) = A$ and $\lim_{x \to x_0} g(x) = B$.

(1) $\lim_{x \to x_0} [f(x) + g(x)] = A + B$

(2) $\lim_{x \to x_0} [f(x) \cdot g(x)] = A \cdot B$

(3)   If $B \neq 0$,   then $\lim_{x \to x_0} \dfrac{f(x)}{g(x)} = \dfrac{A}{B}$.

In many cases, it is necessary to perform algebraic operations on an expression before its limit becomes obvious. We saw this in using the Δ-process.

Example 2 (A nonobvious limit).   Find $\lim_{x \to 4} \dfrac{\sqrt{x} - 2}{x - 4}$.

SOLUTION.

Let $y = \dfrac{\sqrt{x} - 2}{x - 4}$, and consider the behavior of y as the values of x approach 4. The numerator will approach $\sqrt{4} - 2$, or zero. The denominator will also approach zero. When the numerator and denominator of a fraction get close to zero it is not at all clear what value, if any, the fraction is approaching. Note that Part 3 of the Limit Theorem cannot be applied to help us.

An empirical approach to this problem could be attempted with a hand calculator. We could compute the value of y for various values of x near 4. From these values of y, we might be able to guess the limit. Table 2 shows such calculations. These results suggest that y approaches 0.25 as x approaches 4.

Table 2

Value of x	$\sqrt{x} - 2$	$x - 4$	Value of $y = \dfrac{\sqrt{x} - 2}{x - 4}$
3.9	−0.025158234	−0.1	0.251582340
3.99	−0.002501564	−0.01	0.250156400
3.999	−0.000250016	−0.001	0.250016000
4.1	0.024845673	0.1	0.248456730
4.01	0.002498439	0.01	0.249843900
4.001	0.000249984	0.001	0.249984000

An algebraic operation can change this nonobvious limit problem to an obvious one. Note that $x - 4 = (\sqrt{x} - 2)(\sqrt{x} + 2)$. Thus

$$(4) \quad y = \frac{\sqrt{x} - 2}{x - 4} = \frac{\sqrt{x} - 2}{(\sqrt{x} - 2)(\sqrt{x} + 2)} = \frac{1}{\sqrt{x} + 2}.$$

From Equation 4, it is obvious that

$$\lim_{x \to 4} y = \lim_{x \to 4} \frac{1}{\sqrt{x} + 2} = \frac{1}{2 + 2} = \frac{1}{4} = 0.25.$$

Continuous Functions. In Chapter 1, a continuous function was defined as one whose graph had no breaks. A more precise definition can be given using limits.

Definition of Continuous Function

Let x_0 be in the domain of a function f. We say that f is **continuous at x_0** if $\lim\limits_{x \to x_0} f(x) = f(x_0)$. We say that f is **continuous** if f is continuous at x_0 for every value x_0 in the domain of f.

Example 3. Show that the function f defined by

$$f(x) = \begin{cases} \dfrac{x^2 - 16}{x - 4} & \text{if } x \neq 4 \\ 8 & \text{if } x = 4 \end{cases}$$

is continuous at 4.

SOLUTION.
We must show that $\lim\limits_{x \to 4} f(x) = f(4)$. Since $f(4) = 8$, this amounts to showing that

$$\lim_{x \to 4} \frac{x^2 - 16}{x - 4} = 8.$$

The algebraic operations

$$\frac{x^2 - 16}{x - 4} = \frac{(x - 4)(x + 4)}{x - 4} = x + 4$$

make it obvious that

$$\lim_{x \to 4} \frac{x^2 - 16}{x - 4} = \lim_{x \to 4} [x + 4] = 8,$$

as desired.

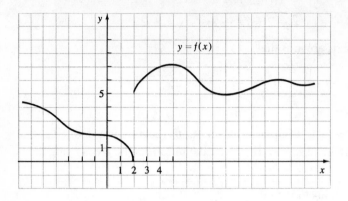

Figure 17

One-sided Limits. Suppose a variable x approaches the value x_0 but always remains greater than x_0. We then say x *is approaching x_0 from the right;* we symbolize this by $x \rightarrow x_0 +$. Similarly, $x \rightarrow x_0 -$ (reads "*x approaches x_0 from the left*") means that the values of x approach x_0, but always remain less than x_0.

Consider the function $y = f(x)$ in Figure 17. As x approaches 2 from the right, the values of y get closer and closer to 5. Therefore

$$\lim_{x \to 2+} f(x) = 5.$$

As x approaches 2 from the left, the values of y get closer and closer to 0. Therefore

$$\lim_{x \to 2-} f(x) = 0.$$

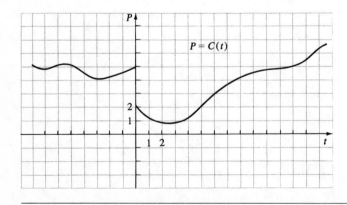

Figure 18

Example 4. Let $P = C(t)$ be the function whose graph is shown in Figure 18. Find each of the following.

a. $\lim_{t \to 0-} C(t)$.

b. $\lim_{t \to 0+} C(t)$.

c. $\lim_{t \to 6-} C(t)$.

d. $\lim_{t \to 6+} C(t)$.

e. $\lim_{t \to 0} C(t)$.

f. $\lim_{t \to 6} C(t)$.

SOLUTION.

a. $\lim_{t \to 0-} C(t) = 5$.

b. $\lim_{t \to 0+} C(t) = 2$.

c. $\lim_{t \to 6-} C(t) = 3$.

d. $\lim_{t \to 6+} C(t) = 3$.

e. As $t \to 0$, the values of $C(t)$ do not approach any unique limit; therefore we say that $\lim_{t \to 0} C(t)$ does not exist.

f. As $t \to 6$, the values of $C(t)$ get closer and closer to 3. Therefore $\lim_{t \to 6} C(t) = 3$.

Infinite Limits. The notation

$$u \to +\infty$$

(read "u tends to plus infinity") means that the values of u become larger and larger, increasing without bound. The notation

$$u \to -\infty$$

(read "u tends to minus infinity") means that the values of u are negative numbers which, in magnitude, become larger and larger without bound. To illustrate the use of these notations, consider the function $y = 1 - x^3$. Figure 19 shows the graph of this function. As x tends to $+\infty$, the values of y decrease without bound. Thus

$$\lim_{x \to +\infty} [1 - x^3] = -\infty.$$

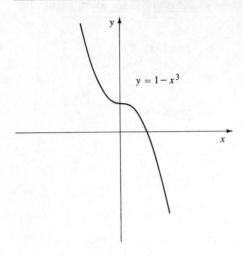

$y = 1 - x^3$

Figure 19

As x tends to $-\infty$, the values of y increase without bound. Therefore

$$\lim_{x \to -\infty} [1 - x^3] = +\infty.$$

Example 5. The graph of $y = 1 + 1/x$ is shown in Figure 20. Find each limit.

a. $\lim_{x \to +\infty} [1 + 1/x]$.

b. $\lim_{x \to -\infty} [1 + 1/x]$.

c. $\lim_{x \to 0+} [1 + 1/x]$.

d. $\lim_{x \to 0-} [1 + 1/x]$.

SOLUTION.

a. Let x get larger and larger without bound. Then the values of $y = 1 + 1/x$ get closer and closer to 1. Thus $\lim_{x \to +\infty} [1 + 1/x] = 1$.

b. If x tends to $-\infty$, then the values of y also approach 1. Therefore $\lim_{x \to -\infty} [1 + 1/x] = 1$.

c. If x is close to, but slightly larger than, zero, then y is a very large positive number. We see that $\lim_{x \to 0+} [1 + 1/x] = +\infty$.

d. As x approaches zero from the left, the values of y tend to $-\infty$. Thus $\lim_{x \to 0-} [1 + 1/x] = -\infty$.

Example 6. Suppose a winery operates at a fixed annual overhead cost of $20,000 and a variable cost of $2 for each gallon of wine produced. Then the average cost $C(x)$ to produce x gallons of wine is

$$C(x) = \frac{\text{total cost}}{\text{number of gallons}} = \frac{2x + 20,000}{x} = 2 + \frac{20,000}{x}.$$

Find each limit.

a. $\lim\limits_{x \to 0+} C(x)$.

b. $\lim\limits_{x \to +\infty} C(x)$.

SOLUTION.

a. If x remains greater than zero while approaching zero, then $20,000/x$ is a large positive number. Thus

$$\lim_{x \to 0+} C(x) = \lim_{x \to 0+}\left[2 + \frac{20,000}{x}\right] = +\infty.$$

We might express this by saying that the average cost becomes outrageously large as the quantity produced tends to zero. Common sense leads to the same conclusion. If the winery produces only 0.01 gallons of wine during the year and tries to sell it at a break-even price, that price would have to be about $20,000, which amounts to a unit price of $2 million per gallon!

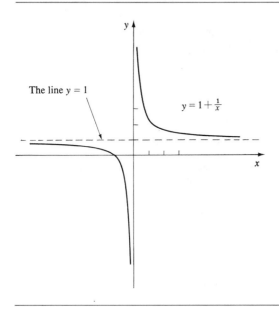

The line $y = 1$

$y = 1 + \frac{1}{x}$

Figure 20

b. Let $x \to +\infty$. Then $20{,}000/x \to 0$ and so

$$\lim_{x \to +\infty} \left[2 + \frac{20{,}000}{x} \right] = 2.$$

The common-sense interpretation is that when the overhead of $20,000 is prorated over a very large output of wine, each gallon bears a negligible share of it. The average cost per gallon is then just the unit cost of $2.

Asymptotes; Limits of Rational Functions. A horizontal line $y = A$ is called a **horizontal asymptote** for a function f if $f(x) \to A$ as $x \to +\infty$ or $x \to -\infty$. In Figure 20, we see that the line $y = 1$ is a horizontal asymptote for the function $f(x) = 1 + 1/x$. In Part b of Example 6, we found that $y = 2$ is a horizontal asymptote for the function $C(x) = 2 + 20{,}000/x$.

A vertical line $x = B$ (that is, a line on which all points have x-coordinate equal to B) is called a **vertical asymptote** for a function f if $f(x) \to +\infty$ or $f(x) \to -\infty$ as $x \to B+$ or $x \to B-$. From Figure 20, we see that the y-axis, $x = 0$, is a vertical asymptote for the function $f(x) = 1 + 1/x$.

The task of drawing graphs of functions is often made easier if we know the horizontal or vertical asymptotes, or the limits as $x \to +\infty$ and $x \to -\infty$. For **rational functions,** functions that are the quotient of two polynomials, this can be done as follows.

Limits at Infinity for Rational Functions

Let

$$(2) \qquad y = \frac{a_m x^m + a_{m-1} x^{m-1} + \cdots + a_1 x + a_0}{b_n x^n + b_{n-1} x^{n-1} + \cdots + b_1 x + b_0}$$

where $a_m \neq 0$ and $b_n \neq 0$. Then

$$(3) \qquad \lim_{x \to +\infty} y = \lim_{x \to +\infty} \frac{a_m x^m}{b_n x^n} \qquad \text{and} \qquad \lim_{x \to -\infty} y = \lim_{x \to -\infty} \frac{a_m x^m}{b_n x^n}.$$

In words, the above rule says that the limit of a rational function can be found by neglecting all terms except the leading ones. The rule is proved by writing Equation (2) in the form

$$y = \frac{a_m x^m}{b_n x^n} \cdot \frac{1 + \dfrac{a_{m-1}}{a_m x} + \dfrac{a_{m-2}}{a_m x^2} + \cdots + \dfrac{a_1}{a_m x^{m-1}} + \dfrac{a_0}{a_m x^m}}{1 + \dfrac{b_{n-1}}{b_n x} + \dfrac{b_{n-2}}{b_n x^2} + \cdots + \dfrac{b_1}{b_n x^{n-1}} + \dfrac{b_0}{b_n x^n}}$$

and observing that the numerator and denominator of the compound fraction on the right tend to 1 as $x \to +\infty$ or $x \to -\infty$.

If, when $x = B$, the denominator of Equation 2 becomes 0 and the numerator does not become 0, then the line $x = B$ must be a vertical asymptote. (If both numerator and denominator become 0 when $x = B$, then the expression should be simplified by dividing numerator and denominator by $x - B$.)

Example 7. Let

$$(4) \quad y = \frac{3x^2 + 6x - 9}{(x - 1)(x - 2)}.$$

Find $\lim\limits_{x \to +\infty} y$, $\lim\limits_{x \to -\infty} y$, and the asymptotes of this function, and sketch the graph.

SOLUTION.
To put Equation 4 in the form of Equation 2, rewrite Equation 4 as

$$y = \frac{3x^2 + 6x - 9}{x^2 - 3x + 2}.$$

Then by Equation 3,

$$(5) \quad \lim_{x \to +\infty} y = \lim_{x \to +\infty} \frac{3x^2}{x^2} = 3 \quad \text{and} \quad \lim_{x \to -\infty} y = \lim \frac{3x^2}{x^2} = 3.$$

Thus $y = 3$ is a horizontal asymptote. To find the vertical asymptotes, we check the values of x that make the denominator of Equation 4 become zero, namely $x = 1$ and $x = 2$. When $x = 2$, the numerator of Equation 4 is different from zero; therefore the vertical line $x = 2$ is a vertical asymptote. When $x = 1$, the numerator of Equation 4 becomes zero. Since $3x^2 + 6x - 9 = 3(x - 1)(x + 3)$, we can divide numerator and denominator of Equation 4 by $x - 1$ to obtain

$$(6) \quad y = \frac{3(x - 1)(x + 3)}{(x - 1)(x - 2)} = \frac{3(x + 3)}{x - 2}.$$

From Equation 6, we see that there are no vertical asymptotes other than $x = 2$. Furthermore, from Equation 5,

$$(7) \quad \lim_{x \to 2^+} y = +\infty \qquad \lim_{x \to 2^-} y = -\infty.$$

The above information (especially Equations 5 and 7), together with plotting a few points, allows us to sketch the general shape of the graph of Equation 4 (see Figure 21).

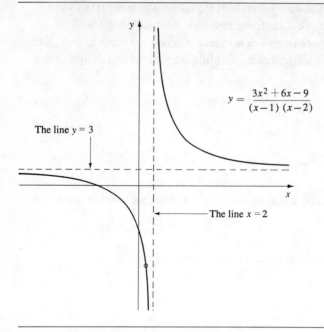

$$y = \frac{3x^2 + 6x - 9}{(x-1)(x-2)}$$

The line $y = 3$

The line $x = 2$

Figure 21

Problems

Find the limits.

1. $\lim\limits_{x \to 3} [2x^2 - 4x - 2]$.

2. $\lim\limits_{x \to 1^-} [4x^2 - x + 1]$.

3. $\lim\limits_{x \to 4} [(x + 1)(6 - x)]$.

4. $\lim\limits_{x \to 1} \left[\dfrac{2x}{5} \cdot \dfrac{3}{x + 1} \right]$.

5. $\lim\limits_{x \to 0} \dfrac{2x^2 - 3x + 4}{4x^2 + x + 2}$.

6. $\lim\limits_{x \to 1} \dfrac{2x - 3}{x - 3}$.

7. $\lim\limits_{x \to 3} \dfrac{x^2 - 9}{x - 3}$.

8. $\lim\limits_{x \to 5} \dfrac{x - 5}{x^2 - 25}$.

9. $\lim\limits_{x \to 0} \dfrac{x^2 - 3x}{x^2 - x}$.

10. $\lim\limits_{x \to 1} \dfrac{\sqrt{x} - 1}{x - 1}$.

11. A function F is defined for all values of x as follows.

$$F(x) = \begin{cases} \dfrac{x^2 - 9}{x - 3} & \text{if } x \neq 3 \\ 6 & \text{if } x = 3. \end{cases}$$

Show that F is continuous at 3.

12. Let $f(x) = 1/x$. Find $\displaystyle\lim_{\Delta x \to 0} \frac{f(x + \Delta x) - f(x)}{\Delta x}$.

13. In Chapter 5, it will be shown that $\displaystyle\lim_{x \to 0+} (1 + x)^{1/x}$ is a certain number between 2 and 3; that number is denoted by e. In terms of this number e, find

$$\lim_{x \to 0+} \frac{(1 + x)^{1/x}}{(2 + x^2)(\sqrt{x} + 4)}.$$

14. Find each limit.
   a. $\displaystyle\lim_{x \to +\infty} [3 + 2/x]$.
   b. $\displaystyle\lim_{x \to -\infty} [3 + 2/x]$.
   c. $\displaystyle\lim_{x \to 0+} [3 + 2/x]$.
   d. $\displaystyle\lim_{x \to 0-} [3 + 2/x]$.

15. Find each limit.
   a. $\displaystyle\lim_{x \to +\infty} \left[2 + \frac{4}{x - 3}\right]$.
   b. $\displaystyle\lim_{x \to -\infty} \left[2 + \frac{4}{x - 3}\right]$.
   c. $\displaystyle\lim_{x \to 3+} \left[2 + \frac{4}{x - 3}\right]$.
   b. $\displaystyle\lim_{x \to 3-} \left[2 + \frac{4}{x - 3}\right]$.

16. For each of the following functions, find the limit as $x \to +\infty$, the limit as $x \to -\infty$, and the asymptotes, and sketch the graphs.
   a. $y = \dfrac{6x + 5}{x - 3}$.
   b. $y = \dfrac{x^2 - 1}{x^3 - 8}$.
   c. $y = \dfrac{x^2 + 1}{x - 1}$.
   d. $y = \dfrac{3x^2 - 3x}{x^2 + 1}$.
   e. $y = \dfrac{3x^2 + 1}{x^2 + x}$.
   f. $y = \dfrac{8x^2 + 1}{4x^2 - 16}$.

17. The electric threshold i in living tissue was found to be $i = a/t + b$, where a and b are positive constants and t is the duration of electrical stimulation.[5] Find $\displaystyle\lim_{t \to 0+} i$ and $\displaystyle\lim_{t \to +\infty} i$.

[5] J. G. Defares, L. N. Sneddon, and M. E. Wise, *An Introduction to the Mathematics of Medicine and Biology,* 2nd. ed. (Amsterdam-London: North Holland, 1973), p. 73.

18. A survey of a printers' union led to the model $q = \dfrac{ax}{ax + 1}$ for the probability q that a worker in a shop with x employees will have his best friend as a coworker[6]; here a is a constant not equal to zero. Find $\lim\limits_{x \to 0+} q$ and $\lim\limits_{x \to +\infty} q$.

Summary

The *derivative* of a function is another name for its *instantaneous rate of change.* An equivalent description is that the derivative of a function at a given value is the *slope of the tangent line* to the graph of the function at the given value.

Three methods have been used to calculate the derivative of a function, say $y = f(x)$.

1. *The Graphical Method.* Sketch the graph of the function and estimate the slope of the tangent line at the desired point.

2. *The Δ-process.* Calculate $\dfrac{\Delta y}{\Delta x}$ and use the equation

$$\frac{dy}{dx} = \lim_{\Delta x \to 0} \frac{\Delta y}{\Delta x}$$

   to find the derivative $\dfrac{dy}{dx}$.

3. *The Differentiation Rules.* If $f(x)$ is a polynomial function then $\dfrac{dy}{dx}$ can be found using the differentiation rules in Section 4.

Miscellaneous Problems

1. Let $y = x^2 - 2x + 2$.

   Use the graphical method to estimate $\dfrac{dy}{dx}$ in each case.

[6]J. S. Coleman, *Introduction to Mathematical Sociology* (London: The Free Press of Glencoe, Collier-Macmillan Ltd. 1964), p. 279.

a. When $x = -1$.   **b.** When $x = 0$.
c. When $x = 1$.   **d.** When $x = 2$.
e. When $x = 4$.

2. Let $s = \frac{1}{4}t^2 - 3$. (a) Use the graphical method to estimate $\dfrac{ds}{dt}$ in each case.

     a. When $t = -2$.   **b.** When $t = 0$.
     c. When $t = 2$.   **d.** When $t = 4$.

3. Let $u = 1/x$.

     a. Use the Δ-process to find $\dfrac{du}{dx}$.

     b. Use Part a and Rule 3 in Section 4 to find $\dfrac{dy}{dx}$ if $y = 5/x$.

4. Let $s = \frac{1}{4}t^2 - 3$. Use the Δ-process to find the exact value of $\dfrac{ds}{dt}$ when

     $t = 4$. Compare with Part d of Problem 2 above.

5. Use the differentiation rules to find each derivative.

     a. $\dfrac{dy}{dx}$ if $y = x^{35}$.   **b.** $\dfrac{dy}{dx}$ if $y = 2x^{35}$.

     c. $\dfrac{dy}{dx}$ if $y = 3x^5 + 1$.   **d.** $\dfrac{dy}{dx}$ if $y = \frac{1}{2}x^6 - 3x + 4$.

     e. $\dfrac{dy}{dx}$ if $y = 3.14x^2 - 2.5x$.   **f.** $\dfrac{dy}{dx}$ if $y = \dfrac{2x^6 + 5x}{3}$.

6. Use the differentiation rules to find each derivative.

     a. $\dfrac{ds}{dt}$ if $s = t^9$.   **b.** $\dfrac{dE}{dr}$ if $E = 3t^9$.

     c. $\dfrac{dw}{du}$ if $w = 9x^4 - 3$.   **d.** $\dfrac{dP}{dy}$ if $P = \frac{1}{3}y^{10} - \frac{1}{3}y^2$.

     e. $\dfrac{dL}{dt}$ if $L = \sqrt{3}t^7 - \pi t + \sqrt{2}$.   **f.** $\dfrac{dA}{dv}$ if $A = \dfrac{7.13t^2 + 4}{10}$.

7. Use the differentiation rules to find each derivative.

     a. $\dfrac{dy}{dx}$ when $x = 1$, if $y = 4x^3 - \frac{2}{3}x^{15} + 3$.

     b. $\dfrac{dC}{dp}$ when $p = 3$, if $C = 4.1p^2 - 9.1p$.

     c. $f'(x)$ if $f(x) = 4x^8 - 31$.
     d. $f'(2)$ if $f(x) = 3x^5 - x^4$.
     e. $G'(3)$ if $G(t) = t^6/27 - 4t$.

8. Use the differentiation rules to find each derivative.

   a. $\dfrac{dy}{dx}$ when $x = 4$, if $y = \dfrac{2x^3}{12} - \dfrac{3x^2}{4}$.

   b. $\dfrac{dR}{dr}$ when $r = 3.1$, if $R = \dfrac{r^2 - 1.7r + 2.3}{1.5}$.

   c. $h'(5)$ if $h(w) = w^3 - 60w + 15$.

9. Differentiate the following functions. (*Hint:* First express each as a polynomial.)

   a. $y = (x - 1)(2x + 1)$.

   b. $y = \dfrac{10x^4}{7} + \dfrac{2x - 1}{4}$.

   c. $f(t) = 2t^3(t + \frac{1}{2})$.

   d. $g(t) = (2t - 3)^2$.

   e. $p = (2q + 1)^3$.

   f. $w = \dfrac{4v^2 + 4v + 1}{2v + 1}$.

10. Find the equation of the line that is tangent to the graph of $y = 5x^2 - 6x + 1$ at the point where the graph crosses the y-axis.

three

Applications of Differentiation

In the preceding chapter, we studied the concept of the derivative, both geometrically and analytically. We also developed basic techniques for computing derivatives. Having sown these seeds, we are now prepared to reap some rewards: we shall examine a number of the applications of differentiation. More advanced applications and techniques will be discussed in Chapter 4.

Section 1 □ **The role of the derivative outside of mathematics.**   Velocity, acceleration, current, marginal cost and other marginal quantities, origins of applied equations.

Section 2 □ **Tests for increasing-decreasing values.**

Section 3 □ **Maxima, minima, and critical points.** Endpoints, nondifferentiability.

Section 4 □ **Higher-order derivatives.**   Concavity, second-derivative tests, points of inflection.

Section 5 □ **Maximum-minimum problems.**

Summary □ **Miscellaneous problems.**

Consider a function $y = f(x)$. The derivative $\left.\dfrac{dy}{dx}\right|_{x=x_0}$ gives the slope of the line tangent to the graph of this function at the point on the graph where $x = x_0$. This derivative is also called the *rate of change* of y with respect to x when $x = x_0$. In areas outside of mathematics the derivative may be known by other names, depending on the meaning of the independent and dependent variables.

Velocity *is the derivative of* **distance** *with respect to* **time.**

For example, suppose a stone is dropped from a tall tower (see Figure 1). It is known that after t seconds, the stone will have fallen a distance s feet, where[1] $s = 16t^2$. Therefore its velocity (or speed) v after t seconds is

$$v = \frac{ds}{dt} = 16(2t) = 32t \text{ feet per second.}$$

Acceleration *is the derivative of* **velocity** *with respect to* **time.**

[1]This equation, like nearly every other one which models a real world situation, is not exact. In this case of a freely-falling body in air, it is nearly impossible to account exactly for effects such as buoyancy, air resistance, etc.

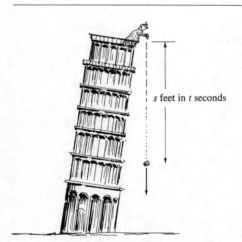

s feet in t seconds

Figure 1

Consider, for example, the falling stone mentioned above. After falling for t seconds, its velocity was $v = 32t$ feet per second. Therefore the acceleration a of the stone after t seconds is

$$a = \frac{dv}{dt} = 32 \text{ feet per second per second.}$$

Current is the derivative of **charge** with respect to **time.**

For example, charged atoms or ions pass through the semipermeable membrane of a nerve cell. Let Q be the total charge that was transported across a piece of membrane after t seconds. Then the current at time t is

$$i = \frac{dQ}{dt}.$$

Marginal cost is the derivative of the supplier's **cost** with respect to the **quantity** supplied.

For example, suppose it costs a fisherman C dollars to catch q tons of tuna. For values of q between 5 and 100, let us assume C is given by the simple formula

(1) $C = 200q^2 + 600q.$

Then the marginal cost at a level of q tons of tuna is

(2) $\dfrac{dC}{dq} = 400q + 600 \text{ dollars per ton.}$

Marginal tax is the derivative of total **tax** with respect to **taxable income.**

For example, the total federal income tax T in dollars on an annual income of x dollars is given in Figure 2.[2] This figure shows that the total tax at an income level of \$32,000 is about \$10,300 (that is, about 32%), yet the marginal tax at that income level is about 0.5 (or 50%).

There are many other uses of derivatives. It would take too long to list them here and, unlike the examples above, most have not earned special names.

[2]Source: Internal Revenue Service, Tax Schedule X.

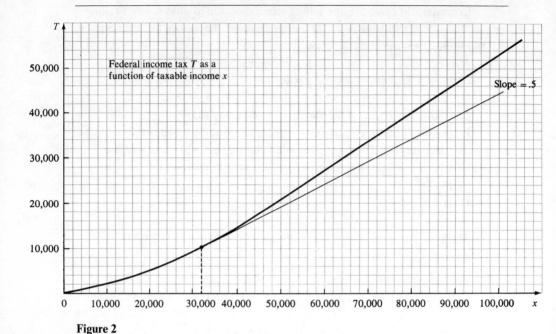

Figure 2

Marginal Quantities in Economics. In business and economics, the word *marginal* indicates differentiation. *Marginal cost* has already been mentioned. Here are some other illustrations.

Utility U of a commodity is a function of the quantity q which is consumed. **Marginal utility** is defined as $\dfrac{dU}{dq}$.

The revenue R received by a manufacturer is a function of the quantity q he produces. **Marginal revenue** is defined as $\dfrac{dR}{dq}$.

Similarly, differentiation is used to define *marginal capital-output ratio, marginal efficiency of investment, marginal product, marginal propensity to consume, marginal productivity, marginal profit,* and so on.

Economists often remark that the word *marginal* in this context is synonomous with *extra.* Let us explain this remark; we use marginal cost as an example. Suppose C is the cost a supplier must pay in order to acquire q units of a commodity. Part a of Figure 3 shows the cost C in dollars for a fisherman to catch q tons of tuna fish.

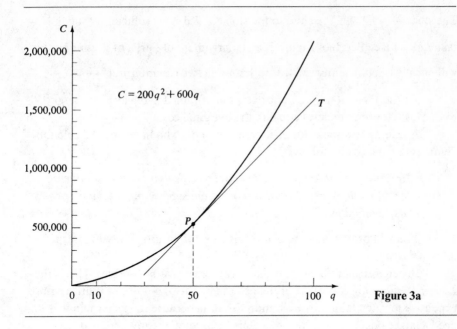

Figure 3a

At the level $q = 50$ tons, the marginal cost $\left. \dfrac{dC}{dq} \right|_{q=50}$ is the slope of the tangent line T in Part a of Figure 3. Let M be a secant line through point P.

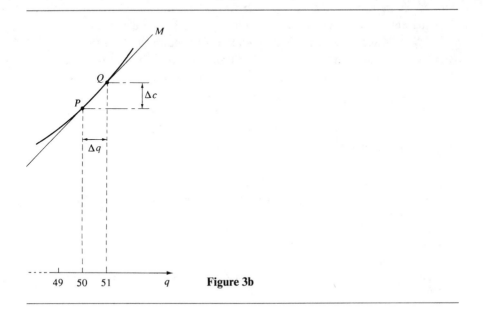

Figure 3b

The slope $\dfrac{\Delta C}{\Delta q}$ of M will be close to the slope of T if Δq is sufficiently small. Now take $\Delta q = 1$ (see Part b of Figure 3, an enlargement of Part a of Figure 3); this will usually be sufficiently small.[3] With $\Delta q = 1$, we can interpret $\dfrac{\Delta C}{\Delta q} = \dfrac{\Delta C}{1}$ $= C_2 - C_1$ as the cost to catch one extra ton of tuna, having already caught 50 tons. This is (approximately) equal to the marginal cost.

A leading textbook[4] for economics students who have not had calculus defines marginal cost as follows.

> *"Marginal cost at any production level q is the extra cost of producing one extra unit more (or less); it comes from subtracting total dollar costs at adjacent outputs."*

You will probably agree that this is a useful, intuitive way to interpret marginal cost.

In general, marginal cost eventually increases as q increases. This is the *Law of Diminishing Returns.* For the fisherman, the one-millionth tuna is more expensive to catch than the one-hundredth tuna because the fisherman will need a larger boat to sail to distant waters he has not already depleted.

From Where Do the Applied Equations Come?   The equations in applied problems usually come from two sources: they may be derived empirically by fitting a standard type of function to a graph of data points, or they may be derived theoretically, often from properties of their derivatives. In either case, the equations are only first approximations to the real-world situation.

We discussed empirically derived linear equations under *Linear Regression* (see Section 8 of Chapter 1). Simple functions other than linear ones may also be used to fit data. In certain areas, there is rarely any need to use complicated functions. For example, Wold and Juréen[5] state that five types of simple functions have been satisfactory fitted to empirical data to obtain demand curves. In the classic treatise on this subject,[6] Schultz introduces the chapter, "The Derivation of Demand Curves from Time Series," with the following.

[3]The reader can check that T has slope 20,600 and M has slope 20,800 in case $C = 200q^2 + 600q$ as above. Thus the slopes differ by less than 1%.

[4]P. A. Samuelson, *Economics,* 8th ed. (New York: McGraw-Hill Publishing Co., 1970) page 429.

[5]H. Wold and L. Juréen, *Demand Analysis, A Study in Econometrics* (New York: John Wiley and Sons, Inc., 1953,) pp. 2–3.

[6]H. Schultz, *The Theory and Measurement of Demand* (Chicago: University of Chicago Press, 1938) p. 61.

"One hypothesis which we must make and which is common to all of the methods used for deriving demand curves is that the unknown theoretical demand curve can be approximated by a more or less simple empirical equation."

The second method, obtaining functions from the expected properties of their derivatives, forms a subject called **differential equations.** A simple example occurs in population studies: the rate of change of certain populations may be expected to be proportional to population size. This leads to an equation for the population function (see Chapter 5, page 204).

The control of messenger RNA in cellular protein synthesis was treated by B. C. G. Goodwin[7] by a combination of these two methods. Differential equations were obtained and then a computer was used to approximate their solutions by means of simple functions.

Finally, it should be emphasized that the concepts we are developing will be useful for understanding given situations, even without introducing mathematical equations. For example, see pages 416–417 of *Economics* (footnote 4 on page 100) where Adam Smith's ancient puzzle "Why is water so very cheap when it is so very valuable?" is solved by using the concept of marginal utility.

Problems

1. The speedometer of a car is broken but the odometer is working. Suppose that the reading on the odometer h hours after beginning a trip is M miles. Interpret the number $\dfrac{dM}{dh}\bigg|_{h=4.5}$

2. In Problem 1, let $\dfrac{dM}{dh}\bigg|_{h=h_0}$ be denoted by $f(h_0)$. Interpret the quantity $f'(4.5)$.

3. Suppose that t seconds after liftoff, a rocket has traveled vertically for h feet. If t and h are related by $h = 0.3t^2 + 0.6t$, find each of the following.

    a. The velocity of the rocket 9 seconds after liftoff
    b. The acceleration of the rocket 9 seconds after liftoff

[7] J. Maynard Smith, *Mathematical Ideas in Biology* (New York: Cambridge University Press, 1971, pp. 170–115.

[8] Samuelson, *Economics,* pp. 416–417.

4. Let $Q = 0.04t^5 + 0.25t^2$ be the total charge in coulombs that have passed through a membrane after t seconds. When $t = 3$, what was the current in amperes across this membrane? (*Note:* An ampere is defined to be one coulomb per second.)

5. Let $C(x)$ be the total cost in dollars to manufacture x radial tires in one month in a certain factory. Suppose the model of this situation is

(1) $C(x) = 0.02x^2 + 10x + 15,000.$

Find the marginal cost of tires at a production level of 1000 tires per month. Interpret this as the cost of making one extra tire beyond 1000, and explain why the manufacturer should increase his production level beyond 1000 if he knows he can sell 1000 tires each month at $65 each.

6. Use Figure 2 (page 98) to estimate the marginal tax at an income level of $100,000. Interpret this as the tax an individual must pay on the last dollar he earned. Such a person is said to be in the "70% tax bracket." What is the average tax he pays on each dollar earned?

7. The tire manufacturer in Problem 5 manufactures x tires each month. Let $p(x)$ be the best price per tire he can receive when he sells x tires each month (as x gets larger, p gets smaller). His monthly revenue $R(x)$ in dollars is then

(2) $R(x) = x \cdot p(x).$

Suppose

(3) $p(x) = 75 - 0.01x.$

Find the marginal revenue when $x = 1000$.

8. The tire manufacturer's monthly profit $P(x)$ (see Problems 5 and 7), which he obtains by making x tires each month, is his revenue minus his cost.

$$P(x) = R(x) - C(x)$$
$$= x \cdot p(x) - C(x) = x[75 - 0.01x] - [0.02x^2 + 10x + 15,000]$$
$$= -0.03x^2 + 65x - 15,000$$

When $x = 1000$, find the profit and the marginal profit.

Section 2 □ Tests for Increasing-Decreasing Values

In this section, we shall be interested in whether the values y of a function $y = f(x)$ are increasing or decreasing as x moves from left to right along an interval on the x-axis. Let us first make some definitions. Suppose I is an interval on the x-axis. Suppose also that the inequality $f(x_1) < f(x_2)$ holds for every pair x_1, x_2 of points in I with $x_1 < x_2$. We then say that f is *increasing on I*. If the inequality $f(x_1) > f(x_2)$ holds for all pairs of points x_1, x_2 in I with $x_1 < x_2$, then we say that f is *decreasing on I*.

To illustrate this terminology, consider the function $y = f(x)$ (see Figure 4). This function f is increasing on the interval from 1 to 10; it is decreasing on the interval from 10 to 20; and it is increasing on the interval from 20 to 30.

In certain applications, we shall want to decide if a given value x_0 is within an interval on which f is increasing. If it is, we shall say that "f is increasing at x_0." To be more precise, the phrase f is *increasing at* x_0 means that there is some interval J on the x-axis, centered at x_0, such that f is increasing on J. (*Note: J* may not be the largest interval on which f is increasing.) Similarly, we say that f is *decreasing at* x_0 if there is an interval J on the x-axis, centered at x_0, such that f is decreasing on J.

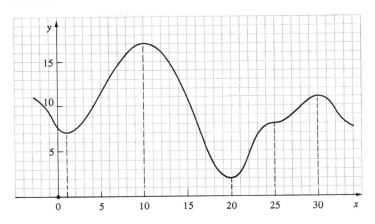

Figure 4

The function *f* in Figure 4 is increasing at 2, at 2.1, at 3, and, in fact, at every value of *x* such that $1 < x < 10$. It is also increasing at every *x* such that $20 < x < 30$. This function *f* is decreasing at every *x* such that $10 < x < 20$ and at every *x* such that $30 < x$. It is important to observe that this function *f* is neither increasing nor decreasing at the values $x = 1$, $x = 10$, $x = 20$, or $x = 30$.

Parts a and b of Figure 5 illustrate the relationship between increasing or decreasing functions and a positive or negative derivative:

---

(1) $\begin{cases} \text{If } f'(x_0) > 0 \text{ then } f \text{ is increasing at } x_0. \\ \text{If } f'(x_0) < 0 \text{ then } f \text{ is decreasing at } x_0. \end{cases}$

Statements 1 may not be new to you, as these facts were observed earlier in the discussions of average and instantaneous rates of change.

Note that if $\dfrac{dy}{dx} = 0$ when $x = x_0$, we are unable to tell, without further information, whether *y* is increasing at x_0, decreasing at x_0, or neither. For example, in Figure 4, $\dfrac{dy}{dx} = 0$ when $x = 10$ (*y* is neither increasing nor decreasing there), and $\dfrac{dy}{dx} = 0$ when $x = 25$ (*y* is increasing here).

Example 1. Let $y = x^2 - 8x$. Is *y* increasing or decreasing when $x = 3$? When $x = 6$?

Figure 5a

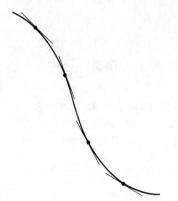

Figure 5b

SOLUTION.

Use the differentiation rules to obtain

(2) $\dfrac{dy}{dx} = 2x - 8.$

When $x = 3$, Equation 2 shows that $\dfrac{dy}{dx} = 6 - 8 = -2$. Since this is negative, y is decreasing when $x = 3$.

When $x = 6$, Equation 2 shows that $\dfrac{dy}{dx} = 12 - 8 = 4$. Since this is positive, y is increasing when $x = 6$.

Example 2. Let $P(q) = 8q - q^2 + 20$ be the profit gained by manufacturing q units of a commodity. Is P increasing or decreasing when $q = 2$? When $q = 5$?

SOLUTION.

The differentiation rules give

(3) $P'(q) = 8 - 2q.$

When $q = 2$, Equation 3 gives $P'(2) = 4$. Since this is positive, P is increasing when $q = 2$. Here we used Equation 1 with P for f and q for x.

When $q = 5$, Equation 3 shows $P'(5) = -2$. Since this is negative, P is decreasing when $q = 5$.

APPLICATIONS OF DIFFERENTIATION

Problems

1. Test if the given function is increasing or decreasing at the given value.

   a. $y = 3x^2 - 20$ at $x = 2$.   b. $y = 3x^2 - 20x$ at $x = 2$.
   c. $y = 3x^2 - 20x$ at $x = 5$.   d. $w = u^6$ at $u = 1$.
   e. $P(q) = q^2 - 3q + 2$ at $q = 1$.   f. $P(q) = q^2 - 3q + 2$ at $q = 2$.
   g. $f(t) = 30 - 5t$ at $t = 6$.   h. $R = x(5 - 2x)$ at $x = 2$.

2. Test if the given function is increasing or decreasing at each of the given values.

   a. $y = 30 - 2x^3$ at $x = 6$; at $x = 7$.
   b. $y = t^3 - 3t^2 + 2$ at $t = 1$; at $t = 3$.
   c. $p = q + 5q^2 - q^3$ at $q = 1$; at $q = 5$.
   d. $w = 3 - v^2 + v$ at $v = -1$; at $v = 1$.
   e. $g(x) = x(x - 2)$ at $x = \frac{1}{2}$; at $x = 5$.
   f. $h(u) = u^{50} - u$ at $u = 1$; at $u = 0$.

3. The marginal cost M is a function of the production level x. Suppose $M = x^2 - 100x + 5000$. Show that M is decreasing when $x < 50$ is increasing when $x > 50$.

4. An aspirin manufacturer finds that his weekly profit P depends on the amount x which he spends each week on advertising. Suppose the formula is $P = 1000 + 2000x - x^2$.

   a. The manufacturer is spending $700 a week on advertising. If he increases this expenditure a little will his profits increase or decrease?
   b. Answer Part a if his weekly expenditure is $1100.

5. In the chemical reaction $H_2 + I_2 \rightleftharpoons 2HI$, one molecule of hydrogen and one molecule of iodine combine to form two molecules of hydrogen iodide and conversely, two hydrogen iodide molecules produce one molecule of hydrogen and one of iodine. Let $x(t)$ be the concentration of hydrogen iodide molecules at time t. (We call $x'(t)$ the **reaction rate**.) Suppose that under certain conditions $x(t) = 0.16t - 0.01t^2$ for values of t between 3 and 14. Is the concentration of HI increasing or decreasing at $t = 6$? At $t = 10$?

6. Let B be the number of bacteria growing in a culture t hours after a dose of penicillin has been introduced. Suppose B can be well approximated by $B(t) = 112 + 24t - t^2$, for values of t between 5 and 20. Is the bacteria population increasing or decreasing when $t = 15$? When $t = 11$?

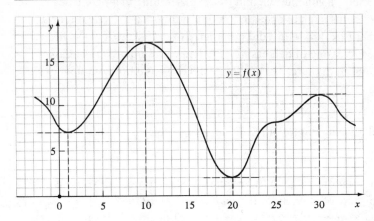

Figure 6

Section 3 □ Maxima, Minima, and Critical Points

An important application of calculus is the use of differentiation to find the *largest* and *smallest values* of a function. These values are called the **absolute maximum** and **absolute minimum** of the function. In Figure 6, for example, the function $y = f(x)$ attains an absolute maximum when $x = 10$; the value of that maximum is $f(10) = 17$. This function attains its absolute minimum when $x = 20$; the value of that minimum is $f(20) = 2$.

A more general notion is that of **relative maximum** (or **minimum**). A relative maximum refers to a situation where $f(x_0)$ is larger than all values $f(x)$ when x is near to x_0. To illustrate, consider Figure 6 again. The function there attains a relative minimum when $x = 1$ (the value of this relative minimum is $f(1) = 7$); it attains a relative maximum when $x = 30$ (the value of this relative maximum is $f(30) = 11$). The absolute maximum at $x = 10$ is, at the same time, a relative maximum. The technical definition of these concepts can be given as follows.

A function $y = f(x)$ is said to attain a *relative maximum* (also called *local maximum*) at x_0 if $f(x_0) \geq f(x)$ for all values of x within some interval on the x-axis centered at x_0. (If $\geq$ is replaced by $\leq$, we obtain the definition of *relative minimum*.)

Consider now a general function $y = f(x)$. Assume that f has a derivative at x_0 and that f attains a relative maximum or minimum at x_0. The important conclusion is that $f'(x_0) = 0$. If it did not, then $f'(x_0) > 0$ or $f'(x_0) < 0$, so that f would be increasing or decreasing at x_0 and could not attain a relative maximum or minimum there. That is, the graph of f has a horizontal tangent line at $x = x_0$.

If f attains a relative maximum or minimum at x_0, then $f'(x_0) = 0$.

This observation will help us find the exact location of all relative maxima and minima of a function. *We shall suppose that the function $y = f(x)$ is defined and has a derivative at all values of x.* To find all relative maxima and minima of f, we usually proceed as follows.

Step 1. Calculate the derivative $f'(x)$.

Step 2. Set the derivative equal to zero and solve for x in $f'(x) = 0$. These values where the derivative is zero are called the **critical points** of f.

Step 3. Examine the critical points found in Step 2 to determine which ones, if any, make f attain a relative maximum or minimum.

Step 3 requires further discussion. How does one tell if a critical point corresponds to a relative maximum, a relative minimum, or neither? In some cases it will be obvious (see Example 1). In some other cases, a very rough sketch of the graph will give the answer (see Example 2). In still other cases, an examination of the sign of the derivative of f will provide the solution (see Example 3).

Example 1. The function $y = x^2 - 6x + 17$ has an absolute minimum. Find it, and find where it occurs.

SOLUTION.

Step 1. $\dfrac{dy}{dx} = 2x - 6.$

Step 2. $2x - 6 = 0; \qquad 2x = 6; \qquad x = 3.$

Step 3. We were told that y has an absolute minimum. It must occur at a critical point. Step 2 shows that y has only one critical point, at $x = 3$. So the absolute minimum must occur at $x = 3$, and its value is

$$y = (3)^2 - 6(3) + 17 = 8.$$

Example 2. Find all relative maxima and minima of $y = x^3 - 6x^2 + 9x + 3$.

SOLUTION.

Step 1. $\dfrac{dy}{dx} = 3x^2 - 12x + 9.$

Step 2. $3x^2 - 12x + 9 = 0;\qquad x^2 - 4x + 3 = 0;\qquad (x - 3)(x - 1) = 0;$ $x = 1, 3.$

Step 3. Let us make a rough sketch of the graph. We plot a few points (see Part a of Figure 7).

x	-1	0	1	2	3	4
y	-13	3	7	5	3	7

Note that we included the critical points $x = 1$ and $x = 3$ found in Step 2, and we indicated on the graph that the function will have a horizontal tangent there since $\dfrac{dy}{dx} = 0$.

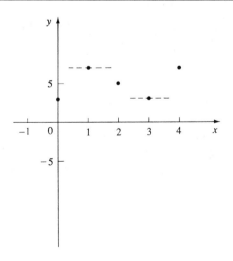

Figure 7a

We now draw a smooth curve through the points in Part a of Figure 7, *remembering that the curve cannot have a horizontal tangent line except at the two indicated points.* The result must be like Part b of Figure 7. (Here is a sample of such reasoning. Begin with x, going from -1 to 1. Some of the time, y must increase. If it also decreased some of the time, then, in changing from decreasing to increasing, it would have to have a horizontal tangent somewhere. Since it does not have a horizontal tangent in this range of x, it must increase all the time in this range.)

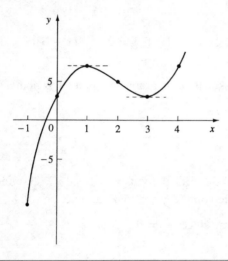

Figure 7b

Now a glance at Part b of Figure 7 allows us to complete the problem. We see that y has a relative maximum when $x = 1$ (its value is $y = 7$), and a relative minimum when $x = 3$ (its value is $y = 3$).

Consider again a general function $y = f(x)$. Suppose x_0 is a critical point of f, so $f'(x_0) = 0$. Can we tell, without sketching a graph, whether f has a relative maximum at x_0, a relative minimum at x_0, or neither? Yes; the sign of $f'(x)$ will provide the answer. As part of Figures 8a, b, c, d show, if $f'(x)$ changes from positive to negative as x passes through x_0 from left to right, then f has a relative maximum at x_0. Similarly, if $f'(x)$ changes from negative to positive, then f has a relative minimum at x_0, while two positive or two negative signs indicate that x_0 is neither a relative maximum nor a relative minimum.

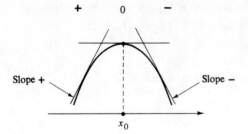

Figure 8a

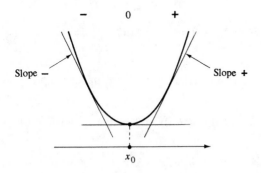

Figure 8b

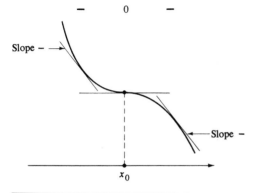

Figure 8c

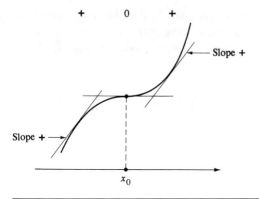

Figure 8d

To test how the sign of f' changes at x_0, pick a point a to the left of x_0 and a point b to the right. Make sure there are no critical points between a and x_0, or between x_0 and b. Then find the sign of $f'(a)$ and $f'(b)$.[8]

Example 3. Find all the relative maxima and minima for the function $y = x^3 - 3x^2 + 3x + 4$.

SOLUTION.

Step 1. $\dfrac{dy}{dx} = 3x^2 - 6x + 3.$

Step 2. $3x^2 - 6x + 3 = 0;$ $x^2 - 2x + 1 = 0;$ $(x - 1)^2 = 0;$

$x = 1.$

Step 3. We pick points to the left of 1 and to the right of 1; suppose we pick 0 and 2.

$$\text{When } x = 0, \qquad \frac{dy}{dx} = 3(0)^2 - 6(0) + 3 = 3 \qquad (+)$$

$$\text{When } x = 2, \qquad \frac{dy}{dx} = 3(2)^2 - 6(2) + 3 = 3 \qquad (+)$$

Since the signs are $+\ +$, Figure 8c applies and shows that the critical point $x = 1$ is neither a maximum nor a minimum. Since there are no other critical points, we conclude that this function has no relative maxima nor minima.

Example 4. Find all relative maxima and minima of $y = 15x^2 - 2x^3$.

SOLUTION.

Step 1. $\dfrac{dy}{dx} = 30x - 6x^2$

Step 2. $30x - 6x^2 = 0;$ $5x - x^2 = 0;$ $x(5 - x) = 0;$ $x = 0,5.$

Step 3. When $x = -1, \dfrac{dy}{dx} = 30(-1) - 6(-1)^2 = -36$ $(-)$

When $x = 1, \dfrac{dy}{dx} = 30(1) - 6(1)^2 = 24$ $(+)$

When $x = 6, \dfrac{dy}{dx} = 30(6) - 6(6)^2 = -36$ $(-)$

---

[8]The theoretical basis of this "obvious" test is an advanced theorem of calculus, the *Intermediate Value Theorem for Derivatives.*

At the critical point $x = 0$, the signs are $-$, $+$ and at the critical point $x = 5$, the signs are $+$, $-$. Therefore y has a relative minimum at $x = 0$ and a relative maximum at $x = 5$. The values of y at these points are:

When $x = 0, y = 0$

When $x = 5, y = 15(5)^2 - 2(5)^2 = 125$

Endpoints. The above discussions concerned functions which were defined and had derivatives for all values of the independent variable. If a function is defined only on an interval which includes its endpoints (see Figure 9), then the endpoints of the interval, as well as the critical values, must be treated as candidates for an absolute maximum or minimum.

Example 5. Find the absolute maximum of the function $y = 2x^3 - 9x^2 + 12x$ if x varies in the interval $0 \leq x \leq 4$.

SOLUTION.

Step 1. $\dfrac{dy}{dx} = 6x^2 - 18x + 12.$

Step 2. $6x^2 - 18x + 12 = 0$; $\quad x^2 - 3x + 2 = 0$;
$(x - 2)(x - 1) = 0$; $\quad x = 1,2.$

Step 3. The absolute maximum will be attained at one of the critical points $x = 1$, $x = 2$, or possibly at one of the endpoints $x = 0$, $x = 4$.

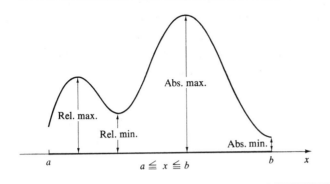

Rel. max.

Rel. min.

Abs. max.

Abs. min.

a

$a \leq x \leq b$

b

x

Figure 9

Therefore we find y at each of these values of x:

When $x = 0, y = 0$

When $x = 1, y = 2(1)^3 - 9(1)^2 + 12(1) = 5$

When $x = 2, y = 2(2)^3 - 9(2)^2 + 12(2) = 4$

When $x = 4, y = 2(4)^3 - 9(4)^2 + 12(4) = 32$.

The largest of these values of y is 32, so this is the absolute maximum of y when x lies in the given interval. (This function is shown in Figure 10.)

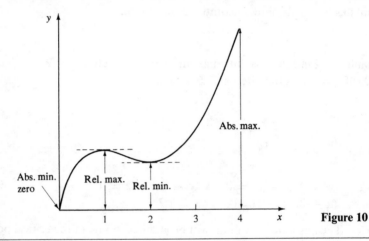

Figure 10

Nondifferentiability. Figures 11 and 12 show functions $y = f(x)$ which do not have a derivative (tangent line) when $x = c$. To find the absolute maximum or minimum of such functions, these exceptional values c, as well as the critical points and endpoints, if any, would have to be treated as possible candidates for an absolute maximum or minimum. These functions do not have simple equations to describe them, and they will not be used in this text.

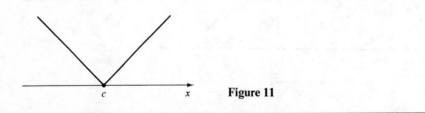

Figure 11

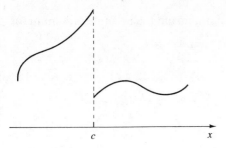

Figure 12

Summary. The relative maxima and minima of a differentiable function are found by examining the critical points.

If a differentiable function has an absolute maximum (or absolute minimum), it can be found by listing all critical points, and all endpoints, if any, and then calculating the value of the function at these points. The largest (or smallest) of these values is the absolute maximum (or minimum). (If the function fails to have a derivative at some points, these points must also be included in the list.)

Problems

In Problems 1–10, find all critical points (save your answers for use in Problems 11–20 and 21–30).

1. $y = 6x - x^2$.

2. $y = 1 + 24x - 3x^2$.

3. $y = 5x^2 - 20x + 46$.

4. $y = 5x^2 - 20x + 26$.

5. $y = 10x - x^2$.

6. $y = x^3 - 15x^2 + 600$.

7. $y = x^3 - 3x$.

8. $y = x^3$.

9. $P = 2q^2 - \frac{1}{9}q^3$.

10. $w = u^3 - 6u^2 + 12u$.

In Problems 11–20, find all relative maxima and minima and where each occurs.

11. $y = 6x - x^2$.

12. $y = 1 + 24x - 3x^2$.

13. $y = 5x^2 - 20x + 46$.

14. $y = 5x^2 - 20x + 26$.

15. $y = 10x - x^2$.

16. $y = x^3 - 15x^2 + 600$.

17. $y = x^3 - 3x$.

18. $y = x^3$.

19. $P = 2q^2 - \frac{1}{3}q^3$. **20.** $w = u^3 - 6u^2 + 12u$.

In Problems 21–30, find the absolute maximum and the absolute minimum for x in the given interval.

21. $y = 6x - x^2, 2 \leq x \leq 5$.

22. $y = 1 + 24x - 3x^2, 0 \leq x \leq 4$.

23. $y = 5x^2 - 20x + 46, 0 \leq x \leq 1$.

24. $y = 5x^2 - 20x + 26, 0 \leq x \leq 1$.

25. $y = 10x - x^2, 0 \leq x \leq 6$.

26. $y = x^3 - 15x^2 + 600, 0 \leq x \leq 5$.

27. $y = x^3 - 3x, -1 \leq x \leq 1$. **28.** $y = x^3, 0 \leq x \leq 2$.

29. $P = 2q^2 - \frac{1}{3}q^3, 0 \leq q \leq 12$.

30. $w = u^3 - 6u^2 + 12u, 0 \leq u \leq 3$.

31. A child operates a lemonade stand. He soon finds that his daily profit P in cents depends on the price x in cents that he charges for each glass of lemonade. Suppose the formula relating P and x is $P = 46x - x^2 - 129$. What price x should he charge in order to earn the most profit?

Section 4 □ Higher-Order Derivatives

In this section, we define higher-order derivatives. These derivatives are applied to problems concerning velocity-acceleration, concavity, maximum-minimum tests, and inflection points.

Suppose we differentiate the function

$$f(x) = 2x^5 - 3x^2 + 1.$$

We obtain

(1) $f'(x) = 10x^4 - 6x$.

The function f' in Equation 1 can also be differentiated. Its derivative, denoted by f'', is

$$f''(x) = 40x^3 - 6.$$

This process can be repeated: $f'''(x) = 120x^2$, and so on.

We call f'' the **second derivative** of f; f''' is called the **third derivative** of f; and so on. In the $\dfrac{dy}{dx}$ notation, the second derivative of y is denoted by $\dfrac{d^2y}{dx^2}$.

If, for example, P is a function of q, then $\dfrac{dP}{dq}$ is the first derivative and $\dfrac{d^2P}{dq^2}$ is the second derivative.

Example 1.   If $f(x) = x^4 - x^3 + 4x^2$ find $f''(x)$ and $f''(3)$.

SOLUTION.

$$f'(x) = 4x^3 - 3x^2 + 8x$$

$$f''(x) = 12x^2 - 6x + 8$$

$$f''(3) = 12(3)^2 - 6(3) + 8 = 108 - 18 + 8 = 98.$$

(*Note:* We must find $f''(x)$ before $f''(3)$ can be calculated.)

Example 2.   If $w = 2u^3 - u^2 + 5u$, find $\dfrac{d^2w}{du^2}$ when $u = 4$.

SOLUTION.

$$\frac{dw}{du} = 6u^2 - 2u + 5$$

$$\frac{d^2w}{du^2} = 12u - 2$$

When $u = 4$, $\dfrac{d^2w}{du^2} = 12(4) - 2 = 46.$

Illustrations.   If we think of the derivative of y as the *rate of change of y*, then the second derivative would be the *rate of change of the rate of change of y*.

There are many examples of this second derivative concept. If $C(t)$ is the cost-of-living index at time t, then $C'(t)$ is a measure of inflation.[9] If $C'(t)$ is

[9]The numerical value $C(t)$ is rarely of interest. Therefore, inflation is often expressed as the ratio $\dfrac{C'(t)}{C(t)}$, written as a percentage. This amounts to adjusting $C(t)$ so that its value is 1 at the time inflation is measured.

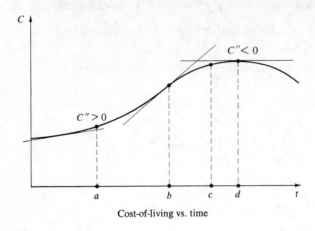

Cost-of-living vs. time   **Figure 13**

large and positive, then prices are rising rapidly at time t; if $C'(t)$ is negative, then prices are falling at that time. If $C''(t)$ is positive, then C' is increasing at t; thus inflation is getting worse (see time $t = a$ in Figure 13). If $C''(t)$ is negative, then inflation is easing off (see time $t = c$ in Figure 13). Note that inflation may be easing off, yet prices continue to rise (see time $t = b$ in Figure 12). In Figure 13, when $t = b$, the politicians will announce that "inflation has been licked" because the inflation has stopped rising. But consumers will not notice a decline in prices until after time $t = d$.

Another application of the second derivative concept is the notion of acceleration. Recall (pages 96–97) that velocity is the rate of change of distance with respect to time, and acceleration is the rate of change of velocity with respect to time. Thus, acceleration is the second derivative of distance with respect to time.

Example 3. A rocket rising vertically reaches a height of h feet after t seconds. Find the velocity and acceleration after 5 seconds if $h = t^3 + \frac{1}{2}t^2$.

SOLUTION.

$$\frac{dh}{dt} = 3t^2 + t \qquad \text{and} \qquad \frac{d^2h}{dt^2} = 6t + 1.$$

Therefore the velocity at time $t = 5$ is

$$3(5)^2 + 5 = 80 \text{ feet per second}$$

and the acceleration at time $t = 5$ is

$$6(5) + 1 = 31 \text{ feet per second per second.}$$

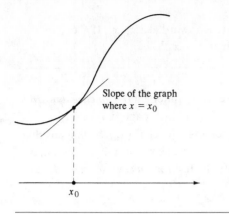

Slope of the graph
where $x = x_0$

x_0

Figure 14

Concavity. Consider a function $y = f(x)$ and its graph. By the **slope of the graph at $x = x_0$**, we mean the slope of the line tangent to the graph at $x = x_0$ (see Figure 14).

If the slope of a graph is increasing as x moves from left to right, then we say that the graph is **concave upward.** In Figure 15, the graph is concave upward at each x to the left of $x = c$. We say the graph is **concave downward** if the slope of the graph is decreasing as x moves from left to right. In Figure 15, the graph is concave downward at each x to the right of $x = c$.

The second derivative is useful for finding where a function is concave upward or concave downward, without the need to draw an accurate graph.

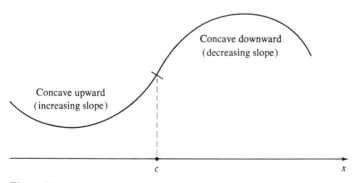

Concave downward
(decreasing slope)

Concave upward
(increasing slope)

c

x

Figure 15

Second-derivative Test for Concavity

If $f''(x_0) > 0$, then the graph of f is concave upward at $x = x_0$. If $f''(x_0) < 0$, then the graph of f is concave downward at $x = x_0$.

The reasoning behind this test is easy to understand. *Concave upward* means that f' is increasing. We know (see Equation 1, page 104) that a function with a positive derivative is increasing. Hence $f''(x_0) > 0$ means that f' is increasing at x_0, and so the graph of f is concave upward there. You may wish to supply similar reasoning to justify the test for *concave downward*.

Example 4. Is the graph of $f(x) = x^3 - 13x^2 + 5x - 1$ concave upward or concave downward when $x = 4$? When $x = 5$?

SOLUTION.

$$f'(x) = 3x^2 - 26x + 5 \quad \text{and} \quad f''(x) = 6x - 26$$
$$f''(4) = 6(4) - 26 = -2 \quad \text{and} \quad f''(5) = 6(5) - 26 = 4$$

By the second-derivative test for concavity, the graph of f is concave downward when $x = 4$ (because $f''(4) < 0$) and is concave upward when $x = 5$ (because $f''(5) > 0$).

The second derivative can be used to test if a critical point of a function is a relative maximum or relative minimum. Figure 16 shows a function f with critical points at $x = a$ and $x = b$. It shows that a relative minimum occurs when the graph is concave upward at the critical point, and that a relative maximum occurs when the graph is concave downward at the critical point.

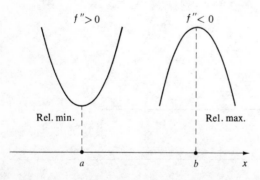

$f'' > 0$

$f'' < 0$

Rel. min.

Rel. max.

a b x

Figure 16

Second-derivative Test for Maxima-minima

Suppose x_0 is a critical point of f (that is, $f'(x_0) = 0$).

1.   If $f''(x_0) > 0$ then f has a relative minimum at x_0.
2.   If $f''(x_0) < 0$ then f has a relative maximum at x_0.
3.   If $f''(x_0) = 0$ this test provides no information.

A device for remembering this test is to draw two curves

and remember: "holds water—positive, spills water—negative." This diagram then suggests Figure 16 and the key points in the test:

f'' is positive means a relative minimum

f'' is negative means a relative maximum

Example 5. Find all the relative maxima and minima of the function $y = \frac{1}{2}x^2 - \frac{1}{3}x^3$.

SOLUTION.

(*Note:* See page 108 for an outline of the steps.)

Step 1. $\dfrac{dy}{dx} = x - x^2$.

Step 2. $x - x^2 = 0$, $x(1 - x) = 0$, $x = 0, 1$.

Step 3.   We now test if $\dfrac{d^2y}{dx^2}$ is positive or negative at the critical values found in Step 2. From Step 1,

$$\frac{d^2y}{dx^2} = 1 - 2x.$$

When $x = 0$, $\dfrac{d^2y}{dx^2} = 1 - 2(0) = 1$

When $x = 1$, $\dfrac{d^2y}{dx^2} = 1 - 2(1) = -1$.

Therefore, by the second-derivative test for max-min, y has a relative minimum when $x = 0$ (because $\dfrac{d^2y}{dx^2}$ is positive when $x = 0$) and y has a relative maximum when $x = 1$ (because $\dfrac{d^2y}{dx^2}$ is negative when $x = 1$).

Points of Inflection. A point on the graph where the concavity changes from upward to downward or from downward to upward is called a **point of inflection**. In Figure 15, there is a point of inflection where $x = c$. In Figure 13, there is a point of inflection where $x = b$. In Figure 13, we see that the curve crosses its tangent line at a point of inflection.

If x_0 is a point of inflection, then f'' is positive on one side of x_0 and negative on the other side. This can happen only if f'' is zero when $x = x_0$.

At a point of inflection the second derivative is zero.

Example 6 (Interpretation of inflection points). Figure 17 shows the graph of an increasing function $y = f(t)$ with an inflection point at t_0. Suppose that y represents the cost-of-living index at time t. Then t_0 is the time when politicians announce "inflation has been licked." Although the cost of living continues to increase after that time, it does so at a slower and slower rate.

Now suppose that y in Figure 17 represents the total number of items a subject has memorized x minutes after being presented with a very long list of items to learn. Psychologists call this graph a **learning curve;** they refer to t_0 as the time of **peak efficiency** because the rate of learning is a maximum then.

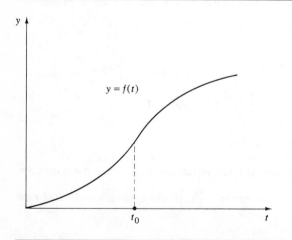

$y = f(t)$

t_0

t

Figure 17

Next, suppose that y in Figure 17 represents the total number of people who have been infected by a certain epidemic disease at time t. Epidemiologists refer to t_0 as the time when the epidemic is "under control"; at this time the rate of spreading of the disease begins to decrease.

Problem 21 in the next problem set illustrates the interpretation of inflection points for cost functions in manufacturing.

Example 7. A learning curve (see Example 6) has the equation

(2) $f(t) = 7.5t + 0.6t^2 - 0.02t^3$

for $0 \leq t \leq 25$. Find the inflection point (that is, the *time of peak efficiency*).

SOLUTION.
Calculate $f''(t)$ from Equation 2 and set it equal to zero. The result is $1.2 - 0.12t = 0$ or $t = 10$. To see that $t = 10$ is actually a point of inflection, we note that

$$f''(t) = 1.2 - 0.12t$$

is negative if $t < 10$ and positive if $t > 10$; therefore $t = 10$ is a point of inflection.

Problems

1. Find $f''(x)$ and $f''(2)$ if $f(x) = x^4 - 2x^3 - 5x$.

2. Find $g''(t)$ and $g''(1)$ if $g(t) = 3t^5 + t^4 - 3t^3$.

3. Find $F''(s)$ and $F''(3)$ if $F(s) = 2s^3 - 10s^2 + s$.

4. Find $f''(x)$ and $f''(0)$ if $f(x) = x^2(x - 5)$.

5. Find $\dfrac{d^2y}{dx^2}$ when $x = 5$ if $y = x^3 + 4x^2 - 3x$.

6. Find $\dfrac{d^2C}{dt^2}$ when $t = 10$ if $C = 0.01t^4 - t + 50$.

7. Find $\dfrac{d^2s}{dt^2}$ when $t = 6$ if $s = 3t^2 - t^3$.

8. Find $\dfrac{d^2w}{du^2}$ when $u = 0$ if $w = u^5 + 4u^2 + 6u$.

In Problems 9–12, a particle moves along a straight line so that the total distance s in feet traveled at time t in seconds is given by the indicated formula. Find the velocity and the acceleration at time $t = 4$.

9. $s = 16t^2$.

10. $s = t^3 + 3t$.

11. $s = 2t^3 + 6t^2 + 2t$.

12. $s = 2 + 5t + 3t^2$.

In Problems 12–17, test whether the graph of the given function is concave upward or concave downward at the indicated value.

12. $y = x^3 - 2x^2 + 3x$ when $x = 5$.

13. $y = x^4 - 2x^3 - 6x^2$ when $x = 2$.

14. $y = x^3 - 10x^2 + 3x$ when $x = 1$.

15. $y = x^2(x - x^2)$ when $x = 3$.

16. $y = (x + 1)(x - 1)$ when $x = 6$.

In Problems 17–20, test whether the given function has a relative maximum or minimum at the given critical point.

17. $y = 12x^2 - x^3$ at $x = 8$.

18. $y = 2x^3 - 9x^2 + 1$ at $x = 3$.

19. $y = 10x - x^2$ at $x = 5$.

20. $y = x^2 - 4x - 1$ at $x = 2$.

21. Let C be the total cost to manufacture q units of a commodity. The first derivative $\dfrac{dC}{dq}$ is the *marginal cost function*. In most businesses, the marginal cost function is U-shaped. Explain why, in such cases, the inflection point for the total cost function occurs when the marginal cost is a minimum.

22. The demand for steel in the United States was estimated by $q = 5 - 0.02p$.[10] The total revenue function is therefore $R = 5p - 0.02p^2$. At what price p is the predicted revenue R a maximum? What is the maximum R?

[10]R. H. Whitmann, *The Statistical Law of Demand for a Producer's Goods as Illustrated by the Demand for Steel*, Econometrica IV (1936): 138–152.

There are problems which ask for "the most profit," "the least cost," "the largest area," "the quickest time," "the shortest route," and so forth. These problems will be solved by finding a function whose maximum (or minimum) leads to the answer. In the following examples, note that the solution consists of two parts.

PART 1. Find a function whose maximum (or minimum) will lead to the answer.

PART 2. Use calculus to find the maximum (or minimum) of this function.

Part 1 is perhaps the more interesting. Creativity and originality are often needed to transform a practical problem into a purely mathematical one; there are no automatic rules for doing it. Part 2 is accomplished by the techniques that were treated in the preceding two sections.

Example 1. A farmer owns the land on one side of a straight river. He wishes to use part of it for a rectangular pasture. He plans to use the river bank for one side of the rectangle, and to build the other three sides with 1200 feet of fencing materials (see Figure 18). What is the largest area his pasture can have?

SOLUTION.
Consider the function $A = f(x)$ where A is the area that the pasture would have if the two sides perpendicular to the river are built to have length x feet (see Figure 18). If we can find the maximum of this function as x varies between 0 and 600, we will have solved the problem.

Figure 18

Let us find a formula for this function. Since the three sides of the pasture have total length 1200 feet, the side parallel to the river has length $1200 - 2x$. Therefore the area $A =$ (length) $\times$ (width) is $A = (1200 - 2x)x$ or

(1) $A = 1200x - 2x^2$.

We have completed Part 1 of the problem. The next part is to use calculus to find the maximum of Equation 1. Therefore differentiate Equation 1, set the derivative equal to zero, and solve for x.

(2) $\dfrac{dA}{dx} = 1200 - 4x,$

(3) $0 = 1200 - 4x,$

$\quad 4x = 1200$

(4) $x = 300$

There are several ways to see that the critical value $x = 300$ actually makes A attain a maximum. We can make a rough sketch of the function in Equation 1 (see Figure 19) to see it; we can note from Equation 2 that A is increasing for

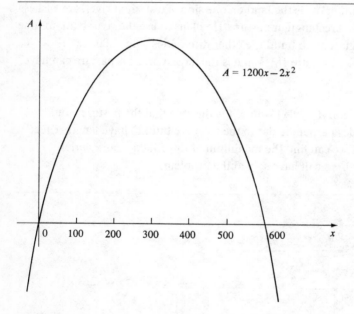

$A = 1200x - 2x^2$

Figure 19

$x < 300$ and is decreasing for $x > 300$; or we can consider x to vary in the interval $0 \leq x \leq 600$ and simply compare the values of A at the endpoints and the critical points.

(5)   When $x = 0, A = 0$

(6)   When $x = 300, A = (1200)(300) - 2(300)^2 = 180,000$

(7)   When $x = 600, A = 0.$

Thus the maximum area occurs when $x = 300$ feet. The value of this maximum area is found by putting $x = 300$ in Equation 1. The result, $A = 180,000$ square feet (see Equation 6), is the largest area this pasture can have.

Example 2 (Monopoly of production). A monopolist maximizes his **profit** P by adjusting the **quantity** q which he produces. The **demand function** $p = f(q)$ determines the best **unit price** p he can charge in order to sell all q units of his product. Usually f is a decreasing function because people buy more when the price is low. His **total cost** $C = h(q)$ to produce q units is an increasing function. His profit P if he produces q units is

(8)
$$\begin{cases} P = (\text{revenue}) - (\text{total cost}) \\ \quad = (\text{quantity produced}) \cdot (\text{unit cost}) - (\text{total cost}) \\ \quad = q \cdot p - C = qf(q) - h(q). \end{cases}$$

The demand function for steel in the United States was estimated to be

(9) $p = f(q) = 250 - 50q$

in suitable units.[11] The total cost to produce steel was estimated to be

(10) $C = h(q) = 182 + 56q$[12].

Assume the existence of a steel monopoly and determine how much steel q it should produce in order to maximize profit.

SOLUTION.
By Equations 8, 9, and 10, we see that the profit is $P = q(250 - 50q) - (182 + 56q)$, or

(11) $P = -50q^2 + 194q - 182.$

[11]Whitmann, *The statistical law of demand for a producers goods*, pp. 138–152.

[12]T. O. Yntema, *United States Steel Corporation*, *TNEC Papers*, Vol. 1 (New York: United States Steel Corporation, 1940.)

Thus

$$\frac{dP}{dq} = -100q + 194,$$

and

$$\frac{d^2P}{dq^2} = -100.$$

We set $\frac{dP}{dq}$ equal to zero and find that $q = 1.94$ is a critical point. Since $\frac{d^2P}{dq^2}$ is negative, this critical point is a relative maximum. There are several ways to see that it is an absolute maximum for all values $q \geqq 0$. For example, the graph of Equation 11 is a parabola opening downward. Or, we can see from $\frac{dP}{dq} = 100(1.94 - q)$ that P is increasing when $q < 1.94$ and decreasing when $P > 1.94$. We can conclude that profit is maximized when $q = 1.94$.

Example 3. A square sheet of tin has sides of length 30 inches. The sheet is to be bent into a rectangular planting box, open on the top, by cutting out small squares from each corner and then bending up the edges (see Figure 20). If the box is to hold the largest possible amount of soil, what size should it be?

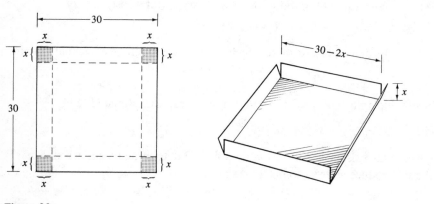

Figure 20

SOLUTION.
Let V be the volume in cubic inches of the box that is obtained when squares of side x are cut from the corners. If we can find the value of x which makes V attain a maximum, we could answer the question. Note that x must be between 0 and 15.

To obtain a formula for V, look again at Figure 20. We see that the box will have length, width, and height equal to $30 - 2x$, $30 - 2x$, and x, respectively. Hence $V = (30 - 2x)^2x$ or, after multiplying,

(12) $V = 900x - 120x^2 + 4x^3$.

Now we use calculus to find the maximum of V for x between 0 and 15. Differentiate Equation 12:

$$\frac{dV}{dx} = 900 - 240x + 12x^2.$$

Next set the derivative equal to zero.

$$0 = 900 - 240x + 12x^2 \quad \text{or} \quad x^2 - 20x + 75 = 0.$$

Finally, solve for x.

$$(x - 15)(x - 5) = 0$$
$$x = 5, 15.$$

Since x varies between 0 and 15, we can find the maximum by testing the endpoints $x = 0$ and $x = 15$ and critical points $x = 5$ and $x = 15$.

When $x = 0$, $V = 0$.

When $x = 5$, $V = 900 \cdot 5 - 120 \cdot 5^2 + 4 \cdot 5^3 = 2000$.

When $x = 15$, $V = 0$.

Thus V is a maximum when $x = 5$; the dimensions of the box will then be 20 inches by 20 inches by 5 inches. (A rough graph of the function V in Equation 12 is shown in Figure 21.)

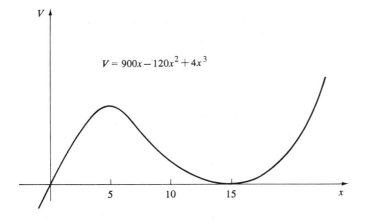

Figure 21

Example 4. At sunrise, ship A is 100 nautical miles due east of ship B. Ship A is steaming due east at 20 knots (a *knot* is a nautical mile per hour). Ship B is steaming due north at 10 knots. When are the ships closest together, and what is their closest distance?

SOLUTION.

Let s be the distance between the ships t hours after sunrise. Then s is a function of t. The minimum of this function will lead to the answer.

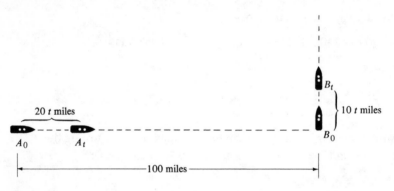

Figure 22

Figure 22 shows the position of the ships at sunrise. A_t and B_t are their positions t hours later. Clearly, the distances A_0A_t and B_0B_t are $20t$ miles and $10t$ miles respectively. We can calculate the other distances (see Figure 23) and thus, by the Pythagorean theorem,

$$(13) \quad s = \sqrt{(10t)^2 + (100 - 20t)^2}.$$

Equation 13 cannot be differentiated by our present techniques (the necessary techniques will be presented in Chapter 4). This difficulty can be

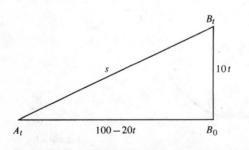

Figure 23

overcome by a trick. Note that the value of t that makes s a minimum will also make s^2 a minimum. So we look for the minimum of this simpler function

$$f(t) = (10t)^2 + (100 - 20t)^2 = 500t^2 - 4000t + 100^2.$$

Differentiate, set the result equal to zero, and solve for t:

$$f'(t) = 2 \cdot 500t - 4000$$
$$0 = 1000t - 4000$$
$$1000t = 4000$$
$$t = 4.$$

The nature of the original problem makes it evident that a minimum exists and hence it must occur when $t = 4$, the only possibility. From Equation 13,

$$\text{when } t = 4, s = \sqrt{40^2 + 20^2} = \sqrt{2000}.$$

Therefore, the ships are closest exactly 4 hours after sunrise at a distance of $\sqrt{2000} \approx 44.72$ nautical miles.

Problems

1. A rancher has 600 yards of fence material. He wishes to fence a rectangular corral and then divide it in half by means of a fence parallel to one of the sides. If he uses all of his fencing, what is the largest corral he can build?

2. A show room is to have three walls of concrete and one of glass. The glass wall costs $100 per foot of length and the concrete wall costs half as much. If $30,000 is budgeted for the cost of four walls, what is the largest area that can be created?

3. A rectangle is inscribed in a semicircle of radius R. The rectangle has one side lying along the diameter (see Figure 24). If the area of the rectangle is to be as large as possible, what should its length and width be? (*Hint:* Consider the area squared, rather than the area, in order to simplify the calculus.)

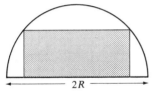

$2R$ **Figure 24**

4. A rectangle is inscribed in an equilateral triangle (see Figure 25). One edge of the rectangle lies along a side of the triangle. What is the maximum area of the rectangle if the triangle has sides of unit length?

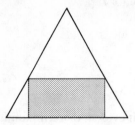

Figure 25

5. A garden has the shape of a rectangle with a semicircle joined to one edge (see Figure 26). There is a path which borders the garden, and which has total length 400 yards. What dimensions give the garden the largest area?

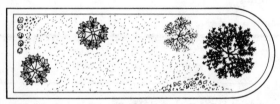

Garden   **Figure 26**

6. A man is walking due west along a highway at 4 miles per hour. A woman is walking due south along a perpendicular highway at 3 miles per hour. At a certain instant, they are each headed toward, and 25 miles away from, the point of intersection of the highways. How close will they come to each other if they continue their present speed and direction? (*Note:* To simplify the calculus, work with the distance squared rather than the distance between them.)

7. The demand function for hamburgers at a town's only restaurant is $p = -0.2q + 270$ for q in the range $100 \leqq q \leqq 1000$; that is, p in cents is the best price the management can charge in order to sell q hamburgers each day. The cost in cents to make q hamburgers is $C = 30q$. To maximize the profit, how many hamburgers should be sold each day and at what price?

8. An airline has a monopoly on flights from city A to city B. The demand function for tickets on this route is $p = 555 - 0.05q$ for q in the range $100 \leq q \leq 10,000$; that is, p in dollars is the best price the airline can charge in order to sell q tickets each week. The total cost to the airlines to transport q people per week is $C = 100,000 + 50q$ dollars. In order to maximize profit, how many passengers per week should the airlines accommodate? What price should the tickets be?

9. The owner of a baseball team knows that he can sell all 20,000 tickets for bleacher seats if he charges 50¢ per ticket. For each 10¢ he charges in excess of that, he estimates that 1000 seats will go unsold. To maximize his receipts, how much should he charge?

10. The city bus company is losing money and so contemplates a fare increase. The present fare is 50¢. Suppose 1,000,000 passengers a day would use the existing buses if the fare were free, but for each 25¢ of fare, 312,500 of these people stop using the bus. What is the optimum fare to maximize the company's revenue?

<hr>

Summary

Velocity, acceleration, current, marginal cost, and marginal tax are just a few examples of the uses of the derivative outside of mathematics. The nonrigorous interpretation of the meaning of *marginal* as *extra* in economics is often very useful as an aid to our intuition.

The first and second derivative can be used to test if a function is increasing, decreasing, concave upward, or concave downward without the need to plot the graph.

If $\dfrac{dy}{dx} > 0$, then y is increasing.

If $\dfrac{dy}{dx} < 0$, then y is decreasing.

If $\dfrac{d^2y}{dx^2} > 0$, then the graph of y is concave upward.

If $\dfrac{d^2y}{dx^2} < 0$, then the graph of y is concave downward.

The first and second derivatives are used to locate and identify maxima and minima of a function. The first derivative is used to find the critical points (Steps 1 and 2 on page 108). Step 3, examining each critical point to determine if the function has a relative maximum or minimum there, can be done by (1) making a rough sketch of the graph, (2) checking the signs $(+ -, - +, + +,$ or $- -)$ of the first derivative, or (3) using the second derivative test for max-min (see page 121).

Absolute maxima and minima can be found by comparing the relative maxima or minima and the values of the function at the endpoints, if any.

Miscellaneous Problems

1. A particle moves along a straight path. After t seconds, it has traveled s feet. If $s = t^3 - 3t^2 + 3t$, find the velocity and the acceleration at time $t = 2$.

2. The cost c in dollars for fuel to run a light generator for eight hours at a capacity x in percentage of full capacity has been estimated by the approximate formula[13] $c = 16.68 + 0.125x + 0.004x^2$. Find the marginal cost $\left(\text{that is } \dfrac{dc}{dx}\right)$ when $x = 50$. Interpret this as an "extra cost."

3. The demand law for cotton during 1915–1929 has been estimated as $q = 6.4 - 0.3p$, where the price p and quantity of cotton q are in convenient units.[14] The revenue $R = p \cdot q$ is therefore given by

$$R = \frac{(6.4 - q)q}{0.3}.$$

   a. Find the marginal revenue $\dfrac{dR}{dq}$ when $q = 3$.
   b. Is the revenue R increasing or decreasing when $q = 3$?
   c. Is the revenue R increasing or decreasing when $q = 4$?

4. Is the function $y = 10 - 3x + 5x^2 - x^3 + x^4 - x^5$ increasing or decreasing when $x = 0$? When $x = 1$?

5. A Chicago meat packing house has a weekly payroll y that is correlated to the weekly total live weight of hogs x in millions of pounds. The relationship between x and y is estimated to be

[13]J. A. Nordin, *Note on a light plant's cost curve,* Econometrica XV (1947): 231ff.
[14]Schultz, *The Theory and Measurement of Demand.*

$y = 0.32 - 0.007x + 0.02x^2$ for a certain range of values x.[15] Find where the function y attains an absolute minimum and where it attains an absolute maximum for x in the interval $0 \leqq x \leqq 1$.

6. Is the graph of the function in Problem 4 concave upward or concave downward when $x = 0$? When $x = 1$?

7. Suppose a particle moves along the x-axis so that its x-coordinate after t minutes is $x = 60t - t^2$. How far to the right of the origin does it get? At what time does it get there?

8. Find all relative maxima and minima of the function $y = 3 - 6x^2 + 2x^3$. Find the point of inflection.

9. Find the absolute minimum and maximum of the function in Problem 8 if x is constrained to lie in the interval $-2 \leqq x \leqq 2$.

10. A quantity x of a drug is administered to a patient. His reaction R (heart rate, temperature, etc., measured in suitable units) may be modeled by the formula $R = Ax^2 - Bx^3$, where A and B are constants.[16] Suppose $A = 150$, $B = 1$, and x lies in the interval $0 \leqq x \leqq 100$. Show each of the following.

    a. R increases for x in this interval.

    b. The maximum value of $\dfrac{dR}{dx}$ for x in this interval is $\dfrac{dR}{dx} = 7500$.

11. Suppose a bacteria culture, when kept at a temperature T in degrees centigrade, will reach a stable size P where $P = 90T^2 - T^3 + 100,000$ if $0 \leqq T \leqq 100$. Find the absolute maximum and absolute minimum of P for this range of T, and find the temperatures at which these extreme values occur.

12. Sketch the graph of $y = 3x^2 - \frac{1}{6}x^4$. Find the critical points, the inflection points, and show where the function is increasing, decreasing, concave upward and concave downward.

13. Sketch the graph of $y = x^3 - 12x$ by finding all critical points, points of inflection, and where the function is increasing, decreasing, concave upward and concave downward.

[15] W. H. Nichols, *Labor Productivity Functions in Meat Packing.* (Chicago: University of Chicago Press, 1948).

[16] K. R. Rebman, H. Slater, and R. M. Thrall, *Some Mathematical Models in Biology.* (Ann Arbor: University of Michigan Press, 1967), p. 59.

four

Further Techniques of Differentiation

In this chapter, you will learn some additional rules for computing derivatives. They will enable you to differentiate functions which cannot be differentiated by the techniques of Chapter 2, such as

$$f(x) = x\sqrt{x}, \qquad g(x) = \frac{2x + 1}{3x + 2}, \qquad \text{and} \qquad h(x) = \sqrt{2x^2 + 1}.$$

Section 1 □ **The Product and Quotient Rules.**

Section 2 □ **Differentiation of Negative and Fractional Powers.**
Review of the algebra of exponents.

Section 3 □ **The Chain Rule.** Rule for differentiating a power of a function.

Section 4 □ **Implicit Differentiation.**

Section 5 □ **Approximation; Differentials.**

Summary □ **Miscellaneous Problems.**

Recall the rule for differentiating a sum (see Rule 5, page 77): the derivative of a sum is the sum of the derivatives. This rule shows, for example, that

$$\text{if} \quad y = x + x^2, \quad \text{then} \quad \frac{dy}{dx} = 1 + 2x.$$

Could a similar rule hold for products? If "the derivative of a product is the product of the derivatives" were a correct rule, then we would have

$$\text{if} \quad y = (x)(x^2), \quad \text{then} \quad \frac{dy}{dx} = (1)(2x).$$

This is certainly false, because the derivative of x^3 is not $2x$. Here is the correct rule for differentiating a product.

The Product Rule
Let u and v be functions of x.

$$(1) \quad \text{If} \quad y = uv, \quad \text{then} \quad \frac{dy}{dx} = (u)\frac{dv}{dx} + (v)\frac{du}{dx}.$$

This rule should be memorized by the words, "the derivative of a product is the first factor times the derivative of the second factor plus the second factor times the derivative of the first factor."

Example 1. Use the product rule to differentiate each product.[1]

a. $y = x^3(x^2 + 3x)$
b. $y = (3x + 1)(x^2 + 2x - 3)$
c. $w = (1 + t + 2t^2)(3t - 2)$

Do not simplify your answers.

SOLUTION.
a. Apply the product rule with $u = x^3$ and $v = x^2 + 3x$. The result is

---

[1]Note that if the factors in these products are multiplied together, then the product rule would not be needed. In this way, the use of the product rule can be avoided in the problems at the end of this section. You should practice the rule at this time, not avoid it—later on, with functions such as $(x + 1)^{1/2}(x + 2)^{1/3}$ or xe^x, it will be unavoidable.

$$\frac{dy}{dx} = x^3(2x + 3) + (x^2 + 3x)(3x^2).$$

In practice, this differentiation is done by mentally repeating the verbal form of the rule:

(2)   Deriv. of Product = (First) · (Deriv. of Second)
                            + (Second) · (Deriv. of First)

Since

$$y = x^3(x^2 + 3x),$$

(2) becomes

$$\frac{dy}{dx} = (x^3)(2x + 3) + (x^2 + 3x)(3x^2).$$

b. From

$$y = (3x + 1)(x^2 + 2x - 3)$$

and (2) we have

$$\frac{dy}{dx} = (3x + 1)(2x + 2) + (x^2 + 2x - 3)(3).$$

c. From

$$w = (1 + t + 2t^2)(3t - 2)$$

and (2) we have

$$\frac{dw}{dt} = (1 + t + 2t^2)(3) + (3t - 2)(1 + 4t).$$

Proof of the Product Rule.   As x changes from x_0 to $x_0 + \Delta x$, let us say that

   y changes from y_0 to $y_0 + \Delta y$,

   u changes from u_0 to $u_0 + \Delta u$,

   v changes from v_0 to $v_0 + \Delta v$.

Then

(3) $\Delta y = (u_0 + \Delta u)(v_0 + \Delta v) - u_0 v_0.$

When Equation (3) is expanded, the terms $u_0 v_0$ cancel and we have

(4) $\Delta y = u_0 \Delta v + v_0 \Delta u.$

Divide both sides of Equation (4) by Δx:

(5) $\dfrac{\Delta y}{\Delta x} = (u_0)\dfrac{\Delta v}{\Delta x} + (v_0)\dfrac{\Delta u}{\Delta x}.$

Now let $\Delta x \to 0$ in Equation (5); the result is $\dfrac{dy}{dx} = (u_0)\dfrac{dv}{dx} + (v_0)\dfrac{du}{dx}.$ Since x_0 can be any value x, we may drop the zero subscripts and we arrive at Rule 1.

The Quotient Rule.   The rule for differentiating a quotient can be found with the help of the product rule (see Problem 24).

The Quotient Rule

Let u and v be functions of x.

(6)   If $y = \dfrac{u}{v}$, then $\dfrac{dy}{dx} = \dfrac{(v)\dfrac{du}{dx} - (u)\dfrac{dv}{dx}}{v^2}$

This rule may be memorized with the phrase, "the derivative of a quotient is the bottom times the derivative of the top minus the top times the derivative of the bottom, all over the bottom squared."

Example 2.   Differentiate; do not simplify your answer.

a. $y = \dfrac{3x + 1}{x^2 - 2}.$

b. $s = \dfrac{t^4}{3t + 1}.$

c. $w = \dfrac{5p^3}{7p^2 - 1}.$

SOLUTION.

In each case, we think

(7)   (Deriv. of quotient)

$$= \dfrac{(\text{Bottom})(\text{Deriv. of Top}) - (\text{Top})(\text{Deriv. of Bottom})}{(\text{Bottom})^2}.$$

The result of using Equation (7) is:

a. $\dfrac{dy}{dx} = \dfrac{(x^2 - 2)(3) - (3x + 1)(2x)}{(x^2 - 2)^2}$.

b. $\dfrac{ds}{dt} = \dfrac{(3t + 1)(4t^3) - (t^4)(3)}{(3t + 1)^2}$.

c. $\dfrac{dw}{dp} = \dfrac{(7p^2 - 1)(15p^2) - (5p^3)(14p)}{(7p^2 - 1)^2}$.

Problems

In Problems 1–14, use the product rule or quotient rule to differentiate the given functions; do not simplify your answer.

1. $y = x^3(5x - 1)$.

2. $p = (2q + 1)(q^4 - q^2)$.

3. $s = (t^3 - 5t^2 + 1)(t^4 + 3)$.

4. $y = \left(\dfrac{1}{2}x^4 - 3\right)(x^5 + 1)$.

5. $y = \dfrac{x^2}{1 + x}$.

6. $w = \dfrac{2u - 1}{4u + 3}$.

7. $f(x) = \dfrac{x^2}{2x + 1}$.

8. $p = \dfrac{2q^3}{3q - 1}$.

9. $P = \dfrac{3t}{1 + 2t}$.

10. $v = \dfrac{2x^2 - 1}{3x + 2}$.

11. $f(x) = 3x(x^2 - 1)$.

12. $g(t) = (3t + 1)(t + 2)$.

13. $H(u) = \dfrac{1}{2u + 1}$.

14. $F(s) = \dfrac{s - 1}{s + 1}$.

In Problems 15–18, find the derivative at the indicated value.

15. $y = (x^3 + 1)(3x^2 + x)$ when $x = 2$.

16. $p = q^3(1 - 7q)$ when $q = -1$.

17. $w = \dfrac{x}{3x + 1}$ when $x = 0$.

18. $v = \dfrac{u}{u^2 + 5}$ when $u = 1$.

19. Is the graph of $y = (x + 1)/(x - 1)$ increasing or decreasing when $x = 5$?

20. Let $C = q/(30 - q^2)$ be the total cost to produce q units of a commodity. Find the marginal cost when the production level is 5 units (that is, find $\dfrac{dC}{dq}$ when $q = 5$).

21. Suppose that t minutes after a glucose injection, the blood sugar level y is given by $y = (t + 100)/(3t + 1)$.
    a. Show that y is decreasing for all $t \geq 0$.
    b. How fast is the blood sugar level falling 33 minutes after the injection?

22. Consider the function

$$f(x) = \frac{1}{Ax - x^2}, \qquad A \text{ a constant.}$$

Show that for x in the interval $0 \leq x \leq A$, f attains its absolute minimum when $x = A/2$, and that the value of this minimum is $2/A$.

23. Find the positive number such that the sum of the number and its reciprocal is a minimum.

24. Use the Product Rule to discover the Quotient Rule as follows: write $y = u/v$ in the form $vy = u$, differentiate both sides with respect to x, and then solve for $\dfrac{dy}{dx}$.

25. There is a function $y = L(x)$ such that $\dfrac{dy}{dx} = \dfrac{1}{x}$ and $L(1) = 3$. Find $f'(1)$ for the function $f(x) = x \cdot L(x)$.

Section 2 □ Differentiation of Negative and Fractional Powers

---

In this section, we shall show how to differentiate functions such as

$$y = \sqrt{x}, \qquad y = \frac{1}{x^3} + 2x^{1.5}, \qquad \text{and} \qquad y = x \sqrt[3]{x}.$$

The technique for differentiating such functions is not difficult, but it requires the ability to use negative and fractional exponents.

Review of the Algebra of Exponents.[2]   What is the definition of a^n? The answer is best given by treating separately the cases that (1) n is a positive integer, (2) n is a positive rational number (that is, a fraction), and (3) n is a negative number.

If n is a positive integer, then a^n means the product of a with itself n times.

$$5^2 = 5 \cdot 5 = 25.$$

$$(-2)^4 = (-2)(-2)(-2)(-2) = 16.$$

$$\left(-\frac{2}{3}\right)^3 = \left(-\frac{2}{3}\right)\left(-\frac{2}{3}\right)\left(-\frac{2}{3}\right) = -\frac{8}{27}.$$

$$(6x)^2 = (6x)(6x) = 6 \cdot 6 \cdot x \cdot x = 36x^2.$$

If $a \neq 0$, then a zero exponent of a is defined by

$$a^0 = 1, \qquad a \neq 0.$$

Laws of Exponents

The following basic laws of exponents hold. (In Law 5, assume temporarily that $m \geq n$.)

LAW 1 $a^{m+n} = a^m a^n.$

LAW 2 $(a^m)^n = a^{mn}.$

LAW 3 $(ab)^n = a^n b^n.$

LAW 4 $\left(\dfrac{a}{b}\right)^n = \dfrac{a^n}{b^n}$

LAW 5 $\dfrac{a^m}{a^n} = a^{m-n}$

If p and q are positive integers, then one defines $a^{p/q}$ to be the qth root (the positive root if there are two) of the number a^p.

(1) $a^{p/q} = \sqrt[q]{a^p}$

We could also define $a^{p/q} = (\sqrt[q]{a})^p$, which is equivalent to Equation 1.

[2]See also the algebra review on pages 359–371.

For example:

$$8^{2/3} = \sqrt[3]{8^2} = \sqrt[3]{64} = 4.$$

$$(-27)^{1/3} = \sqrt[3]{-27} = -3.$$

$$4^{1.5} = 4^{1\frac{1}{2}} = 4^{3/2} = \sqrt{4^3} = \sqrt{64} = 8.$$

$$16^{0.25} = 16^{1/4} = \sqrt[4]{16} = 2.$$

The five Laws of Exponents on page 143 hold also when n is a positive rational number.

Negative exponents are defined by

$$(2) \qquad a^{-n} = \frac{1}{a^n};$$

in Equation 2, n may be any positive rational number, but $a \neq 0$. For example:

$$5^{-2} = \frac{1}{5^2} = \frac{1}{25}.$$

$$4^{-3/2} = \frac{1}{4^{3/2}} = \frac{1}{\sqrt{4^3}} = \frac{1}{\sqrt{64}} = \frac{1}{8}.$$

$$\left(\frac{4}{9}\right)^{-1/2} = \frac{1}{(4/9)^{1/2}} = \frac{1}{\sqrt{4/9}} = \frac{1}{2/3} = \frac{3}{2}.$$

Laws 1–5 continue to hold when n is any rational number—positive, negative, or zero.

Note that a^n now has a meaning when n is any terminating decimal. For example:

$$a^{2.74} = a^{274/100} = \sqrt[100]{a^{274}}.$$

However, we have not yet defined expressions such as

$$5^{\sqrt{2}} \qquad \text{or} \qquad 7^{\pi}.$$

In order to define a^n when n is a real number that may not be rational, the notion of limit must be used. We shall return to this topic in Chapter 5.

Example 1.   Write each expression in the exponent form x^n.

a. $(x^5)(x^3).$

b. $\dfrac{x^2}{x^3}.$

c. $\left(\dfrac{1}{x^4}\right).$

d. $\left(\dfrac{1}{x^4}\right)^3.$

e. $x\sqrt{x}.$

f. $x\sqrt[3]{x}.$

g. $\sqrt{x^{6.4}}.$

h. $\dfrac{x^2}{\sqrt{x}}.$

SOLUTION.

a. $(x^5)(x^3) = x^{5+3} = x^8$.

b. $\dfrac{x^2}{x^3} = x^{2-3} = x^{-1}$.

c. $\dfrac{1}{x^4} = x^{-4}$.

d. $\left(\dfrac{1}{x^4}\right)^3 = \dfrac{1^3}{(x^4)^3} = \dfrac{1}{x^{12}} = x^{-12}$.

e. $x\sqrt{x} = x^1 \cdot x^{1/2} = x^{3/2}$.

f. $x\sqrt[3]{x} = x^1 \cdot x^{1/3} = x^{4/3}$.

g. $\sqrt{x^{6.4}} = (x^{6.4})^{1/2} = x^{(6.4)(1/2)} = x^{3.2}$.

h. $\dfrac{x^2}{\sqrt{x}} = \dfrac{x^2}{x^{1/2}} = x^{2-(1/2)} = x^{3/2}$.

Rule for differentiating x^n. Rule 1 on page 73 holds for any rational number n.

Rule for Differentiating x^n.

(3)   If $y = x^n$ then $\dfrac{dy}{dx} = nx^{n-1}$.

We shall be able to give a general proof of this rule in Chapter 5 when we can make use of logarithms. This rule allows us to differentiate many new functions. In the following examples, note that the first step is to rewrite each algebraic expression so that it is in exponential form.

Example 2. Differentiate each function.

a. $y = \dfrac{1}{x^2}$.

b. $y = \dfrac{3}{x^2} + \dfrac{5}{x} + x^3$.

c. $w = \dfrac{1}{\sqrt{u}}$.

d. $p = 4q^{1.3} - \dfrac{2}{q^{3.1}}$.

SOLUTION.

a. We first write the function in exponent form.

$$y = x^{-2}.$$

Now Rule 3 can be applied with n replaced by -2; the result is

(4) $\quad \dfrac{dy}{dx} = (-2)x^{(-2)-1}.$

Since $(-2) - 1 = -3$, Equation 4 can be written as

$$\frac{dy}{dx} = -2x^{-3} \qquad \text{or} \qquad \frac{dy}{dx} = \frac{-2}{x^3}.$$

b. Write the function in the form

(5) $\; y = 3x^{-2} + 5x^{-1} + x^3.$

Differentiate Equation 5 by using the rule for differentiating sums of functions; use Rule 3 to differentiate each function in the sum. The result is

$$\frac{dy}{dx} = 3(-2)x^{(-2)-1} + 5(-1)x^{(-1)-1} + 3x^2,$$

or

$$\frac{dy}{dx} = -6x^{-3} - 5x^{-2} + 3x^2.$$

If you wish, you can write the answer in the form

$$\frac{dy}{dx} = 3x^2 - \frac{5}{x^2} - \frac{6}{x^3}.$$

c. From $w = u^{-1/2}$, we have

$$\frac{dw}{du} = -\frac{1}{2}u^{(-1/2)-1}$$

or

$$\frac{dw}{du} = -\frac{1}{2}u^{-3/2}.$$

d. From $p = 4q^{1.3} - 2q^{-3.1}$, we obtain

$$\frac{dp}{dq} = 4(1.3)q^{1.3-1} - 2(-3.1)q^{-3.1-1}$$

or

$$\frac{dp}{dq} = 5.2q^{0.3} + 6.2q^{-4.1}.$$

Example 3 (the inventory problem). This example illustrates a famous application of calculus to business problems. A car manufacturer sells 100,000 limousines each year in addition to several million standard-size cars. When the limousine assembly line operates, it produces limousines faster than they can be sold. Therefore, the manufacturer produces them in "runs." He stores the limousines in warehouses and sells them from the warehouses at a uniform rate during the year. Thus the assembly line operates only intermittently.

The manufacturer does not wish to produce too many limousines in each run because then the storage costs would be too high. On the other hand, he does not want to produce too few because that requires many runs and there are large setup costs involved in restarting the assembly line for each run. He must determine the optimum size of the runs in order to minimize the sum of the storage costs and set-up costs.

Suppose that the

(6)   set-up cost = \$64,000 for each run

and that the

(7)   unit storage cost = \$200 per limousine per year.

The only variable is the number of limousines produced in each run; let

(8) x = number of limousines produced in each run.

Since there are to be 100,000 limousines produced in a year, we have

(9) $\dfrac{100,000}{x}$ = number of runs per year.

Our problem is to find x so that

(10) $C(x)$ = yearly set-up costs + yearly storage costs

will be a minimum. From Equations 6 and 9 we see that

(11)   yearly set-up costs = $(64,000)\dfrac{100,000}{x}$.

Storage charges will obviously be least if runs begin when warehouse inventory is zero; the average number of limousines in storage will then be $x/2$ (see Figure 1) and so

(12)   yearly storage costs = (average no. of stored limousines)

$\times$ (unit storage cost)

$= \left(\dfrac{x}{2}\right)(200).$

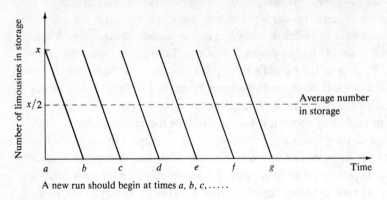

A new run should begin at times $a, b, c, \dots$

Figure 1

Put Equations 11 and 12 into Equation 10.

(13) $\quad C(x) = \dfrac{(64{,}000)(100{,}000)}{x} + 100x = (64 \times 10^8)x^{-1} + 10^2 x.$

To find the minimum of C, set $C'(x)$ equal to zero.

(14) $\quad C'(x) = -(64 \times 10^8)x^{-2} + 10^2$

$\qquad\quad 0 = -(64 \times 10^8)x^{-2} + 10^2$

$\qquad x^2 = 64 \times 10^6$

$\qquad x = \pm 8 \times 10^3$

The negative critical point is of no interest, so $x = 8 \times 10^3 = 8000$ is the only candidate for a minimum. There are several ways to see that this value of x makes $C(x)$ an absolute minimum. For example, Equation 14 shows that C decreases for $0 < x < 8000$ and increases for $x > 8000$. We can conclude that the manufacturer should produce 8000 limousines per run.

Problems

In Problems 1–12, differentiate the given functions.

1. $y = \dfrac{1}{x^3}.$

2. $w = t^{5.1} + 3t^{1.2}.$

3. $f(s) = \sqrt[3]{s}.$

4. $y = \dfrac{7}{\sqrt{x}}.$

5. $C(x) = \dfrac{12}{x} + 5x.$

6. $L(t) = t^2 + \dfrac{1}{t^2}.$

7. $p = \dfrac{7.3}{q^{2.1}}.$

8. $C(p) = \dfrac{2}{\sqrt[5]{p}} + \sqrt[3]{p}.$

9. $y = (5x^{-7} + 1).$

10. $f(x) = (x^{-2.5} - 2).$

11. $C(x) = \dfrac{2}{(x^{2.4} + 1)}.$

12. $y = x^{2.3} + 3x^{1.2} - 1.$

13. If $p = \sqrt[3]{x}$ find $\dfrac{dp}{dx}$ when $x = 8.$

14. If $F(q) = q^2 + \dfrac{3}{q^4}$, find $F'(q)$ and $F'(1).$

15. If $p = \dfrac{8}{q^2}$, find $\dfrac{d^2p}{dq^2}$ when $q = 2.$

16. What positive number x makes y a minimum if $y = x^2 + \dfrac{250}{x}$?

17. Among all positive numbers x, which one makes the expression $52 - x^2 - 16/x$ as large as possible?

18. The Italian economist Vilfredo Pareto found that the distributions of upper class incomes in various countries obey a law of the form $N = A/x^{\alpha}$ where A and α are constants which depend on the time and place, and N is the number of people with incomes of x or higher. In the United States in 1918, the values of A and α were $A = 813 \times 10^9$ and $\alpha = 1.48.$[2]

    For these values of A and α, find $\dfrac{dN}{dx}.$

19. The demand for potatoes in the United States during 1915–29 was estimated as $q = 12.05/p^{0.3}$, where p and q are in suitable units.[3] The *total revenue R* is defined by $R = p \cdot q.$

    a. Find $\dfrac{dq}{dp}.$

    b. Find the *marginal revenue* $\dfrac{dR}{dp}$ in terms of $p.$

[2]V. von Szeliski, *Econometrica* II (1934), p. 215 ff.

[3]H. Schultz, *The Theory and Measurement of Demand* (Chicago: University of Chicago Press, 1938.)

20. For winged animals of fixed biological structure but varying wing span L, the power P required for flying is $P = kL^{3.5}$ and the power that the animal can generate is $P = cL^2$, where k and c are positive constants.[4] Show that the quantity

$$f(L) = cL^2 - kL^{3.5} \ (L \geq 0)$$

has a maximum when $L = (2c/3.5k)^{2/3}$. (*Note:* The animal cannot fly at all if $f(L) < 0$.)

21. A book is to be printed with pages of area 50 square inches. There must be $\frac{1}{2}$-inch margins on the left and right edges of each page, and 1-inch margins at the top and bottom of each page. What page dimensions will allow the greatest printed area?

22. A fishing boat is homeward bound. At a speed of v miles per hour, the fuel costs will be $v/9$ dollars per mile. The refrigeration system costs 25 dollars per hour to operate. What speed should the captain select to minimize total costs?

Section 3 □ The Chain Rule

We have, as yet, no simple method for differentiating a function such as $y = (x^2 + 1)^{15.3}$. However, there is a rule for differentiating such functions. It is called the chain rule because it depends on the idea of forming "chains of functions" (also called "compositions of functions").

This idea can be explained by means of a simple example. Let us find the derivative $\dfrac{dy}{dx}$ of the function

(1) $y = (x^2 + 1)^{15.3}$.

We introduce an auxilary variable, call it u, for example, and use it to write the original function in Equation 1 as a chain of two functions

(2) $y = u^{15.3}$

and

(3) $u = x^2 + 1$.

[4]J. M. Smith, *Mathematical Ideas in Biology* (Cambridge: Cambridge University Press, 1971), p. 14.

Note how Equations 2 and 3 together assign a value of y to each x, and the resulting function y of x is the original function in Equation 1. Now differentiate each function in the chain: from Equation 2, y is a function of u with derivative

(4) $\quad \dfrac{dy}{du} = 15.3u^{14.3}$,

and from Equation 3, u is a function of x with derivative

(5) $\quad \dfrac{du}{dx} = 2x$.

One might guess that the formal identity

(6) $\quad \dfrac{dy}{dx} = \dfrac{dy}{du}\dfrac{du}{dx}$

is actually correct. If so, it can be used together with Equations 4 and 5 to give

(7) $\quad \dfrac{dy}{dx} = (15.3u^{14.3})(2x)$.

When the intermediate variable u is eliminated from this answer—it can be replaced by its expression in Equation 3 in terms of x—the result is

(8) $\quad \dfrac{dy}{dx} = 30.6x(x^2 + 1)^{14.3}$.

The result in Equation 8 is indeed correct. The unproved step, Equation 6, is a theorem in calculus known as the *chain rule*.

The Chain Rule

Suppose we have a chain of functions:

y is a function of u,

and

u is a function of x.

Then y becomes a function of x and its derivative $\dfrac{dy}{dx}$ satisfies

$$\dfrac{dy}{dx} = \dfrac{dy}{du}\dfrac{du}{dx}.$$

To prove the chain rule, let us denote the chain of functions by

$$y = g(u), \quad \text{and} \quad u = h(x)$$

and the function y of x, by

$$y = f(x).$$

Let x change by Δx. Through the relation $u = h(x)$, this change in x causes u to change by an amount which we call Δu. Through the relation $y = g(u)$, the change Δu produces a change in y which we call Δy. Through the relation $y = f(x)$, the change Δx in x produces a change in y; this change must be equal to Δy.

If $\Delta u \neq 0$ for all Δx under consideration, we can write

$$\frac{\Delta y}{\Delta x} = \frac{\Delta y}{\Delta u} \frac{\Delta u}{\Delta x}$$

and from this obtain the chain rule by letting $\Delta x \to 0$. The remaining case ($\Delta u = 0$) is less instructive and will be omitted from the discussion.[5]

Before turning to our first application, differentiating a power of a function, we shall give a few qualitative illustrations of the chain rule.

Example 1. Air pressure p decreases as the altitude a increases. Call this function g.

(9) $p = g(a).$

Suppose a rocket soars vertically upward. Its altitude a at time t is a function we shall call h.

(10) $a = h(t).$

The chain of functions (9) and (10) makes p become a function of t, say

(11) $p = f(t).$

Function (11) gives the atmospheric pressure on the rocket at time t. The chain rule tells us that the derivatives of these three functions are related by the equation

$$\frac{dp}{dt} = \frac{dp}{da} \frac{da}{dt}.$$

[5]It implies $\Delta y = 0$, leads to $\dfrac{dy}{dx} = 0$ and $\dfrac{du}{dx} = 0$, and results in an equation of the form $0 = 0$.

In the above equation, $\dfrac{dp}{dt}$ is the derivative from Equation 11, $\dfrac{dp}{da}$ is the derivative from Equation 9, and $\dfrac{da}{dt}$ is the derivative from Equation 10.

Example 2.   Suppose that a farmer's total profit P for a season depends only on his cost C for irrigation:

$$P = g(C).$$

Suppose that the irrigation cost C depends only on the season's total rainfall r:

$$C = h(r).$$

Then the farmer's profit P is a function of rainfall r, and the rates of change are related by

$$\frac{dP}{dr} = \frac{dP}{dC} \cdot \frac{dC}{dr}.$$

(*Note:* Example 2 is not realistic because P actually depends on more variables than C alone, and C depends on more variables than r alone. In Chapter 9, a chain rule for functions of several variables will be discussed.)

Rule for Differentiating a Power of a Function.   The functions

(14) $y = \sqrt{2x + 1}, \qquad y = \dfrac{1}{x^5 + 1}, \qquad$ and $\qquad y = (x^2 + 1)^{15.3}$

are all of the form

(15) $y = u^n$,

where u is some expression involving x, and n is a rational number. For $y = \sqrt{2x + 1}$, we have $u = 2x + 1$ and $n = \frac{1}{2}$. For $y = 1/(x^5 + 1)$, we have $u = x^5 + 1$ and $n = -1$. For $y = (x^2 + 1)^{15.3}$, we have $u = x^2 + 1$ and $n = 15.3$.

The chain rule shows how to differentiate Equation 15.

$$\frac{dy}{dx} = \frac{dy}{du} \cdot \frac{du}{dx} = nu^{n-1}\frac{du}{dx}.$$

We state this result as a new differentiation rule.

Rule for Differentiating a Power of a Function

The derivative of

(16) $y = u^n$,

where u is a function of x, is

(17) $\dfrac{dy}{dx} = nu^{n-1}\dfrac{du}{dx}$.

That is, the derivative of a quantity to the power n is n times the quantity to the $n - 1$ *times the derivative of the quantity.* We emphasized the last part. Forgetting it is a common error that you should learn to avoid at the start.

Example 3.

a.   The derivative of $y = (5x + 1)^{100}$ is

$$\frac{dy}{dx} = (100)(5x + 1)^{100-1}(5) = 500(5x + 1)^{99}.$$

(*Note:* The answer is *not* $100(5x + 1)^{99}$.)

b.   The derivative of $y = (5x + 1)^{-100}$ is

$$\frac{dy}{dx} = (-100)(5x + 1)^{-100-1}(5) = -500(5x + 1)^{-101}.$$

c.   The derivative of $y = (5x + 1)^{1/2}$ is

$$\frac{dy}{dx} = \tfrac{1}{2}(5x + 1)^{(1/2)-1}(5) = \tfrac{5}{2}(5x + 1)^{-1/2}.$$

d.   The derivative of $y = (x^3 + 6x^2 + 1)^{22.1}$ is

$$\frac{dy}{dx} = 22.1(x^3 + 6x^2 + 1)^{22.1-1}(3x^2 + 12x).$$

Example 4.   Differentiate.

a. $w = (v^3 + v)^{10}$.

b. $p = \dfrac{1}{\sqrt{q^2 + 1}}$.

c. $f(t) = \dfrac{1}{3t^2 - t}$.

SOLUTION.

a. $\dfrac{dw}{dv} = 10(v^3 + v)^9(3v^2 + 1)$.

b. From $p = (q^2 + 1)^{-1/2}$, we obtain

$$\frac{dp}{dq} = \left(-\frac{1}{2}\right)(q^2 + 1)^{(-1/2)-1}(2q) = -q(q^2 + 1)^{-3/2}.$$

c. From $f(t) = (3t^2 - t)^{-1}$, we obtain

$$f'(t) = -1(3t^2 - t)^{-1-1}(6t - 1) = -(6t - 1)(3t^2 - t)^{-2}.$$

Problems

In Problems 1–20, differentiate the given functions.

1. $y = (2x + 1)^5$.

2. $y = (x^4 - 3)^{20}$.

3. $w = (u^2 + 1)^6$.

4. $p = (q^2 - 3q)^4$.

5. $s = (4t - 1)^7$.

6. $y = (5x + 2)^{3.4}$.

7. $w = (2v - 1)^{1/3}$.

8. $F = (t^2 + 1)^{1/2}$.

9. $f(x) = (x^4 - x)^{-3}$.

10. $g(t) = (t^3 + 3)^{-5.4}$.

11. $H(r) = (5 - r^2)^{-4}$.

12. $y = \sqrt[3]{2x + 4}$.

13. $y = \dfrac{1}{\sqrt{5x + 1}}$.

14. $y = \dfrac{1}{5 - 2x}$.

15. $y = \dfrac{1}{(x^3 + 1)^3}$.

16. $w = \dfrac{1}{(1 - 6s)^4}$.

17. $y = \dfrac{5}{1 + x^2}$.

18. $s = 3\sqrt{t^4 + 1}$.

19. $y = \dfrac{15}{2(x^3 + 1)^4}$.

20. $y = 3(8 - x)^{2.5}$.

In Problems 21–24, a chain of functions is given which makes y become a function of x. Find $\dfrac{dy}{dx}$; your answer should not involve the auxiliary variable.

21. $y = \sqrt{u}, u = 1 - x^2$.

22. $y = \dfrac{5}{u^4}, u = 2x^3 + x + 1$.

23. $y = w^{10}, w = x^3 - x$.

24. $y = 3t^{1.2}, t = 9 - x^2$.

25. For the function $y = (1 - x)^5$, compute $\dfrac{dy}{dx}, \dfrac{d^2y}{dx^2}$, and answer the following questions. Is y increasing or decreasing? Is the graph concave upward or concave downward when $x = 0$; when $x = 2$? Find the inflection point.

Section 4 □ Implicit Differentiation

Consider the problem of differentiating the function

(1) $y = \dfrac{1}{x^2}$.

Now that we know the rule for differentiating x^n with negative n, the problem is easy:

(2) $y = x^{-2}$,   so $\dfrac{dy}{dx} = -2x^{-3}$.

Could we have found this answer earlier when we knew only the elementary differentiation rules of Chapter 2? We shall see how this can be done, and in doing so we shall be introduced to the technique of implicit differentiation.

Write Equation 1 in the form

(3) $x^2 y = 1$.

We differentiate each side of Equation 3. The derivative of the left-hand side is indicated by

$$\frac{d}{dx}(x^2 y).$$

To evaluate it, think of $x^2 y$ as the product uv of the two functions of x:

(4) $u = x^2$ and $v = y$.

By the product rule and by Equation 4, we have

(5) $\dfrac{d}{dx}(x^2 y) = \dfrac{d}{dx}(uv) = (u)\dfrac{dv}{dx} + (v)\dfrac{du}{dx} = (x^2)\dfrac{dy}{dx} + y(2x)$.

Equation 5 gives the derivative of the left-hand side of Equation 3. We can equate this to the derivative of the right-hand side, which is $\dfrac{d}{dx}(1)$, or 0, and obtain

(6) $(x^2)\dfrac{dy}{dx} + 2xy = 0$.

Finally, we solve Equation 6 for $\dfrac{dy}{dx}$ and find

(7) $\dfrac{dy}{dx} = -\dfrac{2xy}{x^2}.$

Equation 7 is the answer to our problem.[6] The method used to obtain it is called **implicit differentiation.** This method can be described in general as follows.

Consider an equation which involves an independent and a dependent variable, say x and y. Take the derivative with respect to x of each side of the equation. The result will often be an equation in x, y, and $\dfrac{dy}{dx}$ which can be solved for $\dfrac{dy}{dx}$. Note that the original equation in x and y need not express y *explicitly* in terms of x; it may only indicate an indirect relationship between x and y, so we say it expresses (or defines) y *implicitly* as a function of x. For example,

$$y = \sqrt{x^2 - 4}$$

is an explicit éxpression for a function y, whereas

$$x^2 + y^2 = 4$$

is an implicit expression for this same function (and also for the function $y = -\sqrt{x^2 - 4}$). In Chapter 8, *Differential Equations,* we shall use implicit differentiation to simplify many computations.

The process of taking the derivative with respect to x of each side of a given equation will involve differentiating combinations of x and y as in Equation 5. You should master the following cases and then study Examples 1 and 2.

$$\frac{d}{dx}(1) = 0 \qquad\qquad \frac{d}{dx}(x) = 1$$

$$\frac{d}{dx}(x^2) = 2x \qquad\qquad \frac{d}{dx}(x^n) = nx^{n-1}$$

$$\frac{d}{dx}(y) = \frac{dy}{dx} \qquad\qquad \frac{d}{dx}(y^2) = (2y)\frac{dy}{dx}$$

$$\frac{d}{dx}(y^3) = (3y^2)\frac{dy}{dx} \qquad\qquad \frac{d}{dx}(y^n) = (ny^{n-1})\frac{dy}{dx}$$

[6]Equation 7 is equivalent to the result in Equation 2. To see this, replace y on the right side of Equation 7 with its expression in terms of x from Equation 1.

$$\frac{d}{dx}(xy) = x \cdot \frac{dy}{dx} + y \cdot 1 = (x)\frac{dy}{dx} + y$$

$$\frac{d}{dx}(xy^n) = x \cdot ny^{n-1}\frac{dy}{dx} + y^n \cdot 1 = (nxy^{n-1})\frac{dy}{dx} + y^n$$

$$\frac{d}{dx}(x^2y^n) = (x^2 \cdot ny^{n-1})\frac{dy}{dx} + y^n \cdot 2x = (nx^2y^{n-1})\frac{dy}{dx} + 2xy^n$$

Example 1. Given $3x^4 + y + xy^5 = 2$, find $\frac{dy}{dx}$ by implicit differentiation.

SOLUTION.
We take the derivative with respect to x of each side of the given equation.

$$\frac{d}{dx}(3x^4 + y + xy^5) = \frac{d}{dx}(2).$$

$$3\frac{d}{dx}(x^4) + \frac{d}{dx}(y) + \frac{d}{dx}(xy^5) = 0.$$

$$3 \cdot 4x^3 + \frac{dy}{dx} + x \cdot 5y^4\frac{dy}{dx} + y^5 \cdot 1 = 0.$$

$$\frac{dy}{dx}(1 + 5xy^4) = -12x^3 - y^5.$$

$$\frac{dy}{dx} = -\frac{12x^3 + y^5}{1 + 5xy^4}.$$

Example 2. Find $\frac{dy}{dx}$ when $x = 1$ if it is known that $y = 2$ when $x = 1$, and that x and y are related by the equation $x^3y^3 + y = x + 9$.

SOLUTION.
When each side of the given equation is differentiated with respect to x we obtain

$$(x^3 \cdot 3y^2)\frac{dy}{dx} + y^3 \cdot 3x^2 + \frac{dy}{dx} = 1.$$

$$\frac{dy}{dx}(3x^3y^2 + 1) = 1 - 3x^2y^3$$

(8) $$\frac{dy}{dx} = \frac{1 - 3x^2y^3}{1 + 3x^3y^2}.$$

We now evaluate Equation 8 when $x = 1$ and $y = 2$; the result is

$$\frac{dy}{dx} = \frac{1 - 3 \cdot 1 \cdot 8}{1 + 3 \cdot 1 \cdot 4} = -\frac{23}{13}.$$

Problems

In Problems 1–6, find $\frac{dy}{dx}$ by implicit differentiation.

1. $xy = 4 + x$.

2. $x^2 + xy = 12x - 4$.

3. $xy^2 - 6y = x^3$.

4. $xy = 5$.

5. $x^2y^2 = xy$.

6. $x^2y = 1 + y^3x$.

In Problems 7–10, find the indicated derivative.

7. $xy = 4 + x$; find $\frac{dy}{dx}$ when $x = 2$ and $y = 3$.

8. $x^2 + xy = 12x - 4$; find $\frac{dy}{dx}$ when $x = 4$ and $y = 7$.

9. $xy^2 - 6y = x^3$; find $\frac{dy}{dx}$ when $x = -2$ and $y = 1$.

10. $pq^3 + q = 2p^2$; find $\frac{dq}{dp}$ when $p = 1$ and $q = 1$.

11. The graph of $x^3y^3 + 4y = 3x^2$ is a curve which passes through the point where $x = 2$ and $y = 1$. What is the slope of the curve at that point?

12. In the homohyphallic production function

$$x^n = aL^n + bC^n, \qquad a, b, n \text{ are constants}$$

consider x (the output) as constant, and C (capital) as a function of L (labor). Use implicit differentiation to find $-\frac{dC}{dL}$, which is called the *marginal rate of substitution.*

Consider a function $y = f(x)$ and some fixed value of x, say $x = x_0$. If x changes by an amount Δx, this produces a change in y which, as usual, we denote by Δy. In certain applications, it is handier to have a simple method for estimating Δy approximately rather than to calculate it exactly.

We know that if Δx is small and not equal zero, $\dfrac{\Delta y}{\Delta x}$ will be close to $\dfrac{dy}{dx}$:

(1) $\qquad \dfrac{\Delta y}{\Delta x} \approx \dfrac{dy}{dx} \qquad$ at $x = x_0$.

Therefore

(2) $\qquad \Delta y \approx \left(\dfrac{dy}{dx}\right)\Delta x \qquad$ at $x = x_0$.

Formula 2 gives a handy method for finding an approximate value of Δy when Δx is small:

The change Δy in y, due to a small change Δx in x, is approximately equal to the derivative times Δx.

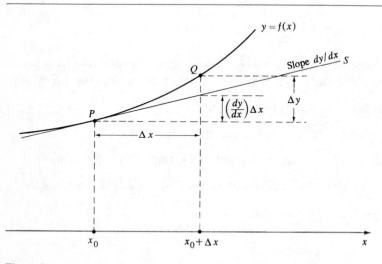

Figure 2

Figure 2 illustrates the geometric interpretation of this method: if Δx is small, the slope $\dfrac{dy}{dx}$ of the tangent line PS is approximately equal to the slope $\dfrac{\Delta y}{\Delta x}$ of the secant line PQ.[7]

Example 1. If $y = x^{7.5}$, how much will y increase when x changes from 1 to 1.003?

SOLUTION.
Since x changes from 1 to 1.003, we have

$$\Delta x = 1.003 - 1 = 0.003.$$

Since $\dfrac{dy}{dx} = 7.5x^{6.5}$, we have

$$\frac{dy}{dx} = 7.5 \quad \text{when} \quad x = 1.$$

Therefore, by Formula 2,

$$\Delta y \approx \frac{dy}{dx}\,\Delta x = (7.5)(0.003) = 0.0225.$$

That is, y will increase by approximately 0.0225.

To find Δy *exactly*, we would need a sophisticated hand calculator to compute $\Delta y = 1.003^{7.5} - 1^{7.5} = 1.0227 - 1 = 0.0227$; note how closely this agrees with the approximation obtained so easily above.

Example 2. Approximate $\sqrt{25.02}$ by introducing the function $y = \sqrt{x}$ and considering the change in y as x changes from 25 to 25.02.

SOLUTION.
Here $\Delta x = 25.02 - 25 = 0.02$. Since $\dfrac{dy}{dx} = \tfrac{1}{2}x^{-1/2}$, we have

$$\frac{dy}{dx} = \frac{1}{10}, \quad \text{when} \quad x = 25.$$

[7]The question of estimating the error in Formula 2 ("How small should Δx be?") belongs to a subject called Taylor Series, of which Formula 2 is a special case. That subject will not be covered in this course.

By Formula 2, we have the approximate equality

(3) $\quad \Delta y = \left(\dfrac{dy}{dx}\right)\Delta x = \left(\dfrac{1}{10}\right)(0.02) = 0.002.$

The meaning of Δy is $\sqrt{25.02} - \sqrt{25}$. Using this and Equation 3, we obtain

$$\sqrt{25.02} = \sqrt{25} + \Delta y = 5 + 0.002 = 5.002.$$

Example 3. The diameter of a ball bearing is measured as 6 centimeters, but the true diameter is 5.98 centimeters. By how much, approximately, was the calculated volume in error? (*Note:* The volume V of a sphere is calculated from the diameter D by the formula $V = \frac{4}{3}\pi r^3$, where $r = D/2$ is the radius.)

SOLUTION.
This problem can be restated as follows. Let

$$V = \frac{4}{3}\pi\left(\frac{D}{2}\right)^3;$$

when D changes from 6 to 5.98, what is the change in V? We answer this using Formula 2.

$$\Delta D = 5.98 - 6 = -0.02,$$

$$\frac{dV}{dD} = \frac{1}{2}\pi D^2 = 18\pi \qquad \text{when} \qquad D = 6.$$

Formula 2 yields the approximations

$$\Delta V = \left(\frac{dV}{dD}\right)\Delta D = (18\pi)(-0.02) = -0.36\pi \approx -1.13.$$

Thus V decreases (because ΔV is negative) by 1.13 cubic centimeters; the calculated volume was in error by about 1.13 cubic centimeters.

Let $y = f(x)$. As in Example 3, consider applications where Δx is a small error in measuring x (Δx = measured value − true value) and Δy is the error in y due to this error in x. Then $\dfrac{\Delta x}{x}$ is called the **relative error** in x; $\dfrac{\Delta y}{y}$ is called the **relative error** in y. Relative error times 100 is called the **percentage error.**

Example 4. A manufacturer selects ball bearings of the desired size by weighing them. The diameter D in centimeters of a steel ball bearing of weight w in grams is $D = 0.63w^{1/3}$. If the percentage error in the weight is 3%, what will be the percentage error in the diameter?

SOLUTION.

We calculate $\dfrac{dD}{dw} = 0.21w^{-2/3}$ and use Formula 2 to obtain the approximate equality

$$\Delta D = 0.21w^{-2/3}\Delta w.$$

Divide by $D = 0.63w^{1/3}$ to obtain

$$\frac{\Delta D}{D} = \frac{0.21w^{-2/3}\Delta w}{0.63w^{1/3}} = \frac{1}{3}\frac{\Delta w}{w}.$$

Thus the relative error in D is one-third of the relative error in w; when $\dfrac{\Delta w}{w}$ is 0.03, $\dfrac{\Delta D}{D}$ will be 0.01 or 1%.

Differentials. Let us write Formula 2 in the form

(4) $\Delta y \approx f'(x_0)\Delta x$ at $x = x_0$.

One might ask, "If the right hand side is not exactly Δy, then exactly what is it?" We can see from Figure 2 that it is the change in y that occurs along the tangent line at x_0 when x changes by Δx. This change is traditionally denoted dy (see Figure 3) and is called the **differential of the function** $y = f(x)$ at x_0; the precise definition is

(5) $dy = f'(x_0)\Delta x$ at $x = x_0$.

With this definition, we can restate Formula 2 or 4 as "when Δx is small, dy is approximately Δy."

It is also customary, in this situation, to use dx as an alternate notation for Δx (see Figure 3):

(6) $dx = \Delta x$.

Now Definition 5 can be written

(7) $dy = f'(x_0)dx$ or $\dfrac{dy}{dx} = f'(x_0)$.

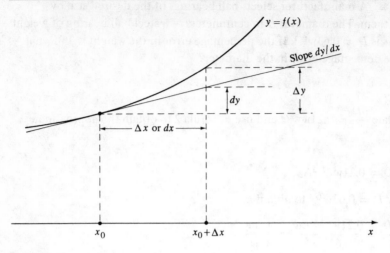

Figure 3

Thus Definitions 5 and 6 allow us to interpret $\dfrac{dy}{dx}$ as an actual fraction, *the ratio of y-change to x-change along the tangent line at x_0.*

Problems

In Problems 1–12, use the method of Example 2 to calculate approximate values of the given quantities. Choose x_0 in each case so that the computations are easy and Δx is small. If you have a hand calculator, compare these approximate answers to the answers determined by use of the calculator.

1. $\sqrt{16.1}$.

2. $\sqrt{15.9}$.

3. $\sqrt{101}$.

4. $\sqrt{99}$.

5. $\sqrt{26.5}$.

6. $\sqrt{23.5}$.

7. $\sqrt[5]{33}$.

8. $\sqrt[5]{31}$.

9. $1.004^{0.351}$.

10. $8.6^{1/3}$.

11. $9.6^{1/2}$.

12. $8.4^{1/2}$.

13. The demand for corn in the United States was estimated as $q = 173p^{-0.49}$, where q is the demand and p is the price in suitable units.[8] Find, approximately, the change in q when p changes from 1 to 1.02. Is the change an increase or decrease?

14. The heat radiation H from an object at absolute temperature T is $H = 10^{-8}T^4$. Find the approximate change in H when T increases from 300 to 302.

15. Find the approximate change in volume and surface area when a ball bearing wears down from a radius of 3 millimeters to 2.98 millimeters. (*Note:* The volume V and surface area A of a sphere of radius r are $V = \frac{4}{3}\pi r^3$ and $A = 4\pi r^2$.)

16. The cost c for fuel to run a light generator at capacity x has been estimated[9] as $c = 16.68 + 0.125x + 0.004x^2$. Find the savings when x is reduced from 100 to 98.

17. When an amount x of a drug is administered to a patient, his reaction R can take the form $R = 150x^2 - x^3$ for $0 \leq x \leq 100$, where R and x are measured in suitable units[10]. Find the change in R when x is increased from 5 to 5.5.

18. The edge of a cube was measured as 2 inches, but the true value was 2.1 inches. Find an approximate value for each.
    a. The error in the volume.
    b. The relative error in the volume.
    c. The percentage error in volume.

19. A bat emits ultrasonic waves which are reflected by a moth back to the bat's ears. The intensity I of the reflected waves is *inversely proportional to* (that is, equal to a constant divided by) r^4, where r is the bat-to-moth distance.[11] Show that a change in r of 1% produces a change in I of about 4%. Is the change an increase or decrease?

[8]H. Schultz, *The Theory and Measurement of Demand,* (Chicago: University of Chicago Press, 1938).

[9]J. A. Nordin, "Note on a Light Plant's Cost Curve," *Econometrica* XV (1947), p. 231ff.

[10]K. R. Rebman, H. Slater, and R. N. Thrall, *Some Mathematical Models in Biology* (Ann Arbor: University of Michigan Report, 1967).

[11]M. Jarman, *Examples of Quantitative Zoology* (London: Edward Arnold, 1970).

Summary

The basic formulas for differentiation are summarized in the following list.

1. **Rule for x^n.** If $y = x^n$, where n is any number, then

$$\frac{dy}{dx} = nx^{n-1}.$$

2. **Rule for constant functions.** If $y = c$, where c is a constant, then

$$\frac{dy}{dx} = 0.$$

3. **Rule for sums and differences.** If $y = u \pm v$, where u and v are functions of x, then

$$\frac{dy}{dx} = \frac{du}{dx} \pm \frac{dv}{dx}.$$

4. **Rule for a constant multiple.** If $y = cu$, where c is a constant and u is a function of x, then

$$\frac{dy}{dx} = c \cdot \frac{du}{dx}.$$

5. **Product rule.** If $y = uv$, where u and v are functions of x, then

$$\frac{dy}{dx} = u \cdot \frac{dv}{dx} + v \cdot \frac{du}{dx}.$$

6. **Quotient rule.** If $y = u/v$, where u and v are functions of x, then

$$\frac{dy}{dx} = \frac{v \cdot \dfrac{du}{dx} - u \cdot \dfrac{dv}{dx}}{v^2}.$$

7. **Rule for powers of a function.** If $y = u^n$, where u is a function of x and n is any number, then

$$\frac{dy}{dx} = (nu^{n-1})\frac{du}{dx}.$$

8. **Chain rule.** If y is a function of u and u is a function of x, then y is a function of x and

$$\frac{dy}{dx} = \frac{dy}{du} \cdot \frac{du}{dx}.$$

The change Δy in $y = f(x)$ when x changes from x_0 to $x_0 + \Delta x$ is approximately $\Delta y \approx f'(x_0)\Delta x$ when Δx is small.

Miscellaneous Problems

In Problems 1–12, differentiate the given functions.

1. $y = (x^2 + 3x + 1)(3x^2 + 5)$.

2. $y = \dfrac{1}{3x^5}$.

3. $w = \dfrac{t}{4t + 1}$.

4. $f(x) = x(\sqrt[3]{x})$.

5. $w = (5t + 1)^6$.

6. $C = (7q^2 + 4)^5$.

7. $f(x) = \sqrt[5]{x}$.

8. $g(x) = \sqrt[5]{3x + 2}$.

9. $g(x) = x^2(3x + 1)^4$.

10. $y = x^2(\sqrt{5x + 1})$.

11. $h(t) = 5t^{2.4} + 3t^{-1.5} + 2$.

12. $f(x) = \sqrt{x} + \dfrac{3}{x\sqrt{x}} + \dfrac{2}{x^3}$.

In Problems 13–15, use implicit differentiation, if necessary, to find the indicated derivatives.

13. Find $\dfrac{dy}{dx}$ when $x = 1$ and $y = 1$ if $y^2 + 2y = x + 2$.

14. Find $\dfrac{dy}{dx}$ when $x = 2$ and $y = 1$ if $y^3 + y = 2x - 2$.

15. Find $\dfrac{dy}{dx}$ when $x = 3$ and $y = 2$ if $y^2 + xy = 3x + 1$.

16. The Thurston learning curve equation in psychology is

$$y = \frac{a(x + b)}{x + c},$$

where a, b, and c are constants.[12] Find $\dfrac{dy}{dx}$ and $\dfrac{d^2y}{dx^2}$.

17. A special case of the Spearman-Brown formula of psychological testing is

[12]L. L. Thurston, "The Learning Curve Equation," *Psychological Bulletin* 14 (1917).

$$R = \frac{2r}{1+r},$$

where r is the reliability (in suitable units, $0 \leq r \leq 1$) of a certain test, and R is the reliability of a test which is similar to, but twice as long as, the first. Find R and $\dfrac{dR}{dr}$ when $r = 0.2$.

18. The value of p which makes

$$(1) \qquad E(p) = p^{10}(1-p)^{70}$$

assume its absolute maximum for $0 \leq p \leq 1$ is called the *maximum likelihood estimate of p*, a concept important in statistics and its applications.[13] Prove that the maximum likelihood estimate of p in Equation 1 is $p = \frac{1}{8}$.

19. For animals of fixed biological structure but of varying length L, the power required to run up a fixed hill at speed v is avL^3 where a is a positive constant.[14] The power that such an animal can generate is bL^2, where b is a positive constant. What value of L makes $g(L) = bL^2 - avL^3$ a maximum for $L \geq 0$? (*Note:* The maximum speed v is proportional to L^{-1}; thus smaller animals can, in general, run uphill faster than large ones. It turns out that in running on level ground, the size L has no effect.[14] Thus if you are tall and wish to compete in a marathon, choose one with few hills!)

20. The demand for corn in the United States during 1915–1929 was $q = 173p^{-0.49}$, where p and q are in suitable units.[15] The *total revenue R* is defined by $R = pq$.

a. Find $\dfrac{dq}{dp}$.

b. Find the *marginal revenue* $\dfrac{dR}{dp}$ expressed as a function of p.

21. The force of gravity r miles from the center of the earth is $F = Gr^{-2}$, where G is a constant. What percentage change in F occurs when one moves from the surface of the earth ($r = 4000$) to a mountain peak four miles high?

22. Let P be the population of a bacteria culture at time t. It is known (we shall study such functions P in the next chapter) that $\dfrac{dP}{dt} = 0.3P$ at any time t. If $P = 800$ when $t = 3$, estimate the population when $t = 3.1$.

[13]Bush and Mosteller, "A Stochastic Model with Applications to Learning," *Annals of Math. Stat.* 24 (1953), pp. 559–685.

[14]J. M. Smith, *Mathematical Ideas in Biology* (Cambridge: Cambridge University Press, 1971), p.14.

[15]H. Schultz, *The Theory and Measurement of Demand* (Chicago: University of Chicago Press, 1938).

five

Exponential and Logarithmic Functions

The functions we have used so far, such as

$$y = \frac{x^3 - 3x + 2}{(x^2 + 5x - 5)^{1/2}}$$

are built up from polynomial functions by a finite number of algebraic operations; such functions are called **algebraic functions.** There are functions which are not of this type; they are called **transcendental functions** (they "transcend" the methods of algebra). Two such functions, the exponential and the logarithmic functions, are the subjects of this chapter. We shall learn their special properties and uses.

Section 1 □ The exponential functions a^x.
Algebraic properties, functional equation.

Section 2 □ Logarithmic functions.   Base 10, other bases, base e, functional equation.

Section 3 □ Differentiation of the exponential function.

Section 4 □ Differentiation of the logarithm function.

Section 5 □ Further applications.   The differential equation of e^x, e as a limit, compound interest.

Summary □ Miscellaneous Problems.

The function

(1) $f(x) = 2^x$

is an example of an **exponential function.** Although it has a superficial resemblance to the polynomial

(2) $g(x) = x^2$,

these two functions f and g are very different. For example, the function g has derivative

$$g'(x) = 2x,$$

but the function f cannot be differentiated by any of the rules we have learned up to now. In particular, the derivative of $f(x) = 2^x$ is *not* $x2^{x-1}$. (The correct

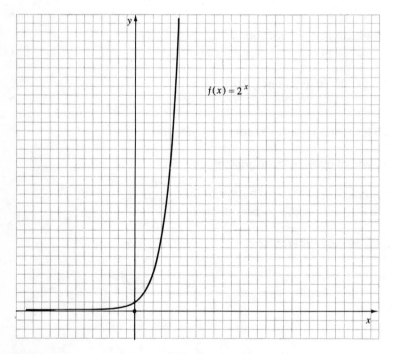

Figure 1

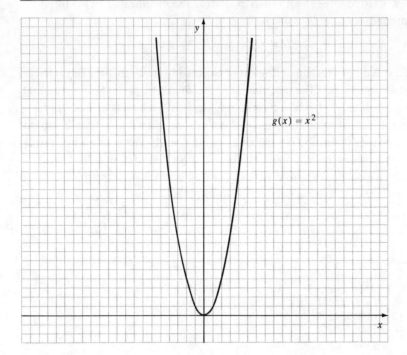

$g(x) = x^2$

Figure 2

way to differentiate $f(x) = 2^x$ will be discussed in Section 3 of this chapter.)

Figures 1 and 2 show the general shape of the functions $f(x) = 2^x$ and $g(x) = x^2$. One important difference between them is that the function $f(x) = 2^x$ eventually grows much faster than $g(x) = x^2$. For example, $f(10) = 2^{10}$ is about 10 times as large as $g(10) = 10^2$, $f(30) = 2^{30}$ is about 1.2 million times as large as $g(30) = 30^2$, and $f(50) = 2^{50}$ is over 400 billion times as large as $g(50) = 50^2$.

A function $y = b^x$, where b is a positive constant is called an **exponential function with base b.** Thus $y = 10^x$ is an exponential function with base 10. A complete discussion of exponential functions would include the rigorous definition of b^x for any real number x.

In this book, we have so far defined b^x only for rational numbers x. This was done by first setting

$$(3) \qquad b^n = \underbrace{b \cdot b \cdot b \cdots b}_{n \text{ times}}$$

when n is a positive integer, and then extending the meaning of b^x to any rational number x by

(4) $b^0 = 1,$ $b^{p/q} = (\sqrt[q]{b})^p,$ and $b^{-x} = 1/b^x.$

We still do not have a definition of expressions such as 10^π. To remedy this, we shall now discuss the meaning of b^x for an arbitrary number x, rational or not. To compute b^x, choose a terminating decimal r which is a good approximation to x. Since a terminating decimal is a rational number, we can compute b^r by our previous definition (the second equation in the list above). The result will be a good approximation to b^x. If a better approximation of b^x is desired, simply choose a decimal approximation to x that is more accurate than r. Table 1 shows this definition being applied to 10^π.

Table 1

$$10^\pi = ?$$
$$10^3 = 1000$$
$$10^{3.1} = 1258.925411$$
$$10^{3.14} = 1380.384265$$
$$10^{3.141} = 1383.566379$$
$$10^{3.1415} = 1385.160186$$
$$10^{3.14159} = 1385.447266$$
$$10^{3.141592} = 1385.453646$$
$$10^{3.1415926} = 1385.455600$$
$$10^{3.14159265} = 1385.455720$$
$$10^{3.141592653} = 1385.455729$$

Figure 3 shows the graphs of $y = b^x$ for various values of the base b, $b > 1$. Note the following properties of these graphs.

If $y = b^x$ where $b > 1$, then

(5) $y > 0,$

(6) $\lim\limits_{x \to -\infty} y = 0,$

(7) $\lim\limits_{x \to +\infty} y = +\infty,$

(8) y is an increasing function.

Property 6 says that if we travel to the left along the graph ($x \to -\infty$), we will get closer and closer to the x-axis. This may be expressed by saying that the x-axis is an **asymptote** of the graph, or that the graph approaches the x-axis

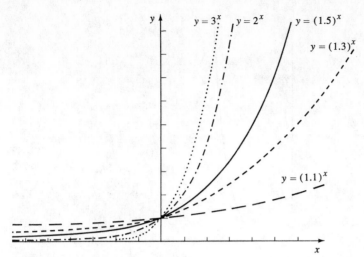

Figure 3

asymptotically. Property 7 says that if we travel to the right along the graph $(x \to +\infty)$, then the value of y will eventually remain larger than any pre-assigned number, no matter how great it is.

Property 8 can be used to advantage in computing b^x for an irrational number x. For example, we can use it to improve Table 1 (see Table 2).

Table 2

$$10^3 = 1,000 < 10^\pi < 10^4 = 10,000$$
$$10^{3.1} = 1258.925411 < 10^\pi < 10^{3.2} = 1584.893193$$
$$10^{3.14} = 1380.384265 < 10^\pi < 10^{3.15} = 1412.537545$$
$$10^{3.141} = 1383.566379 < 10^\pi < 10^{3.142} = 1386.755828$$
$$10^{3.1415} = 1385.160186 < 10^\pi < 10^{3.1416} = 1385.479167$$
$$\vdots$$
$$10^{3.141592653} = 1385.455729 < 10^\pi < 10^{3.141592654} = 1385.455732$$

From the last line of Table 2, we see that

$$10^\pi = 1385.4557 \ldots$$

is accurate to four decimal places.

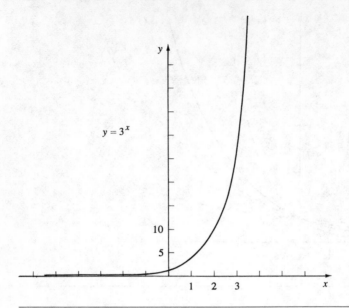

$y = 3^x$

1 2 3 x **Figure 4a**

10

5

Example 1. Graph $y = 3^x$.

SOLUTION.
We see at once that $y = 1$ when $x = 0$. The general shape of the graph can be drawn (see Figure 4a) without plotting any other points. For greater accuracy, a few points (see the table) could be plotted on graph paper and a smooth curve drawn through them.

x	-2	-1	0	1	2	3	4
y	$\frac{1}{9}$	$\frac{1}{3}$	1	3	9	27	81

Example 2. Graph $y = 3^{-x}$.

SOLUTION.
The graph is shown in Figure 4b. This graph can be obtained from Figure 4a, the graph of $y = 3^x$, by interchanging x and $-x$. Note that the equation $y = 3^{-x}$ can also be written as $y = 1/3^x$, or

$$y = \left(\frac{1}{3}\right)^x.$$

Thus it is an exponential function with base $\frac{1}{3}$. Note that since the base is less than 1, Properties 6, 7 and 8 are reversed.

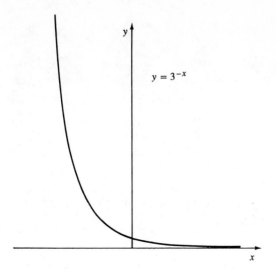

$$y = 3^{-x}$$

x

Figure 4b

Example 3. Graph $y = 2 - 10^{-x}$.

SOLUTION.

A geometric method of obtaining the graph is indicated in Parts a–d of Figure 5. The line $y = 2$ is an *asymptote* of this graph (see Part d of Figure 5). A curve

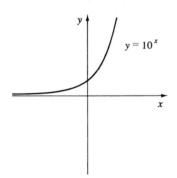

$$y = 10^x$$

x

Figure 5a

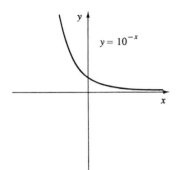

$$y = 10^{-x}$$

x

Figure 5b

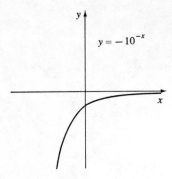

Figure 5c

of this type is often called a *learning curve* in psychology, or a *modified growth curve* in biology, economics, and other areas.

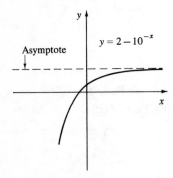

Figure 5d

Example 4. Graph $y = b^{-x^2}$ where $b > 1$.

SOLUTION.
To obtain the general shape of the graph note that

a. If x is large, then $y = \dfrac{1}{b^{(\text{large})}} = \dfrac{1}{\text{large}} = \text{small}$; that is $\lim\limits_{x \to +\infty} y = 0$.

b. If $x = 0, y = 1$.

c. For $x \geqq 0$, as x increases, y decreases.

d. $y > 0$ for all x.

e. The values of y at x and $-x$ are equal. That is, the graph is symmetric about the y-axis.

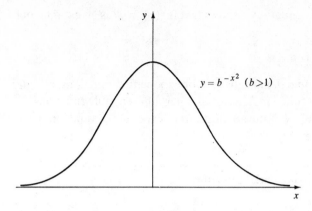

$$y = b^{-x^2} \quad (b>1)$$

Figure 6

Figure 6 shows the graph, a *bell-shaped curve.* This graph is called a *normal curve* in statistics and its applications.

Algebraic Properties of Exponential Functions. The algebraic properties of exponents, previously used only for rational exponents (see page 143). continue to hold when the exponent is any real number. These properties are:

Laws of Exponents

(9) $a^x a^y = a^{x+y}$ and $\dfrac{a^x}{a^y} = a^{x-y}.$

(10) $(ab)^x = a^x b^x$ and $\left(\dfrac{a}{b}\right)^x = \dfrac{a^x}{b^x}.$

(11) $(a^x)^y = a^{xy}.$

For example:

$$(2^{1+\pi})(2^{1-\pi}) = 2^{1+\pi+1-\pi} = 2^2 = 4.$$

$$\frac{\pi^{2+\sqrt{3}}}{\pi^{\sqrt{3}}} = \pi^{2+\sqrt{3}-\sqrt{3}} = \pi^2.$$

$$2^{\sqrt{2}}3^{\sqrt{2}} = (2 \cdot 3)^{\sqrt{2}} = 6^{\sqrt{2}}.$$

$$\frac{6^\pi}{3^\pi} = \left(\frac{6}{3}\right)^\pi = 2^\pi.$$

The first equation, Equation 9, shows that if $f(x) = a^x$ is an exponential function, then

$$f(x_1 + x_2) = f(x_1) \cdot f(x_2).$$

This equation, called the **functional equation** for exponential functions, actually characterizes exponential functions among all other differentiable functions. This fact is the basis for many of the applications of exponential functions (see Problems 8 and 9 on page 179).

Functional Equation for Exponential Functions

　　If

(12)　$f(x_1 + x_2) = f(x_1) \cdot f(x_2)$

then f is an exponential function:[1]

　　　$f(x) = a^x$ for some $a \geq 0$.

═══

Problems

---

1. Find the value of x.

   a. $2^{-3x} = \dfrac{1}{64}$.

   b. $(3^{x+2})^2 = 3^{14}$.

   c. $25^x = 5$.

   d. $(2x)^{\sqrt{2}} = 4$.

   e. $\dfrac{\pi^{5x} \cdot \pi^{x+3}}{\pi^{4x}} = 1$.

   f. $(\sqrt{3})^x = \sqrt[4]{3}$.

2. Sketch the graphs of $y = 4^x$ and $y = 4^{-x}$.

3. Sketch the graphs of $y = 2^x$ and $y = 2^{-x}$.

4. Sketch the modified growth curve $y = 1 - 2^{-x}$.

5. Sketch the normal type curve $y = 2^{-x^2}$.

6. Radioactive substances decay according to a law of the form $y = Ab^{-t}$, where y is the amount of substance remaining at time t; A and b are constants, $A > 0$ and $b > 1$, that depend only on the particular substance (see Problem 9). Show that there is a number h, called the *half-life* of the

---

[1] If Equation 12 holds for all x_1 and x_2, then $f(x) = a^x$ holds for all x. If Equation 12 holds for all positive x_1 and x_2, then $f(x) = a^x$ holds for all $x > 0$.

substance such that during any time interval of length h, the amount of substance remaining at the end of the time interval is one-half of the amount that was present at the beginning of the time interval.

7. Figure 7 shows the graph of $y = 2e^{-t}$, where e is a constant called the *base of natural logarithms* (the value of e is approximately 2.71828182). Determine graphically the half-life of this function y (see Problem 6).

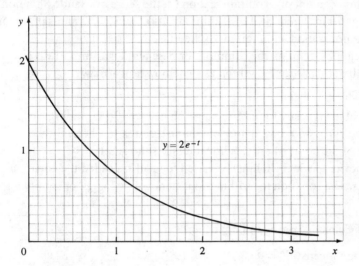

Figure 7

8. A bacteria population, originally of unit size, grows to size $P(t)$ during any length of time t. To investigate this function P, assume that the population size does not affect the growth rate of any subpopulation. Explain why a population of size P_0 will grow to size $P_0 \cdot P(t)$ during a length of time t. Show that this implies $P(t + h) = P(t) \cdot P(h)$. Use Equation 12 to conclude that $P(t) = a^t$ for some $a > 1$.

9. A unit mass of radioactive material decays to size $M(t)$ during any length of time t. This decay process is independent of the surrounding material. Explain why a mass of size M_0 will decay to size $M_0 \cdot M(t)$ during any length of time t. Show that this implies $M(t + h) = M(t) \cdot M(h)$. Use Equation 12 to conclude that $M(t) = a^t$ for some $0 < a < 1$.

Section 2 □ Logarithmic Functions

Logarithm functions, like exponential functions, are a class of transcendental functions. They occur in many mathematical applications; for example, decibels of noise, pH measure of acidity, Richter magnitude of earthquakes, and information content of a communication (in the sense of Claude Shannon) all involve logarithm functions. Logarithms are also useful in business problems that involve amortization and mortgage schedules.

We begin by defining the common (or base ten) logarithm. If s is any positive number, then s can be written as some power of 10, say

$$s = 10^t.$$

In fact, this value t is easily found from a graph of $y = 10^x$. Using Figure 8, locate s on the y-axis and draw a horizontal line through it. Then t is the x-coordinate of the point where this line intersects the graph.

The power to which 10 must be raised to yield s is called the **common logarithm of s.** The common logarithm of s is denoted by $\log_{10}s$, or simply $\log s$.

Definition of Common (Base 10) Logarithm

(2) $t = \log s$ means $10^t = s$

The table on page 386 can be used to find the approximate value of the common logarithm of any number n. This table accepts n to three digits; tables

---

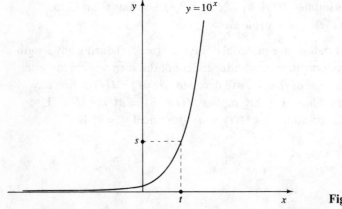

Figure 8

which accept entries to four or five digits can be found in most libraries (note that each additional digit requires a tenfold increase in the length of the table). Any good electronic hand calculator is far more valuable than a table of logarithms. The calculator in Figure 9, for example, accepts entries to ten digits; that is better than a logarithm table with one million pages!

Figure 9

Photo courtesy of The Hewlett-Packard Company

Example 1.

$$\log 1000 = 3 \text{ because } 10^3 = 1000.$$
$$\log 100 = 2 \text{ because } 10^2 = 100.$$
$$\log 10 = 1 \text{ because } 10^1 = 10.$$
$$\log \tfrac{1}{10} = -1 \text{ because } 10^{-1} = \tfrac{1}{10}.$$
$$\log 0.0001 = -4 \text{ because } 10^{-4} = 0.0001.$$

Since $\sqrt{10}$, is approximately equal to 3.16, we can add to this list:

$$\log 3.16 = 0.5 \text{ because } 10^{0.5} = 3.16.$$

We have obtained a new function

$$y = \log x$$

called the **common** (or base ten) **logarithm function.** It is defined for all values of $x > 0$. To sketch its graph, note that Equation 2 can be rephrased as

(3)
$$\begin{cases} (s,t) \text{ lies on the graph of } y = \log x \\ \qquad\qquad \text{means} \\ (t,s) \text{ lies on the graph of } y = 10^x. \end{cases}$$

The graph of $y = 10^x$ was obtained in the previous section. Statement 3 shows how to use it to produce the graph of $y = \log x$. The result is shown in Figure 10; note that the graphs of $y = 10^x$ and $y = \log x$ are mirror images of each other with respect to the line $y = x$. This is expressed by saying $y = 10^x$ and $y = \log x$ are **inverse functions.**

Still another way to express the facts in Statements 2 and 3 is

(4) $10^{\log x} = x,$ and $\log(10^x) = x.$

Other Logarithmic Functions.   The discussion above concerning the common logarithm can be generalized to a base b other than 10. Let b be a given positive number such that $b \neq 1$. Then b can be used as base for a logarithmic function $y = \log_b x$, called the *logarithm to the base b*. The following is the analog of Definition 2.

Definition of Base b Logarithm

(5) $t = \log_b s \text{ means } b^t = s.$

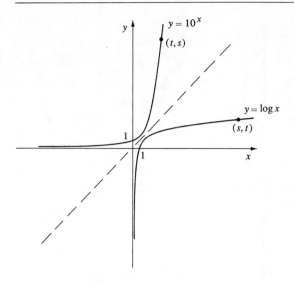

y = 10^x

(t, s)

y = log x

(s, t)

Figure 10

Example 2.

$\log_2 8 = 3$ because $2^3 = 8$

$\log_{16} 2 = \frac{1}{4}$ because $16^{1/4} = 2$

$\log_3 \frac{1}{243} = -5$ because $3^{-5} = \frac{1}{243}$.

Since $\sqrt{2}$ is approximately equal 1.41, we can also write

$\log_2 1.41 = 0.5$ because $2^{0.5} = 1.41$.

The most important base for logarithms is the number e, called the base of natural logarithms.[2] The number e, like the number π, is not a rational number. Its decimal expansion begins with

$e = 2.71828182845904523536 \ldots$

The logarithm function to the base e is called the **natural** or **Napierian**[3] **logarithm.** It is often denoted by ln rather than $\log_e$.

[2] The letter e was first used for this number by the Swiss mathematician Léonhard Euler (1707–1783).

[3] In honor of John Napier (1550–1617), a Scotsman, who spent many years preparing a table of logarithms.

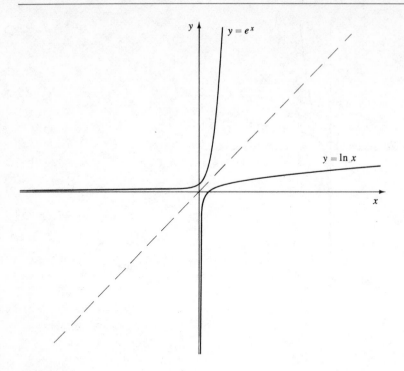

Figure 11

Definition of Natural Logarithm

(6) $t = \ln s$ means $e^t = s$.

The natural logarithm is more useful in calculus and its applications than is the common logarithm.[4] We shall see why that is so in the next section.

The graph of $y = e^x$ and that of its inverse function $y = \ln x$ are shown in Figure 11.

Tables for $y = e^x$ and $y = \ln x$, the natural exponential function and the natural logarithm function, are given on pages 385 and 386. There are, however, no tables available that are as accurate or useful as a good electronic hand calculator such as the one in Figure 9.

The algebraic properties of exponents (see Equations 9, 10, and 11 on page 177) correspond to the following algebraic properties of logarithms.

---

[4]In this book, the natural logarithm function is always denoted by $\ln x$. Some writers prefer to denote it by $\log x$; then they use $\log_{10} x$ to distinguish the common logarithm.

Properties of Logarithms

(7) $\log_b(uv) = \log_b u + \log_b v$

(8) $\log_b \dfrac{u}{v} = \log_b u - \log_b v$

(9) $\log_b(u^k) = k \log_b u$

Example 3. Given that

$$\log 2 = 0.3, \qquad \log 3 = 0.48, \qquad \text{and} \qquad \log 5 = 0.7$$

find $\log \sqrt{24/5}$.

SOLUTION.
We use the properties of logarithms.

$$\log \sqrt{\frac{24}{5}} = \log\left(\frac{24}{5}\right)^{1/2} = \frac{1}{2}\log\frac{24}{5}$$

$$= \frac{1}{2}(\log 24 - \log 5)$$

$$= \frac{1}{2}(\log 8 \cdot 3 - \log 5)$$

$$= \frac{1}{2}(\log 8 + \log 3 - \log 5)$$

$$= \frac{1}{2}(\log 2^3 + \log 3 - \log 5)$$

$$= \frac{1}{2}(3 \log 2 + \log 3 - \log 5)$$

$$= \frac{1}{2}[3(0.3) + 0.48 - 0.7] = 0.34$$

Functional Equation. If $f(x) = \log_b x$, then by Property 7,

$$f(x_1 \cdot x_2) = f(x_1) + f(x_2).$$

This equation expresses an important property of logarithms. In fact, it *characterizes logarithms* among all other differential functions, and is the basis of many applications of logarithms (see Review Problem 8 on page 213).

Functional Equation for Logarithms

If $f \neq 0$ and for all $x_1 > 0$ and $x_2 > 0$,

(10) $f(x_1 \cdot x_2) = f(x_1) + f(x_2),$

then f is a logarithm function.

$$f(x) = \log_b x.$$

Problems

1. Figure 12 shows the graph of $y = 2^x$. Use this figure to sketch the graph of $y = \log_2 x$.

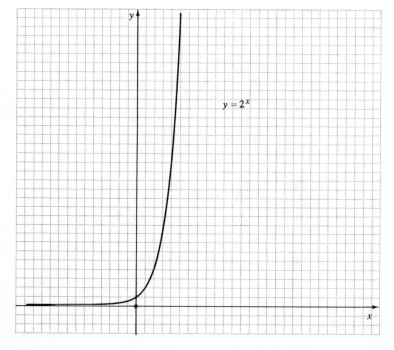

$y = 2^x$

Figure 12

2. Figure 13 shows the graph of $y = 3^x$. Use this figure to sketch the graph of $y = \log_3 x$.

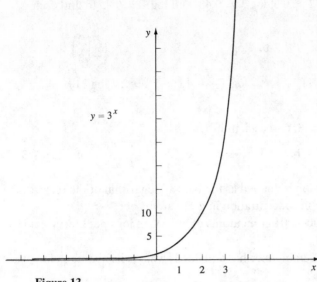

Figure 13

3. The Richter magnitude of an earthquake is defined as the base 10 logarithm of the *strength* of the quake.[5]
    a. Show that an earthquake of magnitude 6 on the Richter scale is 1000 times stronger than an earthquake of magnitude 3.
    b. The Japanese earthquake of 1933 was 4 times stronger than the San Francisco earthquake of 1906. The Japanese earthquake had a magnitude of 8.9 on the Richter scale. What was the magnitude on the Richter scale of the San Francisco earthquake? (*Note:* Use the fact log 4 = 0.6.)

4. In 1972, there was an earthquake in Nicaragua which registered 6.23 on the Richter scale. How many times stronger was the Japanese earthquake of 1933 (see Problem 3)?

5. Use the facts $\ln 2 = 0.69$, $\ln 3 = 1.1$, and $\ln 5 = 1.61$ to find each logarithm.
    a. $\ln 15$.
    b. $\ln 40$.
    c. $\ln \sqrt[3]{5}$.
    d. $\ln 0.4$.
    e. $\ln 30$.
    f. $\ln \dfrac{9}{25}$.

[5]This system, developed by Charles F. Richter in 1935, determines *strength* in terms of the output of a seismograph.

6. Use the facts $\log 2 = 0.3$, $\log 3 = 0.48$, and $\log 7 = 0.85$ to find each logarithm.

    a. $\log 21$.

    b. $\log \dfrac{3}{7}$.

    c. $\log \left(\dfrac{2}{3}\right)^5$.

    d. $\log \sqrt[7]{\dfrac{3}{2}}$.

    e. $\log \left(\dfrac{2^3 \cdot 7^2}{3^4 \cdot 10}\right)$.

    f. $\log 210$.

7. Given that $2^{1.58} = 3$, find each number.

    a. $\log_2 6$.

    b. $\log_2 9$.

    c. $\log_2 \left(\dfrac{3}{2}\right)$.

    d. $\log_2 \sqrt{3}$.

8. The pH of a solution is defined as the base 10 logarithm of the reciprocal of the hydrogen ion concentration in gram atoms per liter.

    a. Water has 0.0000001 gram atoms of hydrogen ions per liter; what is its pH value?

    b. A solution of lemon juice has a pH of 3; what is its hydrogen ion concentration?

9. The altitude h in feet above sea level is related to the air pressure p in inches of mercury by the formula $h = 25{,}000 \ln\left(\dfrac{30}{p}\right)$. What is the air pressure at an altitude of 50,000 feet?

10. The agricultural income in India was studied by Tintner, Narasimhan, and Raghavan.[6] It was found to be

$$y = \frac{4406}{1 + 0.51e^{-0.11t}},$$

    where t is the number of years after 1900. (Functions of this form are called *logistic functions*.) Use natural logarithms to solve this equation for t.

11. An Oriental legend says that the inventor of chess asked a king to reward him by placing some grains of rice on a chess board. He asked the king to put one grain on the first square, double that amount (that is, two grains) on the second, double that amount (that is, four grains) on the third square, and so on. Can you estimate the number N of grains that are placed on the last square—that is, about how many digits are there in the number N? (*Note:* Use these two facts. There are 64 squares in a chess board. The common logarithm of 2 is 0.3.)

[6]G. Tintner and C. B. Milham, *Mathematics and Statistics for Economists,* 2nd ed. (Illinois: Dryden Press), p. 191.

12. The demand q for butter in Stockholm (1925–1937) depended on the price p according to the function[7] $q = \dfrac{38}{p^{1.2}}$. If $Q = \ln q$ and $P = \ln p$, show that the graph of Q as a function of P would be a straight line. What is its slope?

13. British industrial production t years after 1700 was found by Sengupta and Tintner[8] to be $y = \dfrac{573}{1 + 232e^{-0.06t}}$.

    a. Has production been increasing or decreasing?

    b. Use natural logarithms to solve for t as a function of y.

Section 3 □ Differentiation of the Exponential Function

The rules for differentiating exponential and logarithmic functions are simple to state, but the mathematical analysis on which they rest is much less elementary than that used for algebraic functions such as $y = x^n$.

    We begin the discussion with the following observations.

(1)   Let $f(x) = b^x$;   then $f'(x) = b^x \cdot f'(0)$.

The proof of Statement 1 follows from

(2) $\dfrac{f(x + \Delta x) - f(x)}{\Delta x} = \dfrac{b^{x + \Delta x} - b^x}{\Delta x} = b^x \left[\dfrac{b^{0 + \Delta x} - b^0}{\Delta x} \right].$

When $\Delta x \to 0$, the term in brackets approaches $f'(0)$, and the left hand side of Equation 2 approaches $f'(x)$.

    To apply Statement 1, we need to find the value of the constant $f'(0)$; note that it is the slope of the graph of $y = b^x$ when $x = 0$. From Parts a–d of

[7]H. Wold, *Demand Analysis; a Study in Econometrics* (New York: John Wiley and Sons, Inc., 1953), p. 140.

[8]G. Tintner and C. B. Milham, *Mathematics and Statistics for Economists*, 2nd ed. (Illinois: Dryden Press), p. 181.

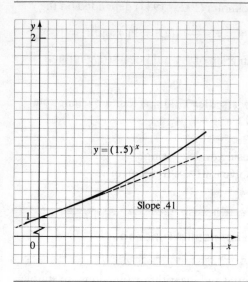

Figure 14a

Figure 14 (see also Figure 3 on page 173), we see that this slope is small if b is close to 1 and is large if b is large. There is a number b for which the graph of $y = b^x$ has slope 1 when $x = 0$. This number is denoted by e. It was introduced

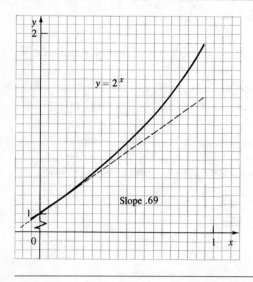

Figure 14b

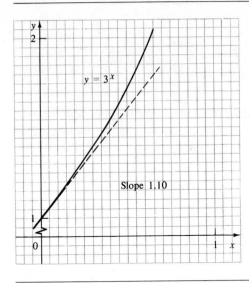

Figure 14c

in the previous section (page 183) and called the *base of natural logarithms.* We now see that its importance is based on its property of yielding the simplest differentiation formula among all exponential functions.

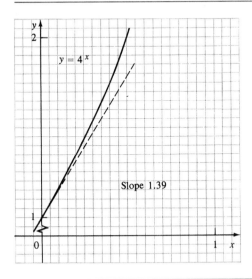

Figure 14d

Rule for Differentiating e^x

(3)   If $y = e^x$,   then $\dfrac{dy}{dx} = e^x$.

This rule says, in words, that the rate of change of the function $y = e^x$ is equal to the value of y. (See Problem 18 on page 195 for an application of this interpretation.) The graph of $y = e^x$ is shown in Figure 15.

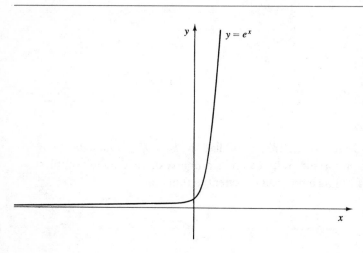

Figure 15

Let u be a function of x, and let $y = e^u$. The chain rule (see page 151) tells us that

$$\frac{dy}{dx} = \frac{dy}{du}\frac{du}{dx},$$

and by Rule (3), $\dfrac{dy}{du} = e^u$. Thus we have the following rule.

Rule for Differentiating $e^{u(x)}$

Let u be a function of x. If

$$y = e^u,$$

then

(4) $\quad \dfrac{dy}{dx} = e^u \dfrac{du}{dx}.$

Rule 4 can be used to differentiate any exponential function $y = b^x$ (see Example 3 below).

Example 1.

a. If $y = e^{2x}$, $\dfrac{dy}{dx}$ can be found by Rule 4 with $u = 2x$. Thus

$$\dfrac{dy}{dx} = e^{2x} \cdot (2) = 2e^{2x}.$$

b. If $f(x) = e^{4x^3}$, then by Rule 4 with $u = 4x^3$, we find

$$f'(x) = e^{4x^3}(12x^2) = 12x^2\, e^{4x^3}.$$

c. If $w = 3e^{5t} + 2e^{-3t}$, then

$$\dfrac{dw}{dt} = 15e^{5t} - 6e^{-3t}.$$

d. If $y_{} = xe^x$, then by the product rule (see page 138),

$$\dfrac{dy}{dx} = xe^x + e^x.$$

e. If $y = \dfrac{e^{3x}}{x^2}$, then by the quotient rule (see page 140),

$$\dfrac{dy}{dx} = \dfrac{3x^2 e^{3x} - 2xe^{3x}}{x^4} = \dfrac{e^{3x}(3x - 2)}{x^3}.$$

Example 2. Use the table on pages 385–386 to find the approximate numerical value of each derivative.

a. $f'(3)$ if $f(t) = e^{3x}$.

b. $g'(2)$ if $g(x) = e^{(x^2)}$.

c. $\dfrac{dy}{dx}$ when $x = \frac{1}{10}$, where $y = x^2 e^x$.

SOLUTION.

a. $f'(t) = 3e^{3t}$, so $f'(3) = 3e^9 = 3(8103.1) = 24309.3$.

b. $g'(x) = 2xe^{x^2}$, so $g'(2) = 4e^4 = 4(54.598) = 218.392$.

c. $\dfrac{dy}{dx} = x^2 e^x + 2xe^x$, so when $x = 0.1$, $\dfrac{dy}{dx} = e^{0.1}(0.01 + 0.2) = 1.1052(0.21)$
$= 0.232092$.

Example 3. Find $\dfrac{dy}{dx}$ if $y = 10^x$.

SOLUTION.
There is a neat trick for solving this problem: it is to write 10^x as e^{kx} for a suitable constant k.

$$y = 10^x = (e^{\ln 10})^x = e^{x \ln 10}.$$

Now use Rule 4 for $u = x \ln 10$.

$$\frac{dy}{dx} = e^{x \ln 10}(\ln 10) = (e^{\ln 10})^x (\ln 10) = 10^x \ln 10.$$

The method of Example 3 can be used to derive the general rule.

Rule for Differentiating b^x

(5) If $y = b^x$ where $b > 0$, then $\dfrac{dy}{dx} = b^x \ln b$.

It is worth noting (see Problem 22) that the function $y = b^x$ can always be written as $y = e^{kx}$ for a suitable constant k.

Problems

In Problems 1–16, differentiate the given functions.

1. $y = 3e^x$.

2. $y = \frac{1}{2}e^x$.

3. $y = x^2 + 5e^x$.

4. $y = e^{3x}$.

5. $y = 4e^{5x}$.

6. $y = \sqrt{x} + 2e^{-x}$.

7. $y = \dfrac{15}{e^{3x}}$.

8. $w = te^{2t}$.

9. $P = q^2 e^{-3q}$.

10. $u = \dfrac{v}{e^v}$.

11. $M(t) = (1 + e^{2t})^4$.

12. $p = \dfrac{e^s}{s}$.

13. $f(x) = \sqrt{x + e^x}$.

14. $y = 2e^{x^2 + 2x - 1}$.

15. $y = (e^x)^2$.

16. $g(t) = (e^t + 1)(e^t - 1)$.

17. The function

$$y = \frac{1}{\sigma\sqrt{2\pi}}(e^{-x^2/2\sigma^2}),$$

is called a *normal* (or *Gaussian*) *distribution*. It is the bell-shaped curve used in statistics and probability. The constant σ is called the *standard deviation*. Show that

$$\frac{d^2y}{dx^2} = \left(\frac{x^2 - \sigma^2}{\sigma^5\sqrt{2\pi}}\right)e^{-x^2/2\sigma^2}.$$

A graph of y is shown in Figure 16. Note that the inflection points are at $x = \pm \sigma$.

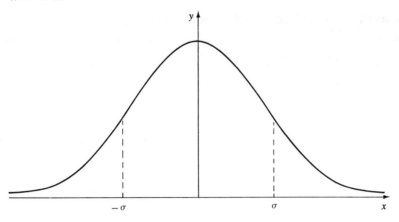

Figure 16

18. Rule 3 says that the graph of $y = e^x$ has slope y at the point (x,y). Use this property to make an approximate sketch of $y = e^x$ as follows. Consider some points $0 = x_0 < x_1 < x_2 < \cdots < x_n$ on the x-axis. Begin by plotting the point $P_0 = (x_0, y_0)$ where $x_0 = 0$ and $y_0 = 1$. Plot $P_1 = (x_1, y_1)$ where y_1 is determined so that $P_0 P_1$ has slope y_0. Then plot $P_2 = (x_2, y_2)$, where y_2 is determined so that $P_1 P_2$ has slope y_1. Continue in this way, and then draw a smooth curve through the points $P_0, P_1, \ldots, P_n$. This curve will approximate $y = e^x$; the smaller the distances $\Delta x = x_{i+1} - x_i$, the better the approximation.

19. The agricultural income y in India t years after 1900 can be approximated by the *logistic curve*[9]

$$y = \frac{4400}{1 + \frac{1}{2}e^{-t/10}}.$$

Find $\dfrac{dy}{dt}$.

20. Let y be the industrial production function of Problem 13 on page 189. Find $\dfrac{dy}{dt}$ when $t = 0$.

21. Find $\dfrac{dy}{dx}$ if $y = 2^x$.

22. Show that $y = b^x$ can be rewritten as $y = e^{kx}$ for a suitable constant k.

Section 4 □ Differentiation of the Logarithm Function

The rule for differentiating $e^{u(x)}$, Rule 4 on page 192, can be used to differentiate the natural logarithm function $y = \ln x$. By definition, this y satisfies

$$e^y = x.$$

Use the method of implicit differentiation to differentiate both sides of this equation (use Rule 4 on page 192 with $u = y$ to differentiate the left-hand side). The result is

$$e^y \frac{dy}{dx} = 1.$$

Since $e^y = x$, the above equation gives $\dfrac{dy}{dx} = \dfrac{1}{x}$.

[9]G. Tintner and C. B. Milham, *Mathematics and Statistics for Economists,* 2nd ed. (Illinois: Dryden Press), p. 191.

Rule for Differentiating ln x

(1)   If $y = \ln x$,   then $\dfrac{dy}{dx} = \dfrac{1}{x}$.

A more general form of Rule 1, which can be obtained from this rule by the chain rule, is the following.

Rule for Differentiating ln $u(x)$

Let u be a function of x. If

$$y = \ln u,$$

then

(2) $\dfrac{dy}{dx} = \dfrac{1}{u} \cdot \dfrac{du}{dx}$.

Example 1.   Differentiate.

a. $y = 5 \ln x$.

b. $y = 2x^2 - 3 \ln x$.

c. $y = \ln (2x - 1)$.

d. $y = x^2 \ln x$.

e. $y = \dfrac{\ln x}{x}$.

f. $y = \dfrac{x}{\ln x}$.

SOLUTION.

a. $\dfrac{dy}{dx} = 5(1/x) = 5/x$.

b. $\dfrac{dy}{dx} = 4x - 3(1/x) = 4x - (3/x)$.

c. Apply Rule 2 with $u = 2x - 1$. Then $\dfrac{dy}{dx} = \left(\dfrac{1}{u}\right)\left(\dfrac{du}{dx}\right)$. Since $\dfrac{1}{u} = \dfrac{1}{2x - 1}$

and $\dfrac{du}{dx} = 2$, the final answer is

$$\dfrac{dy}{dx} = \dfrac{2}{2x - 1}.$$

d. Apply the product rule (see page 138) for $u = x^2$ and $v = \ln x$.

$$\frac{dy}{dx} = u \cdot \frac{dv}{dx} + v \cdot \frac{du}{dx} = (x^2)\left(\frac{1}{x}\right) + (\ln x)(2x)$$

or

$$\frac{dy}{dx} = x + 2x \ln x.$$

e. Apply the quotient rule (see page 140) with $u = \ln x$ and $v = x$.

$$\frac{dy}{dx} = \frac{v \cdot \dfrac{du}{dx} - u \cdot \dfrac{dv}{dx}}{u^2} = \frac{1 - \ln x}{x^2}.$$

f. Apply the quotient rule with $u = x$ and $v = \ln x$.

$$\frac{dy}{dx} = \frac{v \cdot \dfrac{du}{dx} - u \cdot \dfrac{dv}{dx}}{u^2} = \frac{\ln x - 1}{(\ln x)^2}.$$

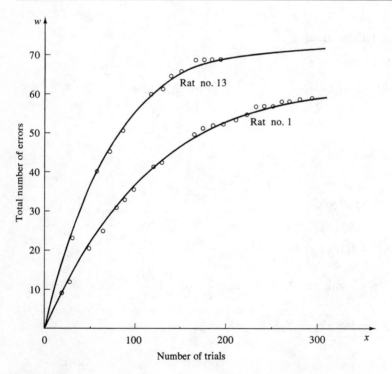

Figure 17

Example 2. A rat is forced to jump toward one of two stimuli, only one of which leads to a reward. The total number of errors w made by the rat after x trials was found to be a function[10] of the form

$$(3) \qquad w = a \ln \frac{b}{M + Ae^{-kx}},$$

where a, b, k, M, and A are constants (see Figure 17).[11] Find $\dfrac{dw}{dx}$, sometimes called *rate of learning* by psychologists.

SOLUTION.
Rewrite Equation 3, using the properties of logarithms, as follows.

$$(4) \qquad w = a \ln b - a \ln (M + Ae^{-kx}).$$

In Equation 4, set $u = M + Ae^{-kx}$ and obtain

$$(5) \qquad \frac{dw}{dx} = 0 - a\left(\frac{1}{u}\right)\left(\frac{du}{dx}\right).$$

From $u = M + Ae^{-kx}$, we obtain

$$\frac{du}{dx} = Ae^{-kx}(-k).$$

When this is substituted in Equation 5, the result is

$$\frac{dw}{dx} = \frac{akAe^{-kx}}{M + Ae^{-kx}}.$$

Logarithmic Differentiation. Sometimes nonlogarithmic functions are differentiated most conveniently by *introducing* logarithms. Consider, for example, the function

$$(6) \qquad y = \frac{\sqrt{x}(2x + 3)^{10}}{(6x + 1)^{1/3}}.$$

An easy way to find $\dfrac{dy}{dx}$ is to take logarithms of each side of Equation 6, simplify by using the properties of logarithms, and then differentiate implicitly. This technique is called **logarithmic differentiation.** To illustrate, take logarithms of both sides of Equation 6 and obtain

[10]N. Rashevsky, *Mathematical Biophysics,* 3rd rev. ed., Vol. 2 (New York: Dover Publications Inc., 1960), p. 138.

[11]Data by H. Gulliksen, *Journal of General Psychology* 11 (1934), p. 395.

$$\ln y = \ln\left(\frac{\sqrt{x}(2x + 3)^{10}}{(6x + 1)^{1/3}}\right),$$

$$= \ln[\sqrt{x}(2x + 3)^{10}] - \ln[(6x + 1)^{1/3}]$$

$$= \ln x^{1/2} + \ln(2x + 3)^{10} - \ln(6x + 1)^{1/3}$$

or

(7) $\ln y = \dfrac{1}{2}\ln x + 10\ln(2x + 3) - \dfrac{1}{3}\ln(6x + 1).$

Differentiate each side of Equation 7:

$$\frac{1}{y}\cdot\frac{dy}{dx} = \frac{1}{2}\left(\frac{1}{x}\right) + 10\left(\frac{1}{2x + 3}\right)(2) - \frac{1}{3}\left(\frac{1}{6x + 1}\right)(6),$$

so

(8) $\dfrac{dy}{dx} = y\left(\dfrac{1}{2x} + \dfrac{20}{2x + 3} - \dfrac{2}{6x + 1}\right).$

In Equation 8, we can replace y by its expression in Equation 6 to obtain an answer in the form

$$\frac{dy}{dx} = \frac{\sqrt{x}(2x + 3)^{10}}{(6x + 1)^{1/3}}\left(\frac{1}{2x} + \frac{20}{2x + 3} - \frac{2}{6x + 1}\right).$$

Example 3.   Use logarithmic differentiation to find $\dfrac{dy}{dx}$ for the function

$$y = \frac{3}{x^3 e^{2x}(3x + 1)^5}.$$

SOLUTION.

$$\ln y = \ln 3 - \ln[x^3 e^{2x}(3x + 1)^5]$$

$$= \ln 3 - \ln x^3 - \ln e^{2x} - \ln(3x + 1)^5$$

or

$$\ln y = \ln 3 - 3\ln x - 2x - 5\ln(3x + 1).$$

Therefore

$$\frac{1}{y}\cdot\frac{dy}{dx} = 0 - \frac{3}{x} - 2 - \frac{15}{3x + 1}$$

or

$$\frac{dy}{dx} = \frac{-3}{x^3 e^{2x}(3x + 1)^5}\left(\frac{3}{x} + 2 + \frac{15}{3x + 1}\right).$$

Example 4. Differentiate $y = x^x$.

SOLUTION.
We will use logarithmic differentiation. We have

(9) $\ln y = \ln(x^x)$

and, since $\ln(x^x) = x \ln x$, Equation 9 simplifies to

(10) $\ln y = x \ln x$.

Differentiate Equation 10 implicitly to obtain

$$\frac{1}{y}\cdot\frac{dy}{dx} = x\left(\frac{1}{x}\right) + \ln x$$

or

(11) $\frac{dy}{dx} = y(1 + \ln x)$.

We can replace y in Equation 11 by x^x to obtain the explicit answer.

$$\frac{dy}{dx} = x^x(1 + \ln x).$$

Problems

1. Differentiate.
    a. $y = \ln x^{10}$.
    c. $s = \ln(2r + 1)$.

    b. $w = 5t^2 - \ln t$.
    d. $y = x \ln(3x - 1)$.

2. Differentiate.
    a. $f(x) = x^3 \ln x$.
    c. $g(t) = t \ln(2t)$.

    b. $y = \ln(2x + 1)$.
    d. $y = [\ln(2x + 1)]^3$.

3. Population growth in an environment with a limited food supply is studied by means of an equation

$$T = a \ln \frac{x}{k - x} + b,$$

where a, k, and b are constants.[12] Find $\dfrac{dT}{dx}$.

4. The diffusion of water through membranes is studied by means of an equation

$$T = a - b - x + a \ln \frac{b}{a - x}$$

where a and b are constants. Show that $\dfrac{dT}{dx} = \dfrac{x}{a - x}$.

5. Use logarithmic differentiation to find each of the following.

a. $\dfrac{dy}{dx}$ if $y = x(3x + 1)^2(x + 6)^{10}$.   b. $\dfrac{dT}{dx}$ if $T = 10^x$.

c. $\dfrac{du}{ds}$ if $u = \dfrac{53.1}{s(s - 1)\sqrt{s - 2}}$.   d. $\dfrac{dy}{dx}$ if $y = x^{2x}$.

e. $\dfrac{dy}{dx}$ if $y = x^{\ln x}$ f. $f'(x)$ if $f(x) = x^{x^3}$.

6. In economics, if $q = f(p)$ is a demand function, then the *elasticity of demand* is

$$\frac{\dfrac{dQ}{dp}}{\dfrac{dP}{dp}},$$

where $Q = \ln q$ and $P = \ln p$.

a. Find the elasticity of demand for the demand function $q = 5 - 2p$.
b. Find the elasticity of demand for the demand function $q = 5 - 2p^2$.
c. Find the elasticity of demand for the demand function $q = e^{-3p}$.
d. Show that for a general demand function $q = f(p)$, the elasticity of demand is equal to $\dfrac{p}{q} \cdot \dfrac{dq}{dp}$.

7. In genetics, the following problem[13] arises: Find a function $n = f(u)$ such that $n = 0$ when $u = 2$ and $\dfrac{dn}{du} = \dfrac{u + 1}{u}$. Show that $n = u - a + \ln \dfrac{u}{a}$ is such a function if $a = 2$.

[12]J. M. Smith, *Mathematical Ideas in Biology* (Cambridge: Cambridge University Press, 1971), p. 41.

[13]J. M. Smith, *Mathematical Ideas in Biology* (Cambridge: Cambridge University Press, 1971), p. 75.

Why is the exponential function so useful in the physical sciences, the life sciences, the social sciences, and business? One reason is reflected in the differential equation of the exponential function; it expresses a growth property that is apparent in a multitude of phenomena. Another reason lies in the characterization of e^x as a limit of quantities that commonly occur in probability, statistics, and compound interest calculations.

The Differential Equation of e^x. The applications we shall treat are based on the following theorem on differential equations.

Differential Equation for Exponential Growth

If y is a function of x such that

(1) $\quad \dfrac{dy}{dx} = ky, \qquad k$ a constant,

for every x, then

(2) $\quad y = Ae^{kx}$

for some constant A.

In applications, one may be searching for a function to describe a real-life situation. From the nature of the situation, it may be apparent that *the function changes at a rate proportional to its value,* a property that is expressed mathematically by Equation 1. We may think of Equation 1 as an equation, involving a derivative, which is to be solved for the unknown function y; such equations are called **differential equations.**

The value of the above theorem on exponential growth is that it tells us almost exactly what the unknown function y is—all we need to do is find the value of the constant A in Equation 2. Half of this theorem can be proved easily; one merely checks that the function y in Equation 2 satisfies the differential equation, Equation 1. The converse is more difficult to prove; it will be omitted in this course.

As an illustration of the theorem on exponential growth, consider a plant or animal species with an unlimited food supply and without predators or competitors. The death rate will then be equal to a constant fraction of the population size. The birth rate, of course, is also equal to a constant fraction of the population size. The rate of change of population size, which is the difference between the birth rate and death rate, is therefore also proportional to population size. If y denotes the population size at time t, this conclusion means $\dfrac{dy}{dt} = ky$ for some constant k. By the theorem, we conclude $y = Ae^{kt}$ for some constant A.

Example 1. A bacteria population grows at a rate (cells per hour) equal to 9% of the population size. There were 1000 cells at time $t = 0$.

a. Find a formula for the number of cells after t hours.

b. How long will it take for the population to reach 13,000 cells? Give an exact answer, and then use a hand calculator, if possible, to find an approximate numerical answer.

SOLUTION.

a. Let y be the number of cells after t hours. We are given that $\dfrac{dy}{dt} = 0.09y$.

Therefore, by the theorem on exponential growth, y is of the form

$$y = Ae^{0.09t}.$$

Since there are 1000 cells at time $t = 0$, we can determine A:

$$1000 = Ae^0 = A,$$

and therefore

$$y = 1000e^{0.09t}$$

is the desired formula.

b. To find when the population reaches 13,000 we must solve the following equation for t.

$$1000e^{0.09t} = 13,000.$$

To solve, we use logarithms to obtain

$$\ln 1000 + \ln e^{0.09t} = \ln 13,000$$
$$0.09t = \ln 13,000 - \ln 1000$$
$$= \ln \frac{13,000}{1000} = \ln 13$$

so
$$t = \frac{1}{0.09} \ln 13.$$

Thus it takes $(\ln 13)/0.09$ hours for the bacteria count to reach 13,000. A hand calculator shows this to be about $28\frac{1}{2}$ hours.

Example 2. Air pressure decreases as altitude increases. The rate of change of pressure with respect to altitude is proportional to the pressure itself. At sea level, the pressure is 30 inches of mercury; at 25,000 feet it is 11.04 inches of mercury. Find a formula for the pressure p at any height h.

SOLUTION.

We are told $\dfrac{dp}{dh} = kp$ for some constant k. Therefore, by the theorem on exponential growth,

(3) $p = Ae^{kh}$.

We wish to determine the constants A and k. We also have the information

(4) $p = 30$ when $h = 0$

and

(5) $p = 11.04$ when $h = 25{,}000$.

By Statement 4 above, $30 = Ae^{k \cdot 0} = A$, so Equation 3 becomes

(6) $p = 30e^{kh}$.

From Statements 5 and 6,

$$11.04 = 30e^{k(25,000)}.$$

This equation can be solved for k using logarithms.

$$\ln 11.04 = \ln 30 + \ln e^{25,000k}$$

$$\ln \frac{11.04}{30} = 25{,}000k$$

so

(7) $k = \dfrac{1}{25{,}000} \ln \dfrac{11.04}{30}.$

A hand calculator shows this value (7) is approximately 0.00004. The desired formula for p is

$$p = 30e^{-0.00004h}.$$

e as a Limit. The number e was chosen so that the corresponding exponential function would have derivative 1 at the origin (see page 190). We now derive an important expression for e; it can be used to compute the decimal expansion of e and it will have interesting applications outside of mathematics.

Let $y = \ln x$. Then $\dfrac{dy}{dx} = \dfrac{1}{x}$, so

(8) $\qquad \dfrac{dy}{dx} = 1 \qquad$ when $\qquad x = 1.$

In Statement 8, we can replace $\dfrac{dy}{dx}$ by its definition as the limit of $\dfrac{\Delta y}{\Delta x}$ as $\Delta x \to 0$.

Now

(9) $\qquad \dfrac{\Delta y}{\Delta x} = \dfrac{\ln(1 + \Delta x) - \ln 1}{\Delta x} = \dfrac{1}{\Delta x}\ln(1 + \Delta x) = \ln(1 + \Delta x)^{1/\Delta x}.$

Hence Statement 8 implies that

$$\ln(1 + \Delta x)^{1/\Delta x} \to 1 \qquad \text{as} \qquad \Delta x \to 0,$$

which in turn implies that

(10) $\quad (1 + \Delta x)^{1/\Delta x} \to e \qquad$ as $\qquad \Delta x \to 0.$

Let $\Delta x = 1/n$; then Statement 10 can be written in the form

(11) $\quad \left(1 + \dfrac{1}{n}\right)^n \to e \qquad$ as $\qquad n \to +\infty.$

If $x \neq 0$ is fixed, then $x/n \to 0$ as $n \to +\infty$. Hence we can let $\Delta x = x/n$ in Statement 10 and obtain

$$\left(1 + \dfrac{x}{n}\right)^{n/x} \to e \qquad \text{as} \qquad n \to +\infty,$$

which implies

(12) $\quad \left(1 + \dfrac{x}{n}\right)^n \to e^x \qquad$ as $\qquad n \to +\infty.$

Statement 12 and its special case Statement 11 are the characterizations of e that we sought.

---

e as a Limit

(13) $\quad \lim_{n \to \infty} \left(1 + \dfrac{1}{n}\right)^n = e,$

(14) $\quad \lim_{n \to \infty} \left(1 + \dfrac{x}{n}\right)^n = e^x.$

Table 3 shows how Equation 13 can be used to calculate the decimal expansion of e. The value for $n = 1{,}000{,}000$ is correct to all nine decimal places.

n	$\left(1 + \dfrac{1}{n}\right)^n$
1	2.0
10	2.593742460 . . .
100	2.704613815 . . .
1,000	2.716923842 . . .
10,000	2.718145918 . . .
100,000	2.718254646 . . .
1,000,000	2.718281828 . . .

Example 3. In an experiment with extrasensory perception, a subject is given n sealed envelopes in random order. Each one contains a number $1, 2, 3, \ldots, n$. The subject uses his psychic powers to try to arrange the envelopes in order. Each envelope that ends up in its correct position constitutes a *match*. For example, if the tenth envelope in the final arrangement has the number 10 inside, that would be a match.

Mathematicians have shown that the probability that the subject scores zero matches is[14]

$$\left(1 - \frac{1}{n}\right)^n + \epsilon_n$$

where $\epsilon_n \to 0$ as $n \to \infty$.

By Equation 14, the probability of zero matches is about e^{-1} and is therefore *almost independent of n if n is large!* So no matter how large the number of envelopes, the subject has a probability of about $1 - e^{-1} \approx 0.63$ of obtaining at least one match.

Compound Interest. Businessmen often use tables to compute interest payments, mortgage payments, and other payment schedules. There are, however, many problems that arise which cannot be solved conveniently by tables. In such cases, it is necessary to know the basic exact and approximate formulas that apply.

[14]M. Dwass, *Probability Theory and Its Applications* (New York: W. A. Benjamin, Inc., 1970). p. 86.

A bank advertises 6% annual interest compounded quarterly on its savings accounts. This means that the interest rate of 0.06 is divided into four equal parts; this rate of interest, $0.06/4 = 0.015$ is paid on the balance each three months and added to the account,.

In general, suppose P dollars are deposited in an account paying $100r$ percent annual interest (thus 6% interest corresponds to $r = 0.06$). Suppose interest is compounded k times a year. The amount A in the account after n years is given by the following law.

Compound Interest Law

(15) $\quad A = P\left(1 + \dfrac{r}{k}\right)^{kn}.$

To prove the compound interest law, let S be the balance at the beginning of any interest period (one-kth of a year). The interest earned during that period is $(r/k)S$, so the balance at the end of the period is $S + (r/k)S$ or

$$S\left(1 + \frac{r}{k}\right).$$

We apply this conclusion to the original amount P and find that after n years (kn interest periods), the balance A is

$$\underbrace{P\left(1 + \frac{r}{k}\right)\left(1 + \frac{r}{k}\right)\cdots\left(1 + \frac{r}{k}\right),}_{kn\text{ factors}}$$

which gives Equation 15.

Not too long ago, banks compounded interest only annually or semi-annually. They lacked the bookkeeping capacity to calculate interest more frequently. If a customer withdrew money during an interest period, he received no interest at all for that period. With the development of electronic computers and data processing systems, banks are now able to offer their customers interest compounded more frequently. An advantage of this practice for the customer is that it makes money grow faster. Table 4 shows the balance A after one year if $1000 is deposited at 6% interest, compounded k times a year:

Table 4

k (frequency of compounding)	$A = 1000\left(1 + \dfrac{0.06}{k}\right)^k$
1 (annually)	$1060.00
2 (semiannually)	1060.90
4 (quarterly)	1061.36
6 (bimonthly)	1061.52
12 (monthly)	1061.68
365 (daily)	1061.83

Example 4. A ten-year account at a savings and loan pays 9% interest compounded daily. How much should be deposited now so that the balance in ten years will be $30,000?

SOLUTION.
Apply the compound interest law in Equation 15 with $A = \$30,000$, $r = 0.09$, $k = 365$, and $n = 10$.

$$30,000 = P\left(1 + \frac{0.09}{365}\right)^{3650}.$$

The desired answer is found by solving this equation for P.

$$(16) \quad P = \frac{30,000}{\left(1 + \dfrac{0.09}{365}\right)^{3650}}.$$

With a hand calculator that has exponentiation capability, we easily obtain, from Equation 16, $P = \$12,198.46$.

We can write the compound interest law in the form

$$(17) \quad A = P\left[\left(1 + \frac{1}{k/r}\right)^{k/r}\right]^{rn}.$$

Since r is very small, k/r is very large. Therefore, by the characterization in Equation 13 of e as a limit (see page 206), the term in square brackets in Equation 17 is approximately e. If we make it e, we obtain

$$(18) \quad A \approx Pe^{rn}$$

as an approximate compound interest law.

Equation 18 can be interpreted as an exact equation for *continuously compounded* interest, a concept introduced whimsically by the Swiss mathematician Jacob Bernoulli (1654–1705), but actually used by some banking establishments today. If we added "compounded continuously" to Table 4 on page 209, we would obtain

$$A = 1000e^{0.06} = 1061.836546 \ldots ,$$

which rounds off to \$1061.84, only one penny more than daily compounding.

The compound interest law is strictly correct only when used at the end of interest periods. The continuous interest formula in Equation 18 can be used for any number n; we emphasize this by replacing n with t.

Continuously-compounded Interest

(19) $A = Pe^{rt}$

Equation 19 gives the amount A in dollars that will result when P dollars are deposited in an account paying $100r$ percent annual interest for t years if interest is *continuously compounded.* Equation 19 is also used to approximate this situation when interest is compounded k times a year for t years.

Example 5.   A \$10,000 investment increases in value to \$12,550 in 307 days. What annual interest rate, compounded continuously, would yield the same result?

SOLUTION.
We apply Equation 19, the formula for continuously-compounded interest, with $P = \$10,000$, $A = \$12,550$, and $t = 307/365$ years. This gives

$$12,550 = 10,000e^{307r/365}.$$

We solve this equation for r, the unknown interest rate, by using logarithms. The result will be

$$r = \frac{365}{307} \ln \frac{12,550}{10,000}.$$

A hand calculator gives the value $0.270047179 \ldots$ for r, so an interest rate of approximately 27% would be the correct answer.

Problems

In the following problems, give your answer in a form that can be computed numerically with a good hand calculator (that is, one with operations $\ln x$, e^x, and y^x, as well as $+$, $-$, $\times$, $\div$). If you have such a calculator, give the numerical answer as well.

1. The population of a colony of Drosophila increases at a rate proportional to the population itself. A population of 100 grew to 1000 in 5.4 days. What was the population after 8 days?

2. A ship collision in a harbor resulted in a spill of contaminants. Their concentration decreased, due to tidal circulation with the sea, at a rate proportional to the concentration itself. The initial concentration was 25 parts per million, and after 24 hours, it was 5 parts per million. When will the concentration reach 0.5 parts per million?

3. The population of the Philippines grows at an annual rate of 0.035 times the population size.[15] The corresponding rates for the United States and England are 0.012 and 0.005, respectively. Assume these rates continue in the future. How long for the population to triple in the Philippines? In the United States? In England?

4. Find the following limits.

   a. $\displaystyle\lim_{n \to \infty} \left(1 + \frac{2}{n}\right)^n$.

   b. $\displaystyle\lim_{t \to \infty} \left(1 + \frac{1}{3t}\right)^t$.

   c. $\displaystyle\lim_{x \to 0} (1 + 3x)^{1/x}$.

   d. $\displaystyle\lim_{m \to \infty} \left(2 + \frac{2}{3m}\right)^{3m}$.

5. A man deposits \$10,000 with a bank that pays 7.5% interest. If interest is compounded semiannually, what will his balance be after 9 years?

6. A man bought some stock for \$1000 in 1930. He sold it in 1978 for \$46,525.
   a. What interest rate, compounded quarterly, would have yielded the same result?
   b. What interest rate compounded continuously would have yielded the same result?

7. A bank advertises that money left in one of its special savings accounts for 18 years will triple in value. What interest does the bank pay (assume continuous compounding)?

8. In 1626, Peter Minuit bought Manhattan Island for \$24. Had that sum been placed in an account earning 6% annual interest compounded continuously, what would the balance have been in 1976?

[15] *The World Almanac* (New York: Doubleday and Co., 1977).

Summary

Perhaps the most important new formulas in this chapter are the *differentiation formulas:*

$$\text{If} \quad y = e^x, \quad \text{then} \quad \frac{dy}{dx} = e^x.$$

$$\text{If} \quad y = e^u, \quad \text{then} \quad \frac{dy}{dx} = (e^u)\frac{du}{dx}.$$

$$\text{If} \quad y = \ln x, \quad \text{then} \quad \frac{dy}{dx} = \frac{1}{x}.$$

$$\text{If} \quad y = \ln u, \quad \text{then} \quad \frac{dy}{dx} = \frac{1}{u} \cdot \frac{du}{dx}.$$

You should also be very familiar with the algebraic properties of exponential and logarithmic functions given in Properties 9, 10, and 11 on page 177 and Properties 7, 8, and 9 on page 185.

The *differential equation for the exponential function* says

$$\text{if} \quad \frac{dy}{dx} = ky, \quad \text{then} \quad y = Ae^{kx}.$$

This fact is central for many applications of the exponential function. Also of interest is the functional equation for e^x (Equation 12, page 178) and the expression of e^x *as a limit:*

$$e^x = \lim_{n \to \infty} \left(1 + \frac{x}{n}\right)^n.$$

Applications to business and finance rest on the *compound interest law*

$$A = P\left(1 + \frac{r}{k}\right)^{kn}$$

and the *continuously-compounded interest law*

$$A = Pe^{rt}.$$

Miscellaneous Problems

1. Sketch the functions $y = \ln x$ and $y = \ln \dfrac{1}{x}$ on the same graph.

2. Sketch the following functions.

    a. $y = e^{2x}$.
             b. $y = e^{-2x}$.

    c. $y = 1 - e^{-2x}$.
             d. $y = \dfrac{1}{1 + e^x}$.

3. Use the facts that $\ln 2 = 0.69$ and $\ln 3 = 1.10$ to find each of the following.

    a. $\ln 36$.
             b. $\ln(4/3)$.

    c. $\ln \sqrt[3]{4/3}$.
             d. $e^{1.79}$.

4. The Richter magnitude of an earthquake is the base 10 logarithm of the strength of the earthquake as shown on a seismograph. On November 6, 1976, an earthquake of Richter magnitude 6.8 rocked the eastern coast of Mindanao. On August 16 of that year, in the same area, an earthquake 16 times stronger had caused tidal waves that killed 8000 people. What was the Richter magnitude of the August 16 earthquake?[16]

5. Differentiate the following functions.

    a. $y = 2x^{1/3} - e^{3x} + \ln x$.
           b. $s = e^{0.06t}$.

    c. $f(x) = 1255e^{-3.2x}$.
           d. $y = \frac{1}{2}(e^x + e^{-x})$.

    e. $w = 6^{2t}$.
             f. $q = \ln(2p + 1)$.

6. Use logarithmic differentiation to find $\dfrac{dS}{dr}$ if $S = r^{1.2}(3r + 2)(r^2 + 1)$.

7. The probability that exactly k calls will come through a telephone company switchboard in a one second time interval is

$$P = \frac{\lambda^k e^{-\lambda}}{1 \cdot 2 \cdot 3 \, \cdots \, k},$$

where λ, the Greek letter *lambda*, is a number which depends on the time of day (the *Poisson probability parameter*). Fix $k = 4$; then find $\dfrac{dP}{d\lambda}$.

8. Let $I(p)$ be the surprise you feel when you win a game in which your probability of winning was p, with $0 < p \le 1$.[17] Thus $I(1) = 0$ (no surprise in winning a certainty) and $I(p) \to +\infty$ as $p \to 0$ (infinite surprise over an impossible win). Argue that $I(p \cdot q) = I(p) + I(q)$ (use the fact that the probability of winning a game with probability p *and a* game with probability q is $p \cdot q$). Use the functional equation (Equation 10 on page 186) to show that $I(p) = \log_b p$, and in this case $0 < b < 1$. If, in addition, $I(1/e) = 1$ (a "normalization"), show that $I(p) = -\ln p$.

---

[16]Give your answer in a form that can be evaluated easily with a hand calculator that has e^x, $\ln x$, and y^x operations. If you have such a calculator, give the numerical answer as well.

[17]Adapted from C. E. Shannon and W. Weaver, *The Mathematical Theory of Communication* (Urbana, Illinois: The University of Illinois Press, 1949).

9. A culture of bacteria grows at a rate proportional to the population. If the population was 800 at time $t = 0$ and 1100 at time $t = 6.4$, find the population at time $t = 10$.[18]

10. Bank A pays 6% interest compounded annually. Bank B pays 5.9% interest compounded continuously. A deposit of $1000 is made in each bank.
    a. What is the balance in bank A after 10 years?
    b. What is the balance in bank B after 10 years?[18]

11. Radium decays at a rate in milligrams per century which is 0.038 times the mass M in milligrams present. What will be the mass 10,000 years from now of a vial of radium of present mass 100 milligrams?

12. A stone, initially at rest, is allowed to fall freely. Let v be the speed after t seconds; let s be the distance fallen after t seconds. Galileo asked if v is proportional to s or to t. He argued that the assumption $v = ks$, k a constant, is impossible.[19] Use the differential equation for exponential growth to confirm that $v = ks$ leads to the impossible conclusion that s is constantly zero or else s was not zero when $t = 0$.

13. In the study of pollution and sewage disposal, the equation $\dfrac{dD}{dt} = -K_2 D$ arises,[20] where D is dissolved oxygen deficit, t is time, and K_2 is a constant called the *reaeration coefficient*. Suppose $D = 2$ when $t = 0$, and $K_2 = 3.2$; express D as a function of t.

[18]Give your answer in a form that can be evaluated easily with a hand calculator that has e^x, ln x, and y^x operations. If you have such a calculator, give the numerical answer as well.

[19]Galileo Galilei, *Dialogues Concerning Two New Sciences* (1636).

[20]See Equation 6, p. 517, of *The McGraw-Hill Encyclopedia of Environmental Science* (1974).

six

Integration

The development of calculus can be traced back to two fundamental problems. One was calculating the tangent line to a given curve—a problem in **differential calculus.** The other was calculating the area surrounded by a given closed curve—a problem in **integral calculus.** There is an important connection between these two problems which is not at all obvious at first; the connection is called the fundamental theorem of calculus (see Section 5, page 250).

In this chapter, we do not follow the historical development. Rather than treat the area problem as a starting point, we treat it as an application of the technique of antidifferentiation (reversing differentiation rules). The significance of the area problem can still be appreciated because it provides an intuitive existence proof for solutions to the problem of reversing differentiation.

Section 1 □ Antidifferentiation.   Uniqueness, side conditions.

Section 2 □ Integration Formulas.   Evaluating an integral by guesswork.

Section 3 □ Techniques of Integration.   Integration by substitution, integration by parts, tables of integrals.

Section 4 □ Applications of Integration.   Integrating rates of change.

Section 5 □ The Area under a Graph.   Signed area, the fundamental theorem of calculus.

Section 6 □ Volumes.

Summary □ Miscellaneous Problems.   Consumer surplus, average value, statistical applications (mean and standard deviation).

In preceding chapters, we studied the differentiation problem:

Given a function, find a formula for its derivative.

We now study the reverse problem:

Given a formula, find a function with that formula as its derivative.

An example of the reverse problem is to find a function $y = f(x)$ so that $\frac{dy}{dx} = x^2$. One answer is $y = x^3/3$ (there are other answers). In this reverse problem, we are given the rate of change of a quantity (in this case $\frac{dy}{dx}$, the rate of change of y) and we are asked to find a formula for the quantity itself (in this case a formula for y at any value of x). The process of passing from the derivative of a function to the function itself is called **integration** or **antidifferentiation.** Differentiation and integration are said to be **inverse operations.**

The following three examples illustrate how to find a quantity by integrating its rate of change.

Example 1. A rocket is fired vertically upward from the ground. Its speed after t seconds is $\sqrt{t}$ feet per second. Find the height h of the rocket after t seconds.

SOLUTION.

If h is the height in feet after t seconds, then the speed after t seconds is $\frac{dh}{dt}$ (see page 96; *speed is the rate of change of distance.*) We are given the speed:

(1) $\quad \frac{dh}{dt} = \sqrt{t}$;

we want to find the function h. You can verify that

(2) $\quad h = \frac{2}{3}t^{3/2} + C$

will satisfy Equation 1 if C is any constant (indeed, from Equation 2, we have $\frac{dh}{dt} = (\frac{2}{3})(\frac{3}{2})t^{(3/2)-1} + 0 = t^{1/2} = \sqrt{t}$). We shall see later in this section that Equation 2 is the most general function h which satisfies Equation 1. In the present problem, we know that $h = 0$ when $t = 0$ because the rocket was fired from ground level. This forces the value of C in Equation 2 to be zero. The answer is $h = (\frac{2}{3})t^{3/2}$.

Example 2.   A patient is injected with a glucose solution containing a radioactive carbon isotope tracer. The radioactive carbon appears instantly in his exhaled breath. A Geiger counter reveals that after t minutes, the metabolized isotope is being released in exhaled breath at a rate of $2.4e^{-0.12t}$ microcuries per minute. How much isotope has been exhaled after 10 minutes?

SOLUTION.

Let q be the total amount in microcuries of isotope exhaled after t minutes. We are given that

$$(3) \qquad \frac{dq}{dt} = 2.4e^{-0.12t}.$$

We wish to find a function q which satisfies Equation 3. We shall learn later that the most general function is

$$(4) \qquad q = -20e^{-0.12t} + C;$$

at this time, you should check that the function q in Equation 4 does indeed satisfy Equation 3.

By the nature of this problem, we must have $q = 0$ when $t = 0$; thus $C = 20$ in Equation 4. Therefore the total amount of exhaled isotope at time t is

$$(5) \qquad q = 20 - 20e^{-0.12t}.$$

The total amount exhaled after 10 minutes is found by setting $t = 10$ in Equation 5. Using a hand calculator, we obtain the answer

$$q = 20 - 20e^{-(0.12)(10)} \approx 13.98 \text{ microcuries.}$$

Uniqueness.   Let $y = f(x)$ be a given function. If F is a function with

$$F'(x) = f(x),$$

then F is called an **antiderivative** or **indefinite integral** of f.

For example, an antiderivative of $f(x) = x^3$ is $F(x) = x^4/4$ because

$$F'(x) = \left(\frac{1}{4}\right)(4)x^3 = x^3 = f(x).$$

Similarly, an antiderivative of x^5 is $x^6/6$.

We have just seen that $F(x) = x^4/4$ is an antiderivative of $f(x) = x^3$. We can easily see that if C is any constant, then

$$F(x) = \frac{1}{4}x^4 + C$$

is also an antiderivative of $f(x) = x^3$. Are there others too? The following theorem says there are not.

Uniqueness of Antiderivatives

If $F(x)$ is an antiderivative of a function $f(x)$ then every antiderivative of $f(x)$ is of the form

(6) $F(x) + C$

where C is a constant.

We can give an instructive and intuitive proof of this uniqueness result. Suppose $F(x)$ and $G(x)$ are antiderivatives of $f(x)$. Then the function $H(x) = G(x) - F(x)$ has derivative zero everywhere:

(7) $H'(x) = G'(x) - F'(x) = f(x) - f(x) = 0.$

Equation 7 says that the rate of change of $H(x)$ is always zero. This suggests that $H(x)$ does not change at all and so is a constant. We obtain the above uniqueness result by writing $H(x) = C$, a constant, and thus $G(x) = F(x) + C$.

Notation.   The symbol

(8) $\int f(x)\,dx$

is used to denote the most general antiderivative of f. The sign $\int$ in Notation 8 is called an **integral sign.** The expression following the integral sign, $f(x)$, is called the **integrand.** The dx symbol has its origin in some ideas of Leibnitz concerning infinitesimals. Such infinitesimals are no longer used in standard calculus,[1] but the dx notation has remained because of its usefulness. One purpose served by dx is to designate the name of the variable (see Problem 7 on page 223). Another purpose will be seen in Section 3, when the methods of substitution and integration by parts are discussed.

---

[1]"Nonstandard Calculus," which uses a new notion of infinitesimals, was developed by Abraham Robinson in the 1960s. That approach is followed in the text *Elementary Calculus* by H. Jerome Keisler (Boston: Prindle, Weber, and Schmidt Inc., 1976).

As an illustration of Notation 8, observe that

$$\int x^2 \, dx = \frac{1}{3}x^3 + C$$

because $(\frac{1}{3})x^3 + C$ is the most general antiderivative of x^2. To check the above equation, differentiate the right-hand side; the result should be equal to the integrand on the left-hand side. All differentiation formulas can be checked this way. An important generalization is the following formula, which can be verified in the same way.

Indefinite integral of x^n

(9) $$\int x^n \, dx = \frac{x^{n+1}}{n+1} + C, \qquad n \ne -1$$

Example 3. Evaluate each integral.

a. $\int x^3 \, dx.$ **b.** $\int x^{12} \, dx.$ **c.** $\int \frac{dx}{x^2}.$ **d.** $\int \sqrt{x} \, dx.$

SOLUTION.

a. Apply Formula 9 for $n = 3$.

$$\int x^3 \, dx = \frac{x^4}{4} + C.$$

b. Apply Formula 9 for $n = 12$.

$$\int x^{12} \, dx = \frac{x^{13}}{13} + C.$$

c. The notation in the problem,

$$\int \frac{dx}{x^2},$$

is an abbreviation for

$$\int \frac{1}{x^2} dx.$$

Now $\int (1/x^2)dx = \int x^{-2}dx$, so we may apply Formula 9 with $n = -2$.

$$\int \frac{dx}{x^2} = \frac{x^{-2+1}}{-2+1} + C = -\frac{1}{x} + C.$$

d. Apply Formula 9 with $n = \frac{1}{2}$.

$$\int \sqrt{x}\,dx = \frac{x^{(1/2)+1}}{(1/2) + 1} + C = \frac{2}{3}x^{3/2} + C.$$

Example 4. Verify that $\int(1 + \ln x)dx = x \ln x + C$.

SOLUTION.

If $F(x) = x \ln x$, then by the product rule,

$$F'(x) = (x)\left(\frac{1}{x}\right) + (\ln x)(1) = 1 + \ln x.$$

Thus $x \ln x$ is an antiderivative of $1 + \ln x$. The most general antiderivative is therefore $x \ln x + C$, which gives the desired equation.

Example 5. Notation 8 is also used with letters other than x as the independent variable. For example:

$$\int t^4 dt = \frac{t^5}{5} + C,$$

$$\int u^{3/5}\,du = \frac{5}{8}u^{8/5} + C,$$

(10) $$\int \frac{dp}{\sqrt{p}} = 2\sqrt{p} + C.$$

Note the notational shorthand used in the last equation: $\int \dfrac{dp}{\sqrt{p}}$ is used for $\int \dfrac{1}{\sqrt{p}}dp$. Note also that equations of this kind can be checked directly by differentiating the right-hand side and comparing it with the integrand.

CHECK OF EQUATION 10.

$$\frac{d}{dp}(2\sqrt{p} + C) = \frac{d}{dp}(2p^{1/2} + C)$$

$$= (2)\left(\frac{1}{2}\right)p^{(1/2)-1} + 0$$

$$= p^{-1/2}$$

$$= \frac{1}{\sqrt{p}} = \text{the integrand.}$$

Example 6. (Finding antiderivatives which satisfy a side condition.) Find the antiderivative which satisfies each side condition.

a. $\dfrac{dy}{dx} = x^3$ and $y = 5$ when $x = 0$.

b. $\dfrac{dy}{dx} = x^{12}$ and $y = 1$ when $x = 1$.

c. $\dfrac{dy}{dx} = \dfrac{1}{x^2}$ and $y = 0$ when $x = \frac{1}{2}$.

d. $\dfrac{dy}{dx} = \sqrt{x}$ and $y = 1$ when $x = 9$.

SOLUTION.
In each case, find the most general antiderivative and then choose a value for the constant C so that the side condition is satisfied. The general antiderivatives were found in Example 3.

a. $y = \dfrac{x^4}{4} + C$ General antiderivative

$5 = \dfrac{0^4}{4} + C,$ so $C = 5$ Impose the side condition and solve for C

$y = \dfrac{x^4}{4} + 5$ Answer

b. $y = \dfrac{x^{13}}{13} + C$ General antiderivative

$1 = \dfrac{1}{13} + C,$ so $C = \dfrac{12}{13}$ Impose the side condition and solve for C

$y = \dfrac{x^3}{13} + \dfrac{12}{13}$ Answer

c. $y = -\dfrac{1}{x} + C$ General antiderivative

$0 = -\dfrac{1}{1/2} + C,$ so $C = 2$ Impose the side condition and solve for C.

$y = -\dfrac{1}{x} + 2$ Answer

d. $y = \dfrac{2}{3}x^{3/2} + C$ General antiderivative

$$1 = \frac{2}{3}(9)^{3/2} + C \qquad \text{so} \qquad C = -17 \qquad \begin{array}{l}\text{Impose the side condition}\\ \text{and solve for } C\end{array}$$

$$y = \frac{2}{3}x^{3/2} - 17 \qquad\qquad\qquad \text{Answer}$$

Problems

1. Evaluate.

   a. $\int x^6\, dx.$ b. $\int q^{1.3}\, dq.$

   c. $\int \frac{dt}{t^3}.$ d. $\int \frac{\sqrt{x}\, dx}{x^2}.$

2. Evaluate.

   a. $\int x\sqrt{x}\, dx.$ b. $\int \frac{du}{u\sqrt{u}}.$

   c. $\int \frac{t}{\sqrt[3]{t}}\, dt.$ d. $\int \frac{p^{1.5}\, dp}{\sqrt{p}}.$

3. Match each entry in the first column with an entry in the second column.

   a. $\int 6e^{-3x}\, dx.$ 1. $2e^{3x} + C.$

   b. $\int -6e^{3x}\, dx.$ 2. $-2e^{-3x} + C.$

   c. $\int 6e^{3x}\, dx.$ 3. $2e^{-3x} + C.$

   d. $\int -6e^{-3x}\, dx.$ 4. $-2e^{3x} + C.$

4. Match each entry in the first column with an entry in the second column.

   a. $\int (3x + 1)^3\, dx.$ 1. $\frac{1}{6}(3x + 1)^2 + C.$

   b. $\int (3x + 1)\, dx.$ 2. $\frac{1}{2}(3x + 1)^2 + C.$

   c. $\int 3(3x + 1)\, dx.$ 3. $\frac{1}{4}(3x + 1)^4 + C.$

   d. $\int 3(3x + 1)^3\, dx.$ 4. $\frac{1}{12}(3x + 1)^4 + C.$

5. Find antiderivatives which satisfy the side conditions (compare with Problem 1).

   a. $\dfrac{dy}{dx} = x^6, \qquad y = 1 \qquad \text{when} \qquad x = 1.$

b. $\dfrac{dp}{dq} = q^{1.3}$, $\quad p = 8.17$ when $\quad q = 0$.

c. $\dfrac{ds}{dt} = t^{-3}$, $\quad s = 8$ $\quad$ when $\quad t = \frac{1}{2}$.

d. $\dfrac{dC}{dx} = \dfrac{\sqrt{x}}{x^2}$, $\quad C = 0$ $\quad$ when $\quad x = 4$.

6. Find antiderivatives which satisfy the side conditions (compare with Problem 2).

a. $\dfrac{dy}{dx} = x\sqrt{x}$, $\quad y = -3$ when $\quad x = 0$.

b. $\dfrac{dv}{du} = \dfrac{1}{u\sqrt{u}}$, $\quad v = 1$ $\quad$ when $\quad u = 4$.

c. $\dfrac{dR}{dt} = \dfrac{t}{\sqrt[3]{t}}$, $\quad R = \frac{3}{5}$ $\quad$ when $\quad t = 0$.

d. $\dfrac{dq}{dp} = \dfrac{p^{1.5}}{\sqrt{p}}$, $\quad q = 1$ $\quad$ when $\quad p = 4$.

7. Evaluate.

a. $\displaystyle\int 2uv^2 \, du.$ $\qquad\qquad\qquad$ **b.** $\displaystyle\int 2uv^2 \, dv.$

Section 2 □ Integration Formulas

In this and the next section we discuss the problem of evaluating indefinite integrals. Although there are many methods for attacking such problems, none of them is guaranteed to succeed. The "method" to be discussed in this section is that of guesswork.

Before we discuss the use of guesswork, we mention two formulas that help to simplify indefinite integrals.

---

(1) $\displaystyle\int [f(x) \pm g(x)] \, dx = \int f(x) \, dx \pm \int g(x) \, dx.$

(2) $\displaystyle\int kf(x) \, dx = k \int f(x) \, dx, \quad k$ a constant.

The verification of these formulas is accomplished by showing that the derivative of the right-hand side of the equation is equal to the integrand on the left-hand side.

Rules 1 and 2, together with the formula

$$(3) \quad \int x^n \, dx = \frac{x^{n+1}}{n+1} + C, \qquad n \neq -1$$

from the previous section, allow us to integrate any polynomial function. For example:

$$\int (5x^3 - 8x) \, dx = \int 5x^3 \, dx - \int 8x \, dx$$

$$= 5 \int x^3 \, dx - 8 \int x \, dx$$

$$= 5\left(\frac{x^4}{4}\right) - 8\left(\frac{x^2}{2}\right) + C,$$

$$(4) \qquad\qquad = \frac{5}{4}x^4 - 4x^2 + C.$$

Note that we used only one arbitrary constant C; we could have introduced a different constant from each integral as in the following.

$$5 \int x^3 \, dx - 8 \int x \, dx = 5\left(\frac{x^4}{4} + C_1\right) - 8\left(\frac{x^2}{2} + C_2\right)$$

$$(5) \qquad\qquad = \frac{5}{4}x^4 - 4x^2 + 5C_1 - 8C_2.$$

Answers 4 and 5 are equivalent; if C_1 and C_2 are arbitrary constants, so is $5C_1 - 8C_2$, and therefore it may be replaced by C. Since Answer 4 is simpler, it is preferable.

Rule 1 can be extended to sums and differences of more than two functions.

$$\int \left(6\sqrt{x} - 5 + \frac{3}{x^2}\right) dx = \int 6\sqrt{x} \, dx - \int 5 \, dx + \int \frac{3}{x^2} \, dx$$

$$= 6 \int x^{1/2} \, dx - 5 \int dx + 3 \int x^{-2} \, dx$$

$$= 6\left(\frac{x^{3/2}}{3/2}\right) - 5x + 3\left(\frac{x^{-1}}{-1}\right) + C$$

$$= 4x^{3/2} - 5x - 3x^{-1} + C.$$

In working from the second to the third line, we used

$$(6) \quad \int dx = x + C_1,$$

where $\int dx$ is an abbreviation for $\int 1 \, dx$.

Two other important integration formulas are the following.

Integral of $\dfrac{1}{x}$

(7) $\quad \displaystyle\int \frac{1}{x}\,dx = \ln x + C.$

Integral of e^{kx}

(8) $\quad \displaystyle\int e^{kx}\,dx = \frac{1}{k}(e^{kx}) + C, \qquad k$ a constant.

Formulas 7 and 8 are verified in the usual way—differentiate the right-hand side and verify that the result equals the integrand on the left-hand side.

Example 1. Evaluate.

a. $\displaystyle\int \left(e^{3x} + \frac{2}{x} \right) dx.$

b. $\displaystyle\int \left(\frac{1}{5t} - 4e^{-3t} \right) dt.$

SOLUTION.

a. $\displaystyle\int \left(e^{3x} + \frac{2}{x} \right) dx = \int e^{3x}\,dx + 2\int \frac{1}{x}\,dx$

$\qquad\qquad\qquad\qquad = \frac{1}{3}e^{3x} + 2 \ln x + C.$

b. $\displaystyle\int \left(\frac{1}{5t} - 4e^{-3t} \right) dt = \frac{1}{5}\int \frac{1}{t}\,dt - 4\int e^{-3t}\,dt$

$\qquad\qquad\qquad\qquad = \frac{1}{5}\ln t - 4\left(\frac{1}{-3}e^{-3t} \right) + C$

$\qquad\qquad\qquad\qquad = \frac{1}{5}\ln t + \frac{4}{3}e^{-3t} + C.$

Evaluating an Integral by Guesswork. To evaluate an integral by guesswork, make a *first guess* at the answer. Differentiate that first guess and compare the result to the original integrand. The chances are that it will not agree perfectly—

if it does you have solved the problem with one guess. Ask yourself whether you can multiply your first guess by a constant, whether you can add a term to it, or whether some other modification of it might lead to a correct answer.

Example 2. Evaluate $\int (3x + 5)^6\, dx$.

SOLUTION.
Try

(9) $(3x + 5)^7$

as a first guess. The derivative of Expression 9 is

(10) $21(3x + 5)^6$,

which is not equal to the original integrand:

(11) $21(3x + 5)^6 \neq (3x + 5)^6$.

We see by Inequality 11 that our first guess was in error by a factor of 21. If we multiply our first guess, $(3x + 5)^7$, by $\frac{1}{21}$, we might obtain the correct answer. Therefore we now guess

(12) $\frac{1}{21}(3x + 5)^7$

as the answer. When we differentiate Expression 12, we obtain $\frac{1}{21} \cdot 7 \cdot 3(3x + 5)^6$, or $(3x + 5)^6$, which is exactly equal to the integral. Thus

(13) $\int (3x + 5)^6\, dx = \frac{1}{21}(3x + 5)^7 + C$.

As a final check, differentiate the right-hand side of Equation 13 and make sure the result is the integrand on the left.

Example 3. Evaluate $\int \dfrac{4x^3 + x - 2}{3x}\, dx$.

SOLUTION.
Before guessing, let us see if algebraic simplifications will help.

$$\int \frac{4x^3 + x - 2}{3x}\, dx = \int \left(\frac{4x^3}{3x} + \frac{x}{3x} - \frac{2}{3x} \right) dx$$

$$= \int \left(\frac{4}{3}x^2 + \frac{1}{3} - \frac{2}{3x} \right) dx$$

$$= \frac{4}{3}\int x^2 \, dx + \frac{1}{3}\int dx - \frac{2}{3}\int \frac{1}{x} \, dx$$

$$= \frac{4}{3}\left(\frac{x^3}{3}\right) + \frac{1}{3}(x) - \frac{2}{3}(\ln x) + C$$

In this case we found the answer without any guesswork.

Example 4. Find $\int x(x^2 - 3)^3 \, dx$.

SOLUTION.

One way to solve this would be to multiply out the integrand, obtaining a polynomial of seventh degree. The multiplications would be tedious, but the resulting polynomial would be easy to integrate. Another way is to guess an answer. A very good first guess would be

(14) $(x^2 - 3)^4$.

Differentiate this, obtaining $4 \cdot 2x(x^2 - 3)^3$, and compare it with the original integrand:

(15) $8x(x^2 - 3)^3 \neq x(x^2 - 3)^3$.

This shows that we should modify the first guess, $(x^2 - 3)^4$, to a new guess,

(16) $\frac{1}{8}(x^2 - 3)^4$.

The derivative of Expression 16 is $\frac{1}{8} \cdot 4 \cdot 2x(x^2 - 3)^3$, or $x(x^2 - 3)^3$. This agrees perfectly with the integrand. The answer is therefore

$$\int x(x^2 - 3)^3 \, dx = \frac{1}{8}(x^2 - 3)^4 + C.$$

Example 5. Find $\int xe^x \, dx$.

SOLUTION.

A good first guess would be

(17) xe^x.

Differentiate Expression 17, obtaining

(18) $xe^x + e^x$,

and compare the result with the integrand:

(19) $xe^x + e^x \neq xe^x$.

We see by Inequality 19 that we have a term e^x to eliminate. We can subtract the term e^x from our first guess; this leads to

(20) $xe^x - e^x$

as a second guess. The derivative of this second guess is

$$xe^x + e^x - e^x,$$

which agrees with the integrand. Therefore

$$\int xe^x dx = xe^x - e^x + C.$$

Example 6. Evaluate $\int \sqrt{5t - 1}\, dt$.

SOLUTION.
Write the integral as $\int (5t - 1)^{1/2} dt$. This suggests

(21) $\dfrac{(5t - 1)^{(1/2)+1}}{(1/2) + 1}$ or $\dfrac{2}{3}(5t - 1)^{3/2},$

as a first guess. Differentiate Expression 21 and compare this result with the desired result:

(22) $5(5t - 1)^{1/2} \neq (5t - 1)^{1/2}.$

The unwanted factor 5 in Inequality 22 suggests that we modify the first guess by putting in a factor of $\frac{1}{5}$. The second guess would then be

(23) $\dfrac{2}{15}(5t - 1)^{3/2}.$

The derivative of Expression 23 is

$$\frac{2}{15}\cdot\frac{3}{2}\cdot 5(5t - 1)^{(3/2)-1} \quad \text{or} \quad (5t - 1)^{1/2},$$

which does agree with $(5t - 1)^{1/2}$. The answer is thus

$$\int \sqrt{5t - 1}\, dt = \frac{2}{15}(5t - 1)^{3/2} + C.$$

Problems

In Problems 1–30, evaluate the indefinite integrals.

1. $\int(2x + 1)dx.$

2. $\int(3x^2 - 2)dx.$

3. $\int(6u - 2)du.$

4. $\int(1 + 8t)dt.$

5. $\int(12x^3 + \frac{1}{2})dx.$

6. $\int(9w^2 - 6w)dw.$

7. $\int(x^2 - 3x + 5)dx.$

8. $\int\left(\frac{6}{x^3} - 1\right)dx.$

9. $\int\left(1.5x^4 - \frac{3}{x^{1.3}}\right)dx.$

10. $\int\left(x^3 - \frac{2}{x}\right)dx.$

11. $\int\frac{4x^3 - 8x^2 + 1}{2x}dx.$

12. $\int\frac{4}{x^2}(x^2 - 2x + 1)dx.$

13. $\int(2x + 1)\left(1 - \frac{1}{x}\right)dx.$

14. $\int(\sqrt{x} - 2e^{5x})dx.$

15. $\int(e^{2x} - 6e^{-3x})dx.$

16. $\int(3e^{-t} + \sqrt[3]{t})dt.$

17. $\int\left(1 + \frac{3}{u}\right)du.$

18. $\int\left(\sqrt{s} - \frac{4}{\sqrt{s}}\right)ds.$

19. $\int 5\left(\frac{1}{3x}\right)dx.$

20. $\int(3u - e^{u/2})du.$

21. $\int(2x + 1)^7 dx.$

22. $\int\left(\frac{x}{3} - 1\right)^8 dx.$

23. $\int x(x^2 + 1)^5 dx.$

24. $\int(8t + 3)^6 dt.$

25. $\int x(2 - x^2)dx.$

26. $\int\left(5 - \frac{1}{2}w\right)^4 dw.$

27. $\int e^{2x}(e^{2x} + 1)^3 dx.$

28. $\int xe^{2x} dx.$

29. $\int\sqrt{5t + 1} dt.$

30. $\int x\sqrt{2x^2 + 1} dx.$

In Problems 31–36, find the antiderivative satisfying the given side condition.

31. $\dfrac{dy}{dx} = \dfrac{3}{x}, y = 7$ when $x = e^2.$

32. $\dfrac{dy}{dx} = 16e^{4x}, y = 0$ when $x = 0.$

33. $\dfrac{dR}{dt} = 2e^{2t} + 1, R = 2$ when $t = 0.$

34. $\dfrac{dp}{dq} = 2 + \dfrac{1}{2\sqrt{q}}, p = 21$ when $q = 9.$

35. $\dfrac{dy}{dx} = 10(2x + 1)^4$, $y = 0$ when $x = 0$.

36. $\dfrac{dw}{dt} = te^t$, $w = 0$ when $t = 0$.

Section 3 □ Techniques of Integration

In the previous section, the problem of evaluating a definite integral was approached by:

1. trying to simplify the integral by algebra or by use of the integration rules (Rules 1 and 2 on page 223);

2. applying, if possible, the integration formulas (Formulas 3, 7, and 8 on pages 224 and 225); and

3. using guesswork to find the solution.

If this routine does not lead to success, there are three more methods to try: *substitution, integration by parts,* and *searching a table of integrals.* We shall consider these methods in the present section.

Integration by Substitution. This method for evaluating

(1) $\int f(x)dx$

has its theoretical basis in the chain rule. To apply it, we select some function of x; call it u. (Some choices for u will lead to success, while some will not. The ability to make a good choice for u should improve with practice.) The function u is substituted into Expression 1 in the hope of arriving at an equation of the form

(2) $f(x)dx = g(u)du,$

which is really an abbreviation for

$$f(x) = g(u)\dfrac{du}{dx}.$$

If Equation 2 can be accomplished and if $\int g(u)du$ can be evaluated, then the method will lead to success. The details of this technique are best explained by means of examples.

Example 1. We shall evaluate

(3) $\int x(x^2 - 3)^3 \, dx$

by substitution. Choose the expression $x^2 - 3$ for u (usually a prominent expression appearing in the integrand is a good choice for u); thus

(4) $u = x^2 - 3,$ and so $\dfrac{du}{dx} = 2x.$

Write Expression 4 in the symbolic form

(5) $u = x^2 - 3,$ $du = 2x \, dx.$

The integrand in Expression 3 is

$$x(x^2 - 3)^3 \, dx.$$

Substitute Expression 5 into this integrand; we obtain

(7) $x(x^2 - 3)^3 \, dx = \dfrac{1}{2}u^3 \, du$

because $\frac{1}{2}du = x \, dx$ and $u^3 = (x^2 - 3)^3$. From Equation 7,

$$\int x(x^2 - 3)^3 \, dx = \int \frac{1}{2}u^3 \, du$$

$$= \frac{1}{2}\int u^3 \, du = \frac{1}{8}u^4 + C.$$

In the last expression, replace u by $x^2 - 3$.

$$\int x(x^2 - 3)^3 \, dx = \frac{1}{8}(x^2 - 3)^4 + C.$$

Example 2. Evaluate $\int \dfrac{3x^2 + 1}{\sqrt{x^3 + x}} dx$ by substitution.

SOLUTION.

Try $u = x^3 + x$. Then $\dfrac{du}{dx} = 3x^2 + 1$, which we write as

(8) $u = x^3 + x,$ $du = (3x^2 + 1)dx.$

From Expression 8,

$$\int \frac{3x^2 + 1}{\sqrt{x^3 + x}} dx = \int \frac{du}{\sqrt{u}} = \int u^{-1/2} \, du$$

$$= 2u^{1/2} + C$$

$$= 2(x^3 + x)^{1/2} + C.$$

Example 3. Evaluate $\int x^3(x^2 + 1)^{2/3}\,dx$ by making the substitution $u = x^2 + 1$.

SOLUTION.

$u = x^2 + 1$, $du = 2x\,dx$.

$$\int x^3(x^2 + 1)^{2/3}\,dx = \frac{1}{2}\int x^2(x^2 + 1)^{2/3}2x\,dx = \frac{1}{2}\int (u - 1)u^{2/3}\,du$$

$$= \frac{1}{2}\int (u^{5/3} - u^{2/3})\,du = \frac{1}{2}\int u^{5/3}\,du - \frac{1}{2}\int u^{2/3}\,du$$

$$= \frac{1}{2}\left(\frac{u^{8/3}}{8/3}\right) - \frac{1}{2}\left(\frac{u^{5/3}}{5/3}\right) + C$$

$$= \frac{3}{16}(x^2 + 1)^{8/3} - \frac{3}{10}(x^2 + 1)^{5/3} + C.$$

Integration by Parts. This method is based on the product rule $(fg)' = f'g + fg'$. From it, we can derive the equation

(9) $\quad \int f(x)g'(x)\,dx = f(x)g(x) - \int g(x)f'(x)\,dx.$

which we abbreviate in the following form.

Formula for Integration by Parts

(10) $\quad \int u\,dv = u \cdot v - \int v\,du.$

To illustrate the meaning of Formula 10, let us write it for the case

(11) $\quad u = \ln x, \qquad v = \frac{1}{2}x^2.$

Then

(12) $\quad du = \frac{1}{x}\,dx, \qquad dv = x\,dx$

and Formula 10, integration by parts, becomes

(13) $\quad \int (\ln x)x\,dx = \frac{1}{2}x^2 \ln x - \int \frac{1}{2}x^2 \cdot \frac{1}{x}\,dx.$

The right-hand side of this expression can be evaluated easily. After doing so, Equation 13 becomes

$$（14）\quad \int x \ln x \, dx = \frac{1}{2}x^2 \ln x - \frac{1}{4}x^2 + C.$$

In practice, the above procedure must be reversed. We would be given $\int x \ln x \, dx$ for evaluation. If we then selected u and v as in Equations 11 and 12, the integration by parts formula would yield Equation 14.

The method can be summarized as follows.

To Evaluate $\int f \, dx$ using Integration by Parts.

Step 1. Select u and dv so that

$$f \, dx = u \, dv.$$

Step 2. Calculate du and v. Record the four equations for u, v, du, and dv.

Step 3. Use the integration by parts formula to write

$$（16）\quad \int f \, dx = \int u \, dv = uv - \int v \, du.$$

Step 4. Try to evaluate $\int v \, du$. If successful, Equation 16 will lead to an evaluation of $\int f \, dx$.

The following three examples will illustrate this technique. They will not explain how to hit upon the successful choice of u and v — you will have to learn that by trial and error, experience, and thinking ahead.

Example 4. Evaluate $\int x e^x \, dx$ using integration by parts.

SOLUTION.
To complete Step 1, we write

$$（17）\quad x e^x \, dx = u \, dv$$

and make a choice for u and dv. The choice $u = x$ and $dv = e^x \, dx$ is a good one, as we shall see.

From $u = x$, we obtain $du = dx$. From $dv = e^x \, dx$, we obtain $v = e^x$. Record these four equations.

$$（18）\quad u = x, \qquad dv = e^x \, dx, \qquad du = dx, \qquad v = e^x.$$

These four equations comprise Step 2. Note that we started with the first two equations (from Step 1), and then we calculated the second two equations.

Step 3 is to use Equations 17 and 18 in the integration-by-parts formula.

$$\int xe^x \, dx = \int u \, dv = u \cdot v - \int v \, du$$
$$= (x)(e^x) - \int e^x \, dx.$$

We can evaluate the last integral; therefore, we have

$$\int xe^x \, dx = xe^x - e^x + C.$$

Example 5. Let us apply integration by parts to $\int x^2 e^x \, dx$.

Step 1. Write $x^2 e^x \, dx = u \, dv$, and choose $u = x^2$, $dv = e^x \, dx$.

Step 2. Calculate du and v; record the results.

$$u = x^2, \qquad dv = e^x \, dx, \qquad du = 2x \, dx, \qquad v = e^x.$$

Step 3. We now have

$$\int x^2 e^x \, dx = \int u \, dv = uv - \int v \, du$$
$$= x^2 e^x - \int e^x \cdot 2x \, dx$$
(19) $$= x^2 e^x - 2 \int xe^x \, dx.$$

Although the last integral is not easy to evaluate, it is simpler than the one we started with. Therefore Expression 19 is a valuable simplification; we should now concentrate on evaluating the last integral, $\int xe^x \, dx$. It can be evaluated by parts, as in the preceding example, or by guessing, as in Example 5 on page 227. Either way gives

$$\int xe^x \, dx = xe^x - e^x + C.$$

When this is put into Expression 19, we obtain

$$\int x^2 e^x \, dx = x^2 e^x - 2xe^x - 2e^x + 2C.$$

(*Note:* The constant $2C$ can be replaced by any other symbol, say K, to denote an arbitrary constant.)

Example 6. Let us try to evaluate $\int x \sqrt{x - 1} \, dx$ using integration by parts. We write

$$x(x - 1)^{1/2} \, dx = u \, dv,$$

and try to make a good choice for u. We shall choose

$$u = x, \qquad \text{and} \qquad dv = (x - 1)^{1/2} \, dx,$$

even though this choice will make the calculation of v somewhat difficult. Step 1 is complete.

We must find v from $\dfrac{dv}{dx} = (x - 1)^{1/2}$. Guesswork (or substitution) will lead to $v = (\tfrac{2}{3})(x - 1)^{3/2}$. Thus Step 2 is summarized by:

$$u = x, \qquad dv = (x - 1)^{1/2}\, dx, \qquad du = dx, \qquad v = \frac{2}{3}(x - 1)^{3/2}.$$

Step 3 results in

$$\int x \sqrt{x - 1}\, dx = \int u\, dv = uv - \int v\, du$$

(20)
$$= (x)\left(\frac{2}{3}\right)(x - 1)^{3/2} - \int \frac{2}{3}(x - 1)^{3/2}\, dx.$$

The final integral can be evaluated by guessing (or substitution) as follows.

$$\int (x - 1)^{3/2}\, dx = \frac{2}{5}(x - 1)^{5/2} + C.$$

When this is put into Expression 20, we obtain the final answer

$$\int x \sqrt{x - 1}\, dx = \frac{2}{3}x(x - 1)^{3/2} - \frac{4}{15}(x - 1)^{5/2} + K.$$

Tables of Integrals. Imagine a list—it would have to be infinitely long—of all the functions we have studied. Imagine also that we differentiated each function in the list and wrote the result next to it.

Each entry in this list would lead to an integration formula. For example, somewhere in the list would be the entry

(21) $\quad y = \ln x - \ln (1 + x), \qquad \dfrac{dy}{dx} = \dfrac{1}{x(1 + x)}.$

It leads to the integration formula

(22) $\quad \displaystyle\int \frac{dx}{x(1 + x)} = \ln x - \ln (1 + x) + C.$

If we were to put all the integration formulas we obtain this way into a "master table of integrals," any time we were faced with the problem of evaluating an integral, we might look through our master table and expect to find the answer.

Of course, no master table exists; there are, however, very comprehensive integral tables which can be found in scientific libraries. Less comprehensive ones appear in the many mathematical handbooks and books of mathematical tables. A smaller integral table is included at the back of this book on pages 388–392.

As with all methods for evaluating integrals, the use of integral tables is somewhat of an art. The integral you wish to evaluate may have to be simplified or transformed by a substitution before it can be found in the table. The examples below illustrate this procedure.

Let us return to the imaginary master table and ask a theoretical question. Since the list was constructed by differentiating all the types of functions we have studied, does it follow that we can use the list to find an antiderivative for every such function?

The answer is no. To see why, suppose we had not studied logarithms Then we would be unable to evaluate

$$\int \frac{1}{x}\, dx,$$

even though the integrand is a function that we studied. This basic situation cannot be remedied by adding $\ln x$ to our list of functions. For example, the integral

$$\int \frac{1}{x^2 + 1}\, dx$$

cannot be evaluated without using the trigonometric functions (in particular, the arctangent function). If you have a theoretical bent, the following question may occur to you: *Does every function have an antiderivative* (even though the antiderivative may not be a function I have studied)? This important question will be answered in Section 5 of this chapter.

Example 7.   Use the table of integrals (pages 388–392) to evaluate

(23) $\int \sqrt{x^2 - 3}\, dx$

SOLUTION.

As we scan the table in search of an integral resembling Integral 23, we might first pause at entry 18, $\int \sqrt{a^2 - x^2}\, dx$. This would be very similar to the given interval when $a^2 = 3$; there is only a difference of sign. However, that sign difference cannot be removed by any obvious means, so we continue to scan the table. When we come to entry 32, we realize success. If we put $a^2 = 3$ in that entry and consistently use the lower sign in each symbol $\pm$, we obtain the answer to our problem.

(24) $\int \sqrt{x^2 - 3}\, dx = \frac{1}{2}[x\sqrt{x^2 - 3} - 3\ln(x + \sqrt{x^3 - 3})] + C.$

Now look at entry 61 in the table. That also could have been used to evaluate Integral 23; we would apply entry 61 by choosing $a = -3$, $b = 0$, and $c = 1$.

(If you actually apply entry 61 this way, you will obtain a result that differs from Expression 24; the difference can be reconciled by the role of the arbitrary constant C.)

Example 8. Suppose we want to find

$$(25) \quad \int \frac{dx}{x^2 \sqrt{3x^2 + 1}}.$$

After scanning the table of integrals, we see that entry 41 on page 390 is almost applicable; it reads as follows.

$$\int \frac{dx}{x^2 \sqrt{x^2 \pm a^2}} = \mp \frac{\sqrt{x^2 \pm a^2}}{a^2 x} + C.$$

To put Integral 25 into this form, we must remove the factor of 3 from the integrand. This can be done algebraically as follows.

$$\int \frac{dx}{x^2 \sqrt{3x^2 + 1}} = \int \frac{dx}{x^2 \sqrt{3\left(x^2 + \dfrac{1}{3}\right)}}$$

$$= \int \frac{dx}{x^2 \sqrt{3}\sqrt{x^2 + \dfrac{1}{3}}}$$

$$= \frac{1}{\sqrt{3}} \int \frac{dx}{x^2 \sqrt{x^2 + \dfrac{1}{3}}}.$$

Entry 41 can be used to evaluate this last integral. We apply it with $a^2 = \frac{1}{3}$, and with the upper sign of the symbol $\pm$. The final result is

$$\int \frac{dx}{x^2 \sqrt{3x^2 + 1}} = \frac{1}{\sqrt{3}}\left(-\frac{\sqrt{x^2 + \frac{1}{3}}}{\frac{1}{3}x} + C\right)$$

$$= -\frac{\sqrt{3}\sqrt{x^2 + \frac{1}{3}}}{x} + \frac{C}{\sqrt{3}}$$

$$= -\frac{\sqrt{3x^2 + 1}}{x} + K.$$

The table of integrals on pages 388–392 makes use of functions that are not discussed in this course (the integrands invariably contain only familiar functions). These functions are arcsin x, arccos x, and arctan x (sometimes written $\sin^{-1}x$, $\cos^{-1}x$, and $\tan^{-1}x$); see, for example, entries 4, 20, and 42. The definitions and graphs of these functions can be found in most textbooks for two-year calculus courses.

Problems

Evaluate the integrals in Problems 1–12 by the method of substitution.

1. $\int \sqrt{3x + 1}\, dx.$

2. $\int (1 - 2t)^{2.4}\, dt.$

3. $\int x\sqrt{x^2 - 1}\, dx.$

4. $\int x^2 \sqrt{1 + x^3}\, dx.$

5. $\int \dfrac{dx}{x + 1}.$

6. $\int \dfrac{dx}{2x - 1}.$

7. $\int \dfrac{dw}{5 - 2w}.$

8. $\int \dfrac{t\, dt}{\sqrt{2t^2 + 1}}.$

9. $\int e^{2x+1}\, dx.$

10. $\int xe^{1-x^2}\, dx.$

11. $\int \dfrac{x}{x - 1}\, dx.$

12. $\int \dfrac{x}{(x - 1)^5}\, dx.$

Use integration by parts to evaluate the integrals in Problems 13–18.

13. $\int \ln x\, dx.$

14. $\int x \ln x\, dx.$

15. $\int x(\log x)^2\, dx.$

16. $\int xe^{kx}\, dx.$

17. $\int x^2 e^{kx}\, dx.$

18. $\int (\ln x)^2\, dx.$

Evaluate the following integrals by any method.

19. $\int x\sqrt{x + 2}\, dx.$

20. $\int (x + 1)(x - 3)\, dx.$

21. $\int (3x - 1)^4\, dx.$

22. $\int x^3(x^4 + 1)^{5/6}\, dx.$

23. $\int xe^{-3x}\, dx.$

24. $\int \dfrac{x^2\, dx}{2x^3 - 1}.$

Section 4 □ Applications of Integration

Examples 1 and 2 on pages 216–217 illustrated a primary use of integration or antidifferentiation:

> *Given the rate of change of a quantity, the quantity itself can be found by integrating the given rate of change.*

In symbols, this can be expressed as

$$\text{(1)} \quad \int \left(\frac{dy}{dx}\right) dx = y + C$$

where $\frac{dy}{dx}$ is the given rate of change of the quantity y. The presence of the arbitrary constant C indicates that y cannot be determined completely from its rate of change—a side condition is needed (see Example 6 on page 221).

In this section, we return to these primary applications; we can now make use of the improved techniques of integration that were learned in the preceding sections.

Example 1. A subject jogs on a computerized treadmill. The treadmill is programmed to start at a slow speed and level angle, and then gradually increase its speed and inclination. Tests verify that after t minutes the subject was using calories at the rate of $2 + 0.86t$ calories per minute. How many calories did the subject burn during the first 20 minutes?

SOLUTION.
Let y be the total number of calories burned after t minutes. We are given the rate of change of y,

$$\text{(2)} \quad \frac{dy}{dt} = 2 + 0.86t,$$

and are asked to find the value of y when $t = 20$.

We first find y at any t.

$$\text{(3)} \quad y = \int (2 + 0.86t)\, dt = 2t + 0.43t^2 + C.$$

The constant C is determined by the side condition

$$y = 0 \quad \text{when} \quad t = 0.$$

From Equation 3, we see that this side condition forces C to be zero. Thus

$$y = 2t + 0.43t^2,$$

$$y = 2(20) + 0.43(20)^2 = 212 \text{ (calories) when } t = 20 \text{ (minutes)}.$$

Example 2. The cost to print one more book when x copies are printed is a decreasing function of x; it is called the *marginal cost* at the production level x. Suppose this marginal cost is $5 + 5e^{-0.0005x}$ dollars per copy. Find the total cost to print 50,000 copies if the constant overhead cost is $1,000. Compare the average cost per book with the marginal cost at this production level.

SOLUTION.

Let y be the total cost in dollars to print x copies. We assume x varies continuously, and that the marginal cost can be approximated by $\dfrac{dy}{dx}$ $\left(\text{recall}\right.$

$\dfrac{dy}{dx} \approx \dfrac{\Delta y}{\Delta x} = \Delta y$ if $\Delta x = 1$; see page 100$\left.\right)$. Thus we have

$$\frac{dy}{dx} = 5 + 5e^{-0.0005x}$$

and wish to find y. The solution is

$$y = \int (5 + 5e^{-0.0005x})\, dx$$
$$= 5\int dx + 5\int e^{-0.0005x}\, dx$$
$$= 5x + 5\frac{1}{-0.0005}e^{-0.0005x} + C$$
$$= 5x - 10{,}000e^{-0.0005x} + C.$$

The side condition, $y = 1000$ when $x = 0$, means $C = 11{,}000$. Therefore

(4) $y = 5x - 10{,}000e^{-0.0005x} + 11{,}000.$

From Equation 4, we see that the cost to print 50,000 copies is

$$y = 5(50{,}000) - 10{,}000e^{-0.0005(50{,}000)} + 11{,}000$$
$$= 261{,}000 - 10{,}000e^{-25} \approx 261{,}000$$

The average cost is therefore $261{,}000 \div 50{,}000 \approx \$5.22$ per copy. The marginal cost at this level (which is interpreted as the cost to print one more copy) is the value of $\dfrac{dy}{dx}$ when $x = 50{,}000$;

$$\frac{dy}{dx} = 5 + 5e^{-0.0005(50{,}000)} \approx 5 \text{ dollars when } x = 50{,}000.$$

Example 3.   In Example 1 (see page 239), the speed of the treadmill belt after t minutes was $380 + 25t$ feet per minute. What distance did the runner cover during the first 20 minutes?

SOLUTION.

Let s be the total distance traveled during the first t minutes. Then $\dfrac{ds}{dt}$ is the speed at time t (recall that *speed* is the *rate of change of distance*) and so

$$\frac{ds}{dt} = 380 + 25t.$$

Therefore

$$s = \int (380 + 25t)\, dt = 380t + \frac{25t^2}{2} + C.$$

The side condition, $s = 0$ when $t = 0$, shows that $C = 0$. Therefore $s = 380t + (\frac{25}{2})t^2$. Put $t = 20$ in this formula to obtain the answer

$$s = 380(20) + \frac{25}{2}(20)^2 = 12{,}600 \text{ feet when } t = 20.$$

Example 4. An oil well produces 400 barrels of oil a month. The current price is \$8 per barrel, and it is expected to rise at a rate of \$.10 per month. If oil is sold as it is produced, what is the total revenue expected during the next 100 months?

SOLUTION.
Let R be the total revenue earned during the next t months. We can find $\frac{dR}{dt}$ as follows.

Let t change by Δt. During that time interval, the well produced $400\Delta t$ barrels of oil; this oil sold at a price that varied from $8 + 0.10t$ (the price at time t) to $8 + 0.10(t + \Delta t)$. Therefore the revenue ΔR during this period satisfied

$$400\Delta t(8 + 0.10t) \le \Delta R \le 400\Delta t[8 + 0.10(t + \Delta t)],$$

and hence

(5) $$400(8 + 0.10t) \le \frac{\Delta R}{\Delta t} \le 400[8 + 0.10(t + \Delta t)].$$

Let $\Delta t \to 0$; then Inequality 5 shows that

(6) $$\frac{dR}{dt} = 400(8 + 0.10t).$$

From Equation 6, we have

(7) $$R = \int 400(8 + 0.10t)\, dt = 400(8t + 0.05t^2) + C.$$

In Equation 7, we must choose $C = 0$ because $R = 0$ when $t = 0$. Thus

$$R = 400(8t + 0.05t^2),$$

so the revenue when $t = 100$ is $400[8(100) + 0.05(100)^2] = 520{,}000$ dollars.

Problems

1. Find the function whose rate of change is given.

    a. $\dfrac{dy}{dx} = 8x^3 - 4x - \dfrac{3}{x^2}, y = 0$ when $x = 1$.

    b. $\dfrac{dy}{dx} = \sqrt{3x + 1}, y = 2$ when $x = 1$.

    c. $\dfrac{dP}{dt} = t\,e^{1-t^2}, P = 0$ when $t = -1$.

    d. $\dfrac{dw}{ds} = \ln s, w = \pi$ when $s = e$.

2. Find the function.

    a. $\dfrac{dy}{dx} = x\sqrt{x^2 - 1}, y = 5$ when $x = 1$.

    b. $\dfrac{dy}{dx} = \dfrac{3}{5 - 2x}, y = 7$ when $x = 2$.

    c. $\dfrac{dQ}{dt} = te^t, Q = 1$ when $t = 0$.

    d. $\dfrac{dp}{dq} = q^2(q^3 + 1)^6, p = 0$ when $q = -1$.

3. At the production level x the marginal cost in a light plant was[2] $\dfrac{dC}{dx} = 0.125 + 0.00878x$. The fixed overhead was 16.68. Find a formula for the total cost C in terms of x.

4. The *marginal productivity* for the Chicago meat packing industry was estimated to be $\dfrac{dy}{dx} = 1.06 - 0.08x$, where y is in millions of pounds of hogs and x is in thousands of man hours.[3] Find y as a function of x if $y = 0.77$ when $x = 3$.

[2] J. A. Nordin, "Note on a Light Plant's Cost Curve," *Econometrica* XV (1947), p. 231 ff.

[3] W. H. Nichols, *Labor Productivity Functions in Meat Packing* (Chicago: University of Chicago Press, 1948).

5. A stone is dropped from a cliff. After t seconds, its speed is $32t$ feet per second. How far did it travel in the first 4 seconds?

6. A bullet is fired vertically upward. Its initial velocity is 1000 feet per second. Let h be the height after t seconds. A law of physics tells us that $\dfrac{d^2h}{dt^2} = -32$. Find h as a function of t.

7. Evaporation rate depends on temperature. Suppose a lake loses water at the rate of $(T + 6)/3$ tons per hour, where T is the temperature. Suppose that T varies according to the law $T = 20 - \frac{1}{4}t^2 + 8t$, where t, the time in hours, varies from 0 to 24. How much water evaporates during the first 3 hours?

8. Suppose the price of gold is $200 per ounce, and that the price steadily rises at the rate of $.16 per day. If a broker sells gold at a steady rate of 5 ounces per day, what will be the total revenue from gold sales after 1000 days?

9. A patient's heart rate t minutes after the completion of exercise is approximately $150 - 3t$ beats per minute, for $0 \le t \le 20$. Each beat pumps 0.07 liters of blood through the heart. How much blood flows through the heart during the first 20 minutes?

10. The integral $\int E'(t)\, e^{\eta t}\, dt$ occurs in the study of memory.[4] Evaluate this antiderivative if $E(t)$, the intensity of a memory trace, is $1 - e^{-t}$ and if $\eta = 0.5$.

Section 5 □ The Area under a Graph

We begin this section with the problem of finding the area of a region with curved boundary. The solution, significant in itself, also has important consequences. In particular, it will show that antiderivatives can be found by calculating areas.

The most convenient way to pose the problem is to consider first the area bounded by the graph of a function, the x-axis, and two vertical lines $x = a$ and $x = b$, with $a < b$. This area, denoted by A in Figure 1, will be referred to simply as the **area under f from a to b.** (The areas of other types of curved regions can usually be reduced to this one; in Figure 2 the area of

[4]N. Rashevsky, *Mathematical Biophysics,* 3rd rev. ed., Vol. 2 (New York: Dover Publications, Inc., 1960), p. 193.

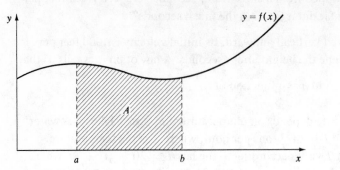

Figure 1

the oval region is the difference between the area under f_1 and the area under f_2, between a and b.)

For simplicity, we shall assume that $f(x) \geqq 0$ for each x. We also assume the graph of f has no breaks or sudden jumps. We shall allow the endpoint b to vary, and calculate the rate of change of A. For that reason we shall replace b by the letter x, which is more suggestive of a variable.

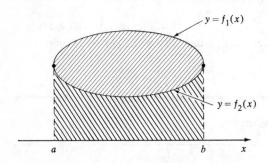

Figure 2

Let A be the area under f between a and x. Then A is a function of x. We shall calculate $\dfrac{dA}{dx}$.

Let x change from x to $x + \Delta x$. The corresponding change in A is denoted ΔA (see Figure 3). The figure shows that the area ΔA is squeezed between the areas of two rectangles with base Δx. The heights of these rectangles are denoted y_{max} and y_{min}, the maximum and minimum values of y which occur between x and $x + \Delta x$. Thus

$$y_{min}\Delta x \leq \Delta A \leqq y_{max}\Delta x$$

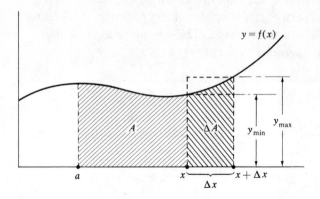

$$y = f(x)$$

$$y_{max}$$
$$y_{min}$$

$$A$$
$$\Delta A$$

$$a$$
$$x$$
$$\Delta x$$
$$x + \Delta x$$

Figure 3

or

(1) $y_{min} \leq \dfrac{\Delta A}{\Delta x} \leq y_{max}.$

As $\Delta x \to 0$, y_{min} and y_{max} both approach $y = f(x)$, as the graph has no jumps. Thus Inequality 1 shows that $\dfrac{dA}{dx} = f(x)$. Hence A is an antiderivative of f. This result can be stated in the following form.

Let A be the area under f between a and x. Then $\dfrac{dA}{dx} = f(x)$. *Equivalently,*

(2) $A = \int f(x)\, dx$

for a suitable choice of the constant on the right side of Equation 2.

Example 1.   Find the area of the region bounded by the graph of $y = \dfrac{1}{2}x$, the x-axis, and the vertical lines $x = 0$ and $x = 8$.

SOLUTION.
The area under this graph between 0 and x is, according to Equation 2,

(3) $A = \displaystyle\int \dfrac{1}{2}x\, dx = \dfrac{1}{4}x^2 + C.$

We can determine the constant C in Equation 3 by observing that $A = 0$ when $x = 0$. This means $C = 0$. Thus $A = \frac{1}{4}x^2$; when $x = 8$, the area is $A = \frac{1}{4}(8)^2 = 16$. (*Note:* The units for this area would be, for example, square inches if the x- and y-axes are scaled in inches.)

We obtained this result by calculus. In this simple example, however, the result could have also been obtained by elementary geometry (see Figure 4). The desired area is that of a right triangle of base 8 and height 4, so the area is $\frac{1}{2} \cdot 8 \cdot 4 = 16$, which agrees with the answer above.

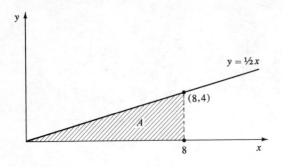

Figure 4

The symbol

(4) $\quad \int_a^b f(x)\, dx$

is called the **definite integral of f from a to b.** It is defined as follows. Let F be *any* antiderivative of f; thus $\dfrac{dF}{dx} = f$. Then

(5) $\quad \int_a^b f(x)\, dx = F(b) - F(a)$.

(*Note:* This value does not depend on the particular antiderivative of f that was chosen. If G is any other antiderivative of f, then, because of uniqueness of antiderivatives, page 218, $G(x) = F(x) + C$, and so $G(b) - G(a) = [F(b) + C] - [F(a) + C] = F(b) - F(a)$.)

Now consider the area A under f between a and x. By Equation 2,

(6) $\quad A = F(x) + C$,

where F is any antiderivative of f. The constant in Equation 6 can be evaluated by observing that $A = 0$ when $x = a$. Therefore $C = -F(a)$. Hence $A = F(x) - F(a)$. The area under f between a and b is therefore $A = F(b) - F(a)$, which is $\int_a^b f(x)\, dx$. We have proved the following result.

Evaluating Areas by Definite Integrals

Let A be the area under f between a and b, with $a < b$. Then

(7) $\quad A = \int_a^b f(x)\, dx$.

The following properties are often useful for simplifying a definite integral in preparation for evaluating it.

$$\int_a^b [f(x) \pm g(x)] \, dx = \int_a^b f(x) \, dx \pm \int_a^b g(x) \, dx$$

$$\int_a^b kf(x) \, dx = k \int_a^b f(x) \, dx, \quad k \text{ a constant.}$$

These properties follow from the corresponding properties for indefinite integrals (see Equations 1 and 2 on page 223).

Another useful device for evaluating definite integrals is the symbolism

$$[g(x)]_a^b = g(b) - g(a).$$

The following examples illustrate this symbol.

$$[x^2]_1^3 = 3^2 - 1^2 = 8$$

$$[5x^3 + 6]_{-1}^2 = [5(2)^3 + 6] - [5(-1)^3 + 6] = 45$$

$$[\ln x]_e^{e^2} = \ln e^2 - \ln e = 2 - 1 = 1.$$

Example 2 illustrates the technique of evaluating a definite integral. Examples 3 and 4 show how to find areas using the definite integral according to Equation 7.

Example 2. Evaluate $\int_1^3 (6x^2 - 2x + 1) \, dx$.

SOLUTION.

$$\int_1^3 (6x^2 - 2x + 1) \, dx = 6 \int_1^3 x^2 \, dx - 2 \int_1^3 x \, dx + \int_1^3 dx$$

$$= 6\left[\frac{x^3}{3}\right]_1^3 - 2\left[\frac{x^2}{2}\right]_1^3 + [x]_1^3$$

$$= 6\left(\frac{27}{3} - \frac{1}{3}\right) - 2\left(\frac{9}{2} - \frac{1}{2}\right) + (3 - 1)$$

$$= 46.$$

Example 3. Find the area A under the graph of $y = e^{2x}$ from 0 to 1. See Figure 5.

SOLUTION.

$$A = \int_0^1 e^{2x} \, dx = \left[\frac{1}{2}e^{2x}\right]_0^1 = \frac{1}{2}e^2 - \frac{1}{2}e^0 = \frac{1}{2}e^2 - \frac{1}{2}.$$

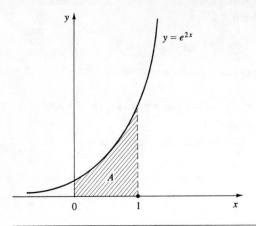

Figure 5

Example 4. Find the area A which lies above the x-axis and below the parabola $y = 5x - x^2 - 4$.

SOLUTION.
Since $y = -(x - 1)(x - 4)$, we see (Figure 6) that the desired area is

$$A = \int_1^4 (5x - x^2 - 4) \, dx.$$

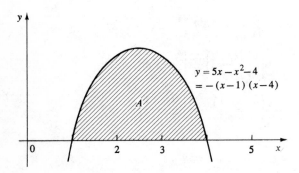

Figure 6

Therefore

$$A = \left[\frac{5}{2}x^2 - \frac{1}{3}x^3 - 4x\right]_1^4$$

$$= \frac{5}{2}(4)^2 - \frac{1}{3}(4)^3 - 4(4) - \left(\frac{5}{2} - \frac{1}{3} - 4\right)$$

$$= \frac{9}{2}.$$

Signed Area. The result on evaluating areas (Equation 7) remains true if $f(x)$ is not always positive. In this case, the "area under f from a to b" must be understood in the following sense: *areas below* the x-axis are counted as *negative numbers*. In Figure 7, for example, the area under f from a to b consists of two regions, labeled II and IV, which lie below the x-axis; therefore

$$\int_a^b f(x)\, dx = \text{area I} - \text{area II} + \text{area III} - \text{area IV}.$$

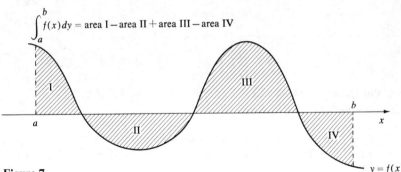

Figure 7

Example 5. Consider $y = x^3$. Then, in the notation of Figure 8, we have

$$-\text{area I} = \int_{-1}^0 x^3\, dx = \left[\frac{x^4}{4}\right]_{-1}^0 = -\frac{1}{4},$$

so area I $= \frac{1}{4}$.

$$\text{area II} = \int_0^2 x^3\, dx = \left[\frac{x^4}{4}\right]_0^2 - \frac{2^4}{4} = \frac{0^4}{4} = 4.$$

Note that

$$\int_{-1}^2 x^3\, dx = \left[\frac{x^4}{4}\right]_{-1}^2 = \frac{15}{4} = -\text{area I} + \text{area II}.$$

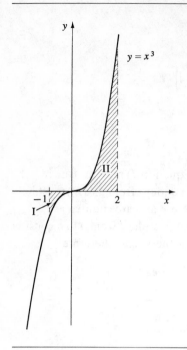

Figure 8

Fundamental Theorem of Calculus. Equation 2 on page 245 may be interpreted in the following manner:

> Given a continuous function f, there is an antiderivative of f; one antiderivative is the function $A = \int_a^x f(t)\, dt$ which is the signed area under f from a to x.

Equation 7 can be given the following interpretation.

> Given a rate of change $\dfrac{dy}{dx} = f'(x)$, the area under this graph from a to b is $f(b) - f(a)$; in symbols

(8) $\displaystyle\int_a^b f'(x)\, dx = f(b) - f(a).$

Either of these interpretations may be called the **fundamental theorem of calculus.** They show the connection between differentiation and the area under a curve.

Example 6. Figure 9 shows the rate at which water flowed down the Nile River during 1955 as measured at Wadi Halfa, the point where the Nile enters Egypt. *The area under this graph is the total volume of water that flowed down the Nile that year.* (To be specific, each square in the grid represents a rectangle of base 10 days and height $33\frac{1}{3}$ millions of cubic meters per day, and therefore of area $333\frac{1}{3}$ million cubic meters. The number of squares, including fractional parts, multiplied by $333\frac{1}{3}$ would give the total water flow down the Nile in one year in millions of cubic meters.)

To see why this is true, let $y = f(x)$ be the total amount of water in millions of cubic meters that flowed past Wadi Halfa during the first x days.

Then $\dfrac{dy}{dx} = f'(x)$ is the rate in millions of cubic meters per day after x days; it is the function graphed in Figure 9. The area under this graph is

$$\int_0^{360} f'(x)\, dx = f(360) - f(0),$$

the total flow up to day 360 minus the total flow (namely zero) up to day zero.

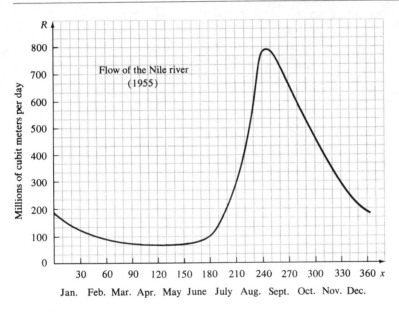

Figure 9

Example 7. A Rolls Royce Silver Shadow accelerates from a complete stop. Its speed $v(t)$ after t seconds is shown in Figure 10.[5] The distance $s(t)$ traveled during the first t seconds is the area under this graph from zero to t. The reason is the relationship

$$\text{area} = \int_0^t v(t)\, dt = s(t) - s(0) = s(t),$$

because $v(t) = s'(t)$.

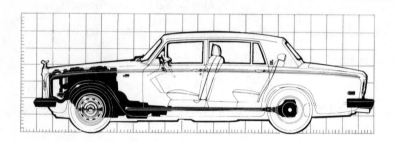

"R & T Road Test ROLLS-ROYCE SILVER SHADOW"   Scale: 10″ divisions

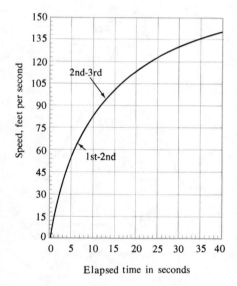

Figure 10

[5]Adapted from *Road and Track*, September, 1976, p. 45.

Problems

Evaluate.

1. $\int_0^2 5x^4 \, dx.$

2. $\int_1^3 27x^2 \, dx.$

3. $\int_0^1 e^{4x} \, dx.$

4. $\int_1^e \dfrac{dx}{5x}.$

5. $\int_{-1}^{e-2} \dfrac{dx}{3(x+2)}.$

6. $\int_1^4 9\sqrt{x} \, dx.$

7. $\int_{-1/2}^0 (2t+1)^{10} \, dt.$

8. $\int_4^6 (4x-1) \, dx.$

9. $\int_1^3 (30x^2 - 6x + 5) \, dx.$

10. $\int_0^1 e^{2t+1} \, dt.$

11. Sketch the region R under the graph of $y = 6x^2 + x$ between 1 and 3. Use calculus to find the area of R.

12. Sketch the region R under the graph of $y = \sqrt{x}$ between 0 and 36. Find the area of R.

13. Sketch the graph of $y = x(x-1)(x-2)$. There are only two finite regions bounded by this curve and by the x-axis. Find the area of each.

14. Table 1 gives the *rate E* in kilojoules per second, or *kilowatts*, at which electric power is consumed in a cafe at various times t in hours. Sketch the graph of this consumption rate as a function of t (use graph paper if possible). Estimate the area under this curve, and thereby find the total power in kilowatt-hours consumed during the 24-hour period. If the electric company charges 15¢ per kilowatt-hour, what is the cost for electricity during this 24-hour period?

Table 1

t	0	1	2	3	4	5	6	7	8	9	10	11	12
E	2	1	.4	.6	1	2	2.9	4.1	3	2	2.1	3	4.1

t	13	14	15	16	17	18	19	20	21	22	23	24
E	4.1	3	2.8	4	5	5.1	4.5	3.9	3.5	2.9	2.1	2

15. Table 2 gives the marginal cost $\dfrac{dC}{dx}$ in dollars per pound to produce x pounds of candy bars. The fixed overhead is \$1000 (that is, $C = 1000$ when $x = 0$). Sketch the graph of $\dfrac{dC}{dx}$ as a function of x (use graph paper if possible). By estimating the area under this graph, find the total cost to produce 20,000 pounds of candy.

Table 2

x	2000	4000	6000	8000	10,000	12,000	14,000	16,000	18,000	20,000
$\dfrac{dC}{dx}$	3.0	2.0	1.5	1	0.8	0.6	0.54	0.52	0.51	0.50

16. Table 6 on page 9 gives the flow F in millions of cubic meters per day of the Blue Nile at Khartoum[6] at time t in days since January 1. Draw the graph (use graph paper if possible). By estimating an area under the graph, find the total volume of water that flowed past Khartoum during the period May 1–December 31.

17. Table 3 gives a patient's heart rate H in beats per minute t minutes after completion of vigorous exercise. Sketch the graph of H as a function of t (use graph paper if possible). By estimating the area under the graph, approximate the total number of heart beats between times $t = 0$ and $t = 20$. If each beat pumps 0.07 liters of blood through the heart, how many liters were pumped during this period from $t = 0$ to $t = 20$?

Table 3

t	0	2	4	6	8	10	12	14	16	18	20
H	150	143.6	137.2	130.8	124.4	118.0	111.6	105.2	98.8	92.4	86.0

18. Figure 10 on page 252 shows the speed v in feet per second of a Rolls Royce Silver Shadow t seconds after beginning an acceleration test (adapted from *Road and Track* magazine, September 1976, page 45). Find the total distance traveled during the first 30 seconds.

[6]Encyclopaedia Britannica, 1957, p. 453.

An instructive application of integration is its use for finding volumes. The method we develop will make it easy to discover such formulas as those for the volume of a sphere or a cone.

Consider the solid shown in Figure 11. Let V be the volume of the portion between the two parallel planes p_0 and p_x; the plane p_x is at the distance x from the fixed plane p_0. Thus V is a function. We shall show that $\dfrac{dV}{dx} = A(x)$, where $A(x)$ is the cross-sectional area of the slice of the solid cut by the plane p_x.

Let x change by Δx. The corresponding change in volume is denoted ΔV. Let A_{max} and A_{min} be the maximum and minimum of the cross-sectional areas cut by p_x as x changes by Δx. It is geometrically plausible, and we shall accept the fact, that

(1) $A_{min}\Delta x \leqq \Delta V \leqq A_{max}\Delta x.$

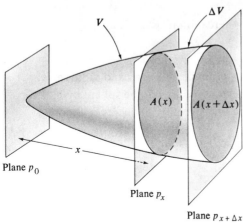

ΔV

V

$A(x)$

$A(x+\Delta x)$

x

Plane p_0

Plane p_x

Plane $p_{x+\Delta x}$ **Figure 11**

(This is especially clear for "expanding" cross sections as in Figure 11; it is also true in general.)

From Inequality 1, we obtain

(2) $A_{min} \leq \dfrac{\Delta V}{\Delta x} \leq A_{max}.$

Let $\Delta x \to 0$. Then A_{min} and A_{max} each approach $A(x)$ if, as we shall always assume, $A(x)$ is a continuous function of x. Therefore Inequality 2 shows $\dfrac{\Delta V}{\Delta x} \to A(x)$ as $\Delta x \to 0$, and so

(3) $\quad \dfrac{dV}{dx} = A(x).$

Equation 3 shows that

(4) $\quad V = \int A(x)\, dx$

if the constant in Equation 4 is properly chosen.

The results Equations 3 and 4 can be restated, using definite integrals, as follows.

Volume by Slicing
The volume of the solid between the planes p_a and p_b $(a < b)$ is

(5) $\quad \displaystyle\int_a^b A(x)\, dx,$

where $A(x)$ is the cross sectional area cut from the solid by p_x.

Example 1.   Find the volume of a sphere of radius R.

SOLUTION.
In Figure 12, p_x is the plane which is perpendicular to a fixed radius at a distance x from the center. The cross section cut from the sphere by p_x is a circle. Its radius, denoted by r, can be found by the Pythagorean theorem, $x^2 + r^2 = R^2$. Thus

$$A(x) = \pi r^2 = \pi(R^2 - x^2).$$

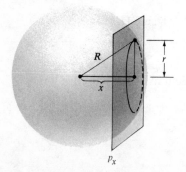

p_x **Figure 12**

The volume V of the sphere is twice the volume of the portion between p_0 and p_R; hence

$$\frac{1}{2} V = \int_0^R A(x)\, dx = \int_0^R \pi(R^2 - x^2)\, dx$$

$$= \pi R^2 \int_0^R dx - \pi \int_0^2 x^2\, dx$$

$$= \pi R^2 [x]_0^R - \pi \left[\frac{x^3}{3}\right]_0^R$$

$$= \pi R^2 \cdot R - \pi \cdot \frac{R^3}{3} = \frac{2}{3}\pi R^3.$$

We solve for V to obtain the result

$$V = \frac{4}{3}\pi R^3.$$

Example 2. The pyramid of Khufu (the *Great Pyramid*) near Cairo, Egypt, is about 480 feet tall. A horizontal cross section x feet from the top is, approximately, a square of sides $1.56x$ feet. Find the volume.

SOLUTION.
Let p_x be the horizontal plane x feet from the top. It cuts a square cross section from the pyramid (see Figure 13) whose area $A(x)$ is

$$A(x) = (1.56x)^2 = (1.56)^2 x^2$$

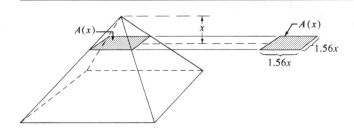

Figure 13

The volume V of the pyramid is therefore

$$V = \int_0^{480} (1.56)^2 x^2\, dx$$

$$= (1.56)^2 \int_0^{480} x^2\, dx$$

$$= (1.56)^2 \left[\frac{x^3}{3} \right]_0^{480}$$

$$= \frac{(1.56)^2 (480)^3}{3} \approx 89{,}712{,}230 \text{ cubic feet.}$$

Problems

1. Figure 14 shows a boat's hull, 20 feet in length. A vertical plane perpendicular to the keel and x feet from the bow cuts, from the interior of the hull, a cross section in the form of a semicircle of radius $\sqrt{x}$ feet. Find the volume interior to the hull.

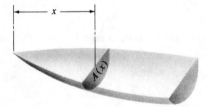

Figure 14

2. A steeple on a tower is in the shape of a pyramid. It is 60 feet tall; the base is a square with sides of 30 feet. Find the volume inside the steeple.

3. A spire (see Figure 15) is one meter in height. Any cross section parallel to the base and x feet from the top is a circle of radius $e^x - 1$ meters. Find the volume inside the spire.

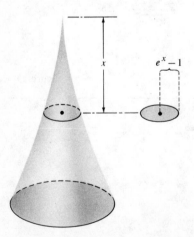

Figure 15

4. A structure (see Figure 16) has a semicircular base of radius $\overline{OA} = \overline{OB}$ $= 100$ meters. A vertical cross section perpendicular to AB and x meters from O is an isosceles right triangle. Find the volume enclosed.

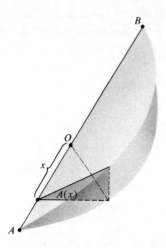

Figure 16

5. A right circular cone (see Figure 17) has height h. The radius of its base is r. Use calculus to derive the formula $V = \frac{1}{3}\pi r^2 h$ for its volume V.

Figure 17

6. Find the volume of a pumpkin 10 inches tall if a horizontal cross section x inches from the bottom is a circle of radius $2\sqrt{60x - 6x^2}$ inches.

7. Find the volume of a football 10 inches long if a cross section perpendicular to the axis and x inches from the center is a circle of radius $\frac{3}{5}\sqrt{25 - x^2}$ inches.

8. A doughnut 2 inches thick lies flat on a table. A horizontal cross section x inches above the table is a circular ring of inner radius $3 - 2x + x^2$ inches and outer radius $3 + 2x - x^2$ inches. Find the volume of the doughnut.

Summary

The integral notation $\int f(x)\,dx$ denotes all functions $F(x)$ such that $F'(x) = f(x)$; $\int_a^b f(x)\,dx$ means $F(b) - F(a)$. The basic formulas that help to evaluate integrals are as follows.

$$\int x^n\,dx = \frac{x^{n+1}}{n+1} + C, \qquad n \neq -1$$

$$\int dx = x + C,$$

$$\int [f(x) \pm g(x)]\,dx = \int f(x)\,dx \pm \int g(x)\,dx,$$

$$\int kf(x)\,dx = k\int f(x)\,dx,$$

$$\int e^{kx}\,dx = \frac{1}{k}e^{kx} + C, \int \frac{1}{x}\,dx = \ln x + C.$$

There is no automatic process for evaluating integrals. The methods studied in this chapter were guesswork, substitution, integration by parts, and the use of integral tables.

Integrals are typically used to find a formula for a function when only its rate of change is known. Finding areas and volumes are problems of this type.

Miscellaneous Problems

1. Evaluate.

   a. $\displaystyle\int \frac{dx}{x^2}.$

   b. $\displaystyle\int_{1/2}^{1} \frac{dx}{x^2}.$

   c. $\displaystyle\int (2x - 1)^8\,dx.$

   d. $\displaystyle\int_0^1 (2x - 1)^8\,dx.$

   e. $\displaystyle\int (t + 1)(t - 1)\,dt.$

   f. $\displaystyle\int_0^3 (t + 1)(t - 1)\,dt.$

   g. $\displaystyle\int u\sqrt{u^2 + 16}\,du.$

   h. $\displaystyle\int_0^3 u\sqrt{u^2 + 16}\,du.$

   i. $\displaystyle\int \frac{x + 1}{x}\,dx.$

   j. $\displaystyle\int_1^e \frac{x + 1}{x}\,dx.$

   k. $\displaystyle\int e^{3x}\,dx.$

   l. $\displaystyle\int_0^1 e^{3x}\,dx.$

   m. $\displaystyle\int xe^x\,dx.$

   n. $\displaystyle\int_0^1 xe^x\,dx.$

2. Let C be the cost to make x units of a commodity. Suppose the marginal cost is given by $\dfrac{dC}{dx} = 2.7x^{1.7} - 0.02x + 20.01$. Find C as a function of x if the overhead is $C = 6.3$ when $x = 0$.

3. Let R be the world rate of petroleum consumption at time t; t is the elapsed time in years measured from 1900, and R is measured in millions of barrels per year. The table below gives some approximate values of R and t. Plot R as a function of t, using graph paper if possible. Estimate a suitable area in order to find the total amount of petroleum consumed between 1900 and 1970.

t	0	10	20	30	40	50	60	70
R	2	50	70	150	220	350	700	1700

4. Let R and t have the same meanings as in Problem 3 above. Suppose this function is described by the equation $R = 8e^{0.08t}$. According to this formula, what amount of petroleum will be consumed between 1900 and 2000?

5. Let A be the rate in micrograms per hour at which corn roots absorb phosphorus from a certain soil. A decreases with time because the diffusion of phosphorus through soil moisture lags behind its uptake through root surfaces, resulting in a decreasing concentration of phosphorus at root surfaces. For Tucumcari fine sandy loam at a fixed moisture level, some approximate[7] values of A and t are given in the table below. Plot A as a function of t. By estimating areas, approximate the total amount of phosphorus absorbed during the fifty hours $t = 0$ to $t = 50$.

t	0	10	20	30	40	50
A	1	0.2	0.18	0.16	0.15	0.15

6. Let A and t have the same meanings as in Problem 5 above. Suppose A can be described by the equation $A = e^{-0.09t}$. Use this formula to predict the total amount of phosphorus absorbed during the first 50 hours.

7. The *marginal propensity to consume* in the United States during 1922–1941 was estimated to be $\dfrac{dc}{dy} = 0.672$, where c is consumption expendi-

[7]Adapted from *The McGraw-Hill Encyclopedia of Environmental Science* (1974), p. 548.

ture in dollars and y is deflated disposable income.[8] Find c as a function of y if $c = 119.82$ when $y = 10$.

8. A stone is hurled vertically upward. Its initial speed was 70 feet per second (about 48 miles per hour). The height y in feet after t seconds obeys the law of gravity $\dfrac{d^2y}{dt^2} = -32$. Find y as a function of t, and find the maximum height reached by the stone.

9. Sketch the graph of $y = 10x - x^2$. Find the area of the finite region R which is bounded by this graph and the x-axis.

10. A mine produces 1000 troy ounces of silver each day. Suppose the market price for an ounce of silver will rise constantly at the annual rate of one dollar per year. The current price is 4 dollars per ounce. Let A be the total revenue after t years if the mine sells its silver as it is produced.

   a. Show that $\dfrac{dA}{dt} = 365,000(4 + t)$.

   b. Find A when $t = 10$.

11. Let $p = f(q)$ be a decreasing demand curve (see Figure 18); p is the price which must be charged for a commodity so that the demand for it will be q. The *consumers' surplus* at the price p_0 is defined to be the shaded region in Figure 18. (This area represents the money saved by consumers in an economy where everyone pays the same price, as compared to a

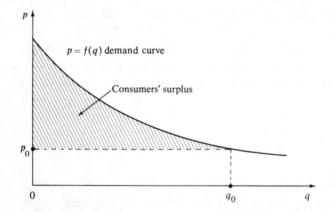

$p = f(q)$ demand curve

Consumers' surplus

Figure 18

perfect monopoly where each consumer could be made to pay a separate price corresponding to his own demand, called "perfect price discrimination.") If the demand function is $p = 49 - q^2$, find the consumers' surplus when $p_0 = 13$.

[8]T. Haavelmo, "Methods of Measuring the Marginal Propensity to Consume," *J. Amer. Stat. Assoc.* XLII (1974), pp. 105–122.

12. The rate r in micrograms per minute at which the adrenal cortex secrets cortisol in response to ACTH concentration changes has been studied by teams of physiologists and electrical engineers working together, using mathematical methods.[9] If $r = 2.6t^{0.3} + e^{0.01t} - 1$ t minutes after a jump increase in ACTH concentration, how many micrograms were secreted between $t = 0$ and $t = 10$?

Problems 13–19 concern *average value*, which is defined in Problem 13.

13. Let $y = f(x)$ be a function for $a \leq x \leq b$. The *average value* of y on this interval, denoted $\bar{y}$, is defined by

$$\bar{y} = \frac{1}{b - a} \int_a^b f(x)\, dx.$$

Figure 19 shows a thin "wave box"; the surface of the water is the graph of $y = f(x)$. Explain why $\bar{y}$ can be interpreted as the water level when the surface is calm.

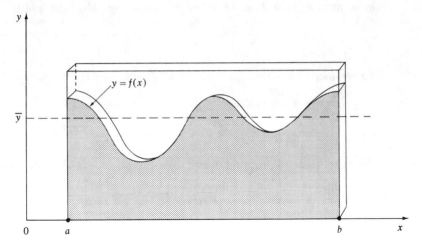

Figure 19

14. The assets A (in millions of dollars) of a bank on day t were $A = 100 + 15\sqrt{t}$. Find the average value $\bar{A}$ of A during the period $t = 0$ to $t = 100$.

15. In Problem 3, find, approximately, the average value $\bar{R}$ of R during $0 \leq t \leq 70$.

[9]C. C. Li and J. Urquhart, "Modeling of Adrenocortical Secretory Dynamics," *Concepts and Models of Biomathematics,* ed. F. Heinmets (New York: Marcel Dekker, Inc., 1969), p. 287.

16. In Problem 4, find the average value $\bar{R}$ during $0 \leq t \leq 100$.

17. A certain city with population 20,000 uses water at time t of the day at the rate $R = 44 + 24t - t^2$, where R is in thousands of gallons per hour and t is the number of hours past midnight $(0 \leq t \leq 24)$. Is this city's water usage above or below the national average of 6.42 gallons per person per hour[10]?

18. An object moves in a straight line with speed $v = 3t^2 + 4t$ at time t, with t in hours and v in miles per hour.
  a. Find the total distance traveled during $0 \leq t \leq 3$.
  b. Find $\bar{v}$, the average value of v during $0 \leq t \leq 3$.

19. In Problem 9, find $\bar{y}$ for $0 \leq x \leq 10$.

20. Let x be a numerical attribute of individuals in a population (for example, x might be height, weight, or IQ). Suppose the values of x always lie between A and B. Suppose f is a function such that for any a and b with $A \leq a < b \leq B$, $\int_a^b f(x)\, dx$ is the fraction of the population for which x lies between a and b. Then f is called the *distribution of x*. The *mean* $\bar{x}$ of x is

$$\bar{x} = \int_A^B x f(x)\, dx.$$

The *variance V* of x is

$$V = \int_A^B (x - \bar{x})^2 f(x)\, dx,$$

and the *standard deviation* σ of x is $\sigma = \sqrt{V}$. If $f(x) = \dfrac{3x(2 - x)}{4}$ and if x lies between 0 and 2, find the mean, variance, and standard deviation.

[10]H. E. Babbitt and E. R. Baumann, *Sewerage and Sewage Treatment,* 8th ed. (New York: Wiley, 1958).

21. Approximately how far can a Lancia Beta Scorpion[11] travel in 40 seconds from a standing start (see Figure 20)?

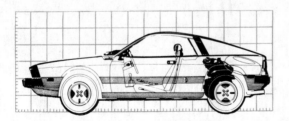

"R & T Road Test LANCIA BETA SCORPION"   Scale: 10″ divisions

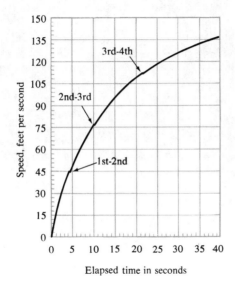

Elapsed time in seconds

Figure 20

[11]Adapted from *Road and Track*, September 1976, page 51.

seven

Functions of Several Variables

The preceding chapters dealt with the methods of calculus for functions of one variable. In the present chapter, some of those methods will be extended to functions of several variables.

Section 1 □ **Basic concepts.**   Notation, graphing.

Section 2 □ **Partial derivatives.**   Increasing or decreasing.

Section 3 □ **The total differential.**   Approximate changes.

Section 4 □ **Higher-order partial derivatives.**

Section 5 □ **Maxima and minima.**   Second derivative test.

Section 6 □ **Maxima and minima with constraints.**   Lagrange multipliers.

Section 7 □ **Least squares curve fitting.**   Square-sum error, least squares criterion, least squares regression line.

Summary □ **Miscellaneous problems.**

We write $z = f(x, y)$ to show that the value of z depends on the values of x and y. In this notation, f denotes a rule that determines the value of z from the two input values x and y; f is called a **function of two variables.** For example, if f is the rule

$$f(x, y) = 2x + 3y,$$

then when x has value 10 and y has value 5, z will have value

$$f(10, 5) = 2 \cdot 10 + 3 \cdot 5 = 35.$$

Similarly, when $x = 1$ and $y = 6$, then z will be

$$f(1, 6) = 2 \cdot 1 + 3 \cdot 6 = 20.$$

Functions of two, three, or more variables occur frequently in mathematics and its applications. In economics, for example, it is sometimes assumed for simplicity that the demand D for a commodity is a function only of the price P of that commodity. We could denote the function by g and write

$$D = g(P).$$

More sophisticated economic models take account of the fact that the prices of competing commodities influence the demand for each other. Thus the demand D for beef actually depends not only the price P of beef but also on the prices Q, R, and S of pork, mutton, and fish. In symbols,

(1) $D = h(P, Q, R, S).$

Thus D is a function of four variables in this model.

Example 1. Suppose the demand function in Equation 1 is given by

$$h(P, Q, R, S) = 2P + Q + 3R + \frac{1}{2}S.$$

Find D when $P = 5$, $Q = 7$, $R = 2$, and $S = 2$.

SOLUTION.

$$D = h(5, 7, 2, 2) = 2 \cdot 5 + 7 + 3 \cdot 2 + \frac{1}{2} \cdot 2 = 24.$$

Example 2. Let $f(x, y) = 2x - xy + 10y$. Calculate $f(5, 6)$ and $f(6, 5)$ to illustrate that $f(5, 6) \neq f(6, 5)$.

SOLUTION.
$$f(5, 6) = 2 \cdot 5 - 5 \cdot 6 + 10 \cdot 6 = 10 - 30 + 60 = 40.$$
$$f(6, 5) = 2 \cdot 6 - 6 \cdot 5 + 10 \cdot 5 = 12 - 30 + 50 = 32.$$

Example 3. The solid rectangular box in Figure 1 has length x, width y, and height z (the units are feet). Let V be its volume, S its surface area, and E the total length of all the edges (V is in cubic feet, S is in square feet, and E is in feet). Then V, S, and E are functions of the three variables x, y, z; find their formulas.

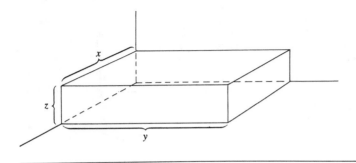

Figure 1

SOLUTION.
$$V = xyz$$
$$S = 2xy + 2yz + 2xz$$
$$E = 4x + 4y + 4z.$$

Functions of Two Variables. In the preceding chapters, we studied functions of one variable. The geometric representation of such functions by means of a graph was a valuable tool. There is a geometric way to represent functions of two variables as surfaces in two-dimensional space. We shall explain the basic idea behind this kind of representation, but we shall not treat surface sketching techniques. There is no convenient way to visualize functions of more than two variables; they must be studied exclusively by analytical methods.

Consider a function of two variables

(2) $z = f(x, y)$.

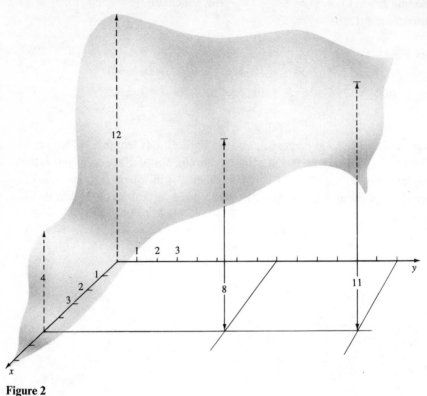

Figure 2

To represent this function geometrically, consider the xy-plane to lie horizontally in three dimensional space. Form a surface whose height above the point (x, y) is $f(x, y)$; if $f(x, y)$ is negative, interpret this to mean that the surface is at the distance $|f(x, y)|$ *below* (x, y).

One can see that the function $f(x, y)$ represented by the surface in Figure 2 satisfies

$$f(0, 0) = 12, \quad f(5, 0) = 4, \quad f(5, 8) = 8, \quad \text{and} \quad f(5, 14) = 11.$$

It is sometimes possible to sketch the graphs of simple equations without great effort. The graph of $z = \sqrt{x^2 + y^2}$ is a cone, as shown in Figure 3. The graph of $z = x^2 + y^2$ is a dish-shaped surface called a **paraboloid;** it is shown in Figure 4. Other equations may require more care to graph. Figure 5 shows part of the graph of $z = y^2 - x^2$ as drawn by a computer.

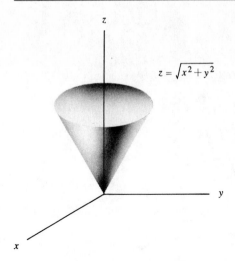

$$z = \sqrt{x^2 + y^2}$$

Figure 3

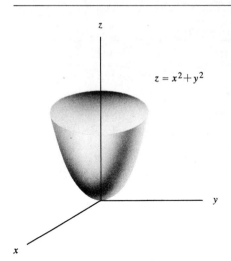

$$z = x^2 + y^2$$

Figure 4

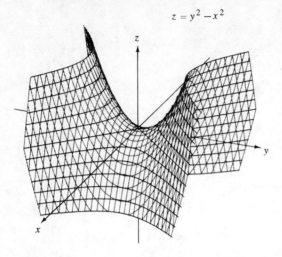

$$z = y^2 - x^2$$

Figure 5

Problems

1. Let $z = p(x, y)$, where $p(x, y) = 2x + 4xy + y$.
   a. What is the value of z when $x = 1$ and $y = 3$?
   b. What is the value of z when $y = 1$ and $x = 3$?
   c. Which is larger, $p(5, 0)$ or $p(0, 5)$?
   d. Find at least four sets of values for (x, y) which make $z = 0$.

2. Let x, y, and z be the price of beef, pork, and mutton, respectively. The demand D for beef has been estimated[1] to be, in suitable units, $D = 63.3 - 1.9x + 0.2y + 0.5z$.
   a. If the price of beef is fixed at 10 and the price of pork is fixed at 5, compare the demands D for beef when the price of mutton is 2 and then changes to 4.
   b. If x and y are fixed, does D increase or decrease when z increases?
   c. If y and z are fixed, does D increase or decrease when x increases?

3. Let $Q = f(L, C)$, where Q is the production output when L dollars are spent for labor and C dollars for fixed costs (capital). We call f a *production function.* If it is of the form

(3) $f(L, C) = aL^b C^{1-b}$,   where a, b are constants,

---

[1]H. Schultz, *Statistical Laws of Demand and Supply* (Chicago: University of Chicago Press, 1928).

then economists call f a *Cobb-Douglas production function*. The production function for New South Wales has been estimated[2] to be approximately $Q = 1.14\,L^{4/5}\,C^{1/5}$, measured in suitable units. Find Q when $L = 32$ and $C = 243$ (*Note:* $32 = 2^5$ and $243 = 3^5$)

4. A production function (see Problem 3) of the form

(4) $f(L, C) = (a_1 L^b + a_2 C^b)^{1/b}$,   where a_1, a_2, and b are constants,

is called a *homohyphallic production function*. Show that both types of production functions, in Equations 3 and 4, have the property that when labor and capital inputs are doubled, the production output is doubled (in symbols, $f(2L, 2C) = 2f(L, C)$).

5. The Olympic Games R-rating[3] of a sailboat is $R = g(L, V, A)$ where L is the length in meters, A is the sail area in square meters, V is volume in cubic meters of water which the boat displaces, and

$$g(L, V, A) = 0.9\left(\frac{L\sqrt{A}}{12\sqrt[3]{V}} + \frac{L + \sqrt{A}}{4} \right).$$

What is the R-rating of a sailboat with $L = 12$, $A = 64$, and $V = 8$?

In Problems 6–11, find a formula for each function f.

6. $A = f(x, y)$, where A is the area in square feet of a rectangular lot whose width is x feet and whose depth is y feet.

7. $I = f(s, t)$, where I is the income in dollars of a salesman who sells s computers and t terminals if the commission on computers is $2000 each and the commission on terminals is $60 each.

8. $S = f(L, F)$, where S is the area in square feet of a sail which has the shape of a right triangle with legs of L feet (length of the luff) and F feet (length of the foot).

9. $W = f(x, y, z)$, where W is the weight in pounds of an adobe brick whose length, width, and height are x inches, y inches, and z inches, respectively, if the density of adobe is 0.24 pounds per cubic inch.

10. $D = f(r, t)$, where D is the distance in miles traveled by a snail in t hours if the snail crawls at a constant rate of r inches per minute.

[2]P. H. Douglas and G. T. Gunn, "The Production Function for Australian Manufacturing," *Quarterly Journal of Economics*, LVI (1941), pp. 108–129.

[3]C. A. Marchaj, *Sailing Theory and Practice* (New York: Dodd, Mead and Company, 1964).

11. $T = f(r_1, r_2)$, where T is the time in years to deplete a certain oil reserve with two oil wells if the first well alone can deplete it in r_1 years and the second well alone can deplete it in r_2 years.

Section 2 □ Partial Derivatives

Suppose the quantity z depends on two variables x and y; in symbols,

$$z = f(x, y).$$

It is often of interest to consider the rate at which z changes when x and y change. This rather complicated notion is best studied by considering the two special cases: the rate of change of z with respect to a changing x when y remains fixed, and the rate of change of z with respect to a changing y when x remains fixed. The first of these is called the **partial derivative of z** (or of f) **with respect to x** and is denoted by

$$\frac{\partial z}{\partial x} \quad \text{or} \quad \frac{\partial f}{\partial x} \quad \text{or} \quad f_x.$$

The second is called the **partial derivative of z** (or of f) **with respect to y** and is denoted by

$$\frac{\partial z}{\partial y} \quad \text{or} \quad \frac{\partial f}{\partial y} \quad \text{or} \quad f_y.$$

These partial derivatives may be found by the ordinary rules of differentiation; it is only necessary to keep clearly in mind which variables are being held constant. We find $\dfrac{\partial z}{\partial x}$ by thinking of y as a constant and then calculating $\dfrac{dz}{dx}$. We find $\dfrac{\partial z}{\partial y}$ by thinking of x as a constant and then calculating $\dfrac{dz}{dy}$.

Example 1. Find $\dfrac{\partial z}{\partial x}$ and $\dfrac{\partial z}{\partial y}$ if

(1) $z = 4x^2 + 6x^3y^5.$

SOLUTION.

To find $\dfrac{\partial z}{\partial x}$, we think of y as a constant in Equation 1 and then calculate $\dfrac{dz}{dx}$. To help remember that y is constant we can temporarily replace y by y_0 in

Equation 1 to emphasize that y is a constant. Then Equation 1 becomes

(2) $z = 4x^2 + 6x^3 y_0^5$.

We calculate $\dfrac{dz}{dx}$ from Equation 2; the result is

(3) $8x + 18x^2 y_0^5$

because y_0^5 is a constant. If we now return y_0 to its original notation y, then Equation 3 gives $\dfrac{\partial z}{\partial x}$:

(4) $\dfrac{\partial z}{\partial x} = 8x + 18x^2 y^5$.

After a little practice you should be able to omit the intermediate step of replacing y by y_0; just look at Equation 1, think of y as a constant, differentiate with respect to x, and obtain Equation 4 at once.

Let us use the shorter method to find $\dfrac{\partial z}{\partial y}$. We look at Equation 1, think of x as a constant, and differentiate with respect to y. We obtain

$$\frac{\partial z}{\partial y} = 30x^3 y^4.$$

The notion of partial derivative also applies to functions of more than two variables. If, for example, z is a function of x, y, u, v, and t, then $\dfrac{\partial z}{\partial u}$ would be found by thinking of all variables except u as constant, and then calculating $\dfrac{dz}{du}$.

Example 2.   Let

(5) $p = 3r^4 s + t^3 r - 8rst^2$.

Find $\dfrac{\partial p}{\partial r}, \dfrac{\partial p}{\partial s}$, and $\dfrac{\partial p}{\partial t}$.

SOLUTION.

To find $\dfrac{\partial p}{\partial r}$, think of s and t as constant in Equation 5 and calculate $\dfrac{dp}{dr}$:

$$\frac{\partial p}{\partial r} = 12r^3 s + t^3 - 8st^2.$$

To find $\dfrac{\partial p}{\partial s}$, think of r and t as constant in Equation 5 and then calculate $\dfrac{dp}{ds}$. The result is

$$\frac{\partial p}{\partial s} = 3r^4 - 8rt^2.$$

To find $\dfrac{\partial p}{\partial t}$, think of r and s as constant in Equation 5 and then calculate $\dfrac{dp}{dt}$. The result is

$$\frac{\partial p}{\partial t} = 3t^2 r - 16rst.$$

Example 3. Let $f(u, v, w) = 5u^2 v^3 w^4$. Find the partial derivatives f_u, f_v and f_w.

SOLUTION.

The partial derivative with respect to u is f_u; this means we differentiate with respect to u while thinking of v and w as constant.

$$f_u = 10uv^3 w^4$$

Similarly,

$$f_v = 15u^2 v^2 w^4,$$
$$f_w = 20u^2 v^3 w^3.$$

$\left(\text{Note: } f_u \text{ has the same meaning as } \dfrac{\partial f}{\partial u}, f_v \text{ is the same as } \dfrac{\partial f}{\partial v}, \text{ and } f_w \text{ is the same as } \dfrac{\partial f}{\partial w}.\right)$

Consider a function of several variables, say $z = f(x, y)$. The value of any of the partial derivatives of z depends, in general, on the values of x and y. The value of $\dfrac{\partial z}{\partial x}$ when x has the value a and y has the value b is denoted by

$$\frac{\partial z}{\partial x} \quad \text{when} \quad (x, y) = (a, b)$$

or

$$\frac{\partial z}{\partial x} \quad \text{when} \quad x = a, \quad y = b$$

or

$$f_x(a, b).$$

This value of the partial derivative is found by first calculating $\dfrac{\partial z}{\partial x}$ as in the above examples, and then replacing x and y by a and b respectively.

Example 4. If $T = 5r^2s^3$ find $\dfrac{\partial T}{\partial r}$ when $r = 7$ and $s = 2$. Find $\dfrac{\partial T}{\partial s}$ when $r = 3$ and $s = 1$.

SOLUTION.

We find that $\dfrac{\partial T}{\partial r} = 10rs^3$ and therefore

$$\dfrac{\partial T}{\partial r} = 10(7)(2)^3 = 560 \qquad \text{when} \qquad r = 7 \qquad \text{and} \qquad s = 2.$$

We also find that $\dfrac{\partial T}{\partial s} = 15r^2s^2$ and therefore

$$\dfrac{\partial T}{\partial s} = 15(3)^2(1)^2 = 135 \qquad \text{when} \qquad r = 3 \qquad \text{and} \qquad s = 1.$$

Example 5. If $f(x,y,z) = x^3z + 2yz + z^2$, find $f_x(2,3,4), f_y(3,0,6)$, and $f_z(1,7,-3)$.

SOLUTION.
We find that $f_x = 3x^2z$, and therefore

$$f_x(2, 3, 4) = 3(2)^2(4) = 48.$$

We find that $f_y = 2z$, and therefore

$$f_y(3, 0, 6) = 2(6) = 12.$$

We also find that $f_z = x^3 + 2y + 2z$, and therefore

$$f_z(1, 7, -3) = (1)^3 + 2(7) + 2(-3) = 9.$$

Increasing or Decreasing. Our knowledge of derivatives for functions of one variable can be applied to functions of several variables. For example, if $z = f(x, y)$ and if $\dfrac{\partial z}{\partial x}$ is positive when $x = a$ and $y = b$, we can conclude that when y is held fixed at b and x increases from a, z will increase.

Similarly, if $\dfrac{\partial z}{\partial x} < 0$ when $(x, y) = (a, b)$, then z will decrease as x increases from a with y held fixed at b. The rate of increase or decrease is measured by the magnitude of $\dfrac{\partial z}{\partial x}$ when $(x, y) = (a, b)$. In the following example, we calculate partial derivatives in an applied situation and interpret the results in this way.

Example 6 (Marginal productivity). The costs of production can be separated into labor cost L (salaries for employees) and capital cost C (cost of land, buildings, machines, and so forth). Let $f(L, C)$ be the amount of commodities produced when the amount spent on labor is L and the amount spent on capital is C. Then f is called the *production function*, f_L is called the *marginal productivity of labor*, and f_C is called the *marginal productivity of capital*.

The production function[4] for Australia during the 1930s was (in suitable units)

(6) $\qquad f(L, C) = L^{0.64} C^{0.36}$.

Find the marginal productivity of labor and of capital when $L = 1.7$ and $C = 1$. In order to increase production, should the government encourage increased spending on labor or on capital? (*Note:* To carry out the calculations, you will need the values $(1.7)^{-0.36} \approx .83$ and $(1.7)^{0.64} \approx 1.40$, which were obtained with a hand calculator.)

SOLUTION.
We compute the partial derivatives from Equation 6 and find that $f_L = 0.64 L^{-0.36} C^{0.36}$ and $f_C = 0.36 L^{0.64} C^{-0.64}$. Therefore the marginal productivities when $(L, C) = (1.7, 1)$ are as follows.

$$f_L(1.7, 1) = (0.64)(1.7)^{-0.36}(1)^{0.36} \approx 0.53$$
$$f_C(1.7, 1) = (0.36)(1.7)^{0.64}(1)^{-0.64} \approx 0.50$$

Since productivity f increases faster with respect to L than with respect to C, it would be better, initially, to increase spending on labor than on capital (with additional calculus techniques, we shall be able to compute a way to divide increased spending between labor and capital so that production will increase faster than it would by increasing labor alone.)

[4]P. H. Douglas and G. T. Gunn, "The Production Function for Australian Manufacturing," *Quarterly Journal of Economics,* LVI (1941), pp. 108–129.

Problems

1. If $z = 4x^2y^4$, find $\dfrac{\partial z}{\partial x}$ and $\dfrac{\partial z}{\partial y}$.

2. If $f(x,y) = 2x^2 - 3y^5 + xy^2$, find f_x and f_y.

3. If $w = ve^{3u}$, find $\dfrac{\partial w}{\partial u}$ and $\dfrac{\partial w}{\partial v}$.

4. If $h(x,y,z) = x^{0.6}y^{0.4} + x^{0.2}z^{0.8}$, find h_x, h_y, and h_z.

5. If $Q = r^2t + \dfrac{s^2}{r^4}$, find $\dfrac{\partial Q}{\partial r}$, $\dfrac{\partial Q}{\partial s}$, and $\dfrac{\partial Q}{\partial t}$.

6. If $E = s\sqrt{1 + t^2}$, find $\dfrac{\partial E}{\partial s}$ and $\dfrac{\partial E}{\partial t}$.

7. Find $\dfrac{\partial S}{\partial m}$ and $\dfrac{\partial S}{\partial b}$ if $S = (12m + 5b)^2$.

8. Find f_x, f_y, and f_z if $f(x,y,z) = x^2zy + xy^2z^3 + \ln z$.

9. Find $\dfrac{\partial P}{\partial L}$ and $\dfrac{\partial P}{\partial C}$ if $P = 5L(2 + 3C)$.

10. Find $\dfrac{\partial z}{\partial x}$ and $\dfrac{\partial z}{\partial y}$ if $z = e^{2x+3y}$.

11. Find $\dfrac{\partial w}{\partial x}$ and $\dfrac{\partial w}{\partial y}$ if $w = e^x \ln y$.

12. Find Q_u and Q_v if $Q = 4u^{1/2}v^{-2/3}$.

13. Find g_s and g_t if $g(r,s,t,u) = \ln(r^{1.5}s^{2.3}t^{0.4}u^{3.1})$.

14. Find $\dfrac{\partial E}{\partial m}$ and $\dfrac{\partial E}{\partial b}$ if $E = (2m + b)^2 + (3m - b)^2$.

15. Find F_x and F_y if $F(x,y,z) = x + z\sqrt{3 - 2y}$.

16. Find $\dfrac{\partial z}{\partial x}$ and $\dfrac{\partial z}{\partial y}$ if $z = 4 - x^3$.

17. In Problem 1, find $\dfrac{\partial z}{\partial x}$ when $x = 12$ and $y = 1$.

18. In Problem 2, find $f_x(3,1)$ and $f_y(3,1)$.

19. In Problem 3, evaluate $\dfrac{\partial w}{\partial u}$ and $\dfrac{\partial w}{\partial v}$ when $(u,v) = (1,2)$.

20. The formula $Q = cw^{10}D^{-10}$ is used in studying the erosion of river beds[5]; here Q, w, and D designate the average river discharge, width, and depth respectively, and c is a positive constant. Find $\dfrac{\partial Q}{\partial w}$ and $\dfrac{\partial Q}{\partial D}$ when $x = 20$ and $y = 20$.

21. Poiseuille's formula[6] for flow within a blood vessel is of the form[7] $v = pr^4/L$ where v is the flow rate, r and L are the radius and length of the vessel, and p is the pressure difference at the ends of the vessel. Calculate $\dfrac{\partial v}{\partial p}, \dfrac{\partial v}{\partial r}$, and $\dfrac{\partial v}{\partial L}$.

22. Blood flow resistance R through a blood vessel is defined as $R = p/v$, where p is the difference in blood pressure at the ends of the vessel and v is the flow rate of the blood.[7] Use Poiseuille's formula from Problem 21 to express the blood flow resistance R as a function of r and L. Find $\dfrac{\partial R}{\partial L}$ and $\dfrac{\partial R}{\partial r}$ when $r = 0.3$ and $L = 2$.

23. The production function[8] for the United States in 1919 was $f(L,C) = L^{0.76}C^{0.24}$, where L is labor and C is capital in suitable units. Find the marginal productivities f_L and f_C. When $L = 80$ and $C = 20$, which will produce a faster increase in productivity, an increase in L or an increase in C? (*Note:* You may need the following numerical results. $20^{0.24} \approx 2.05$, $20^{-0.76} \approx 0.10$, $80^{0.76} \approx 27.95$, and $80^{-0.24} \approx 0.35$.)

24. Suppose $D_A = f(p_A, p_B, p_C)$, where D_A is the demand for commodity A and p_A, p_B, and p_C are the prices of the commodities A, B, and C. Let X and Y stand for any of the letters A, B, or C. The *partial elasticity of demand for commodity X with respect to the price of Y* is defined as

$$\frac{\partial D_X}{\partial p_Y} \frac{p_Y}{D_X}.$$

Discuss the economists' interpretation of this as "the percent of increase in D_X that occurs when the price of Y increases by 1% and the prices of the other commodities remain fixed."

[5] *The McGraw-Hill Encyclopedia of Environmental Science* (1974), pp. 603–604.

[6] Jean Louis Poiseuille (1799–1869) was a French physician who discovered several laws of flow by empirical techniques; the laws were later derived by theoretical methods.

[7] K. R. Rebman, H. Slater, and R. M. Thrall, *Some Mathematical Models in Biology* (Ann Arbor: University of Michigan Report, 1967), p. 14ff.

[8] P. H. Douglas and G. T. Gunn, "The Production Function for American Manufacturing in 1919," *American Economic Review*, XXXI (1941), pp. 67–80.

25. Let p_x, p_y, and p_z be the prices of beef, pork, and mutton, respectively. The demand D_x for beef has been estimated[9] to be $D_x = 63.3 - 1.9p_x + 0.2p_y + 0.5p_z$. Find the partial elasticity of demand for beef with respect to the price of mutton (see Problem 24 above) when $p_x = 10$, $p_y = 5$, and $p_z = 6$.

Section 3 □ The Total Differential

Suppose $z = f(x,y)$ is a function of two variables. We shall show how partial derivatives can be used to estimate the change produced in z by small changes in x and y. This is the analog in several variables of the topic "Approximation; Differentials" treated in Section 5 of Chapter 4.

Let z_0 be the value of z when x has the value x_0 and y has the value y_0; thus $z_0 = f(x_0,y_0)$. Now suppose x and y change by amounts Δx and Δy. The new value of z will be $f(x_0 + \Delta x, y_0 + \Delta y)$. Therefore the exact change in z, denoted by Δz, will be

(1) $\Delta z = f(x_0 + \Delta x, y_0 + \Delta y) - f(x_0, y_0)$.

In many theoretical as well as practical situations, the exact value for Δz in Equation 1 is less useful than the following approximate value:

(2) $\Delta z \approx f_x(x_0, y_0)\Delta x + f_y(x_0, y_0)\Delta y$.

The approximation in Equation 2 increases in accuracy when Δx and Δy are small. The right hand side of Equation 2 is often denoted by dz and is called the **total differential** of z.

Equation 2 is often written in the shorter version

(3) $\Delta z \approx \dfrac{\partial z}{\partial x}\Delta x + \dfrac{\partial z}{\partial y}\Delta y$.

When using the abbreviated form in Equation 3, one must keep in mind that the partial derivatives $\dfrac{\partial z}{\partial x}$ and $\dfrac{\partial z}{\partial y}$ are to be evaluated at or near $(x, y) = (x_0, y_0)$.

In the case of a function of three variables, say $w = f(x,y,z)$, the formula corresponding to Equation 3 is

(4) $\Delta w \approx \dfrac{\partial w}{\partial x}\Delta x + \dfrac{\partial w}{\partial y}\Delta y + \dfrac{\partial w}{\partial z}\Delta z$.

---

[9]H. Schultz, *Statistical Laws of Demand and Supply,* (Chicago: University of Chicago Press, 1928).

In general, the approximate change in the dependent variable is equal to the sum of each partial derivative with respect to each of the independent variables times the change in that independent variable.

The precise statement and proof of Equation 2 are too technical to present here. Let us note, however, that $f_x(x_0,y_0)\Delta x$ can be interpreted as the approximate change in z when y is fixed at y_0 and x changes from x_0 to $x_0 + \Delta x$ (in symbols, $f_x(x_0,y_0)\Delta x \approx f(x_0 + \Delta x, y_0) - f(x_0, y_0)$). This follows from the definition of the derivative

$$f_x(x_0,y_0) = \lim_{\Delta x \to 0} \frac{f(x_0 + \Delta x, y_0) - f(x_0, y_0)}{\Delta x}$$

$$\approx \frac{f(x_0 + \Delta x, y_0) - f(x_0, y_0)}{\Delta x}.$$

Similarly, $f_y(x_0,y_0)\Delta y$ is approximately the change in z when x is fixed at x_0 and y changes from y_0 to $y_0 + \Delta y$. Thus Equation 2 asserts that the sum of these two changes is approximately equal to the actual change in z as (x_0,y_0) changes to $(x_0 + \Delta x, y_0 + \Delta y)$.

Example 1. Let $z = x^{1/2}y^{1/3}$. Suppose x changes from 25 to 25.2 and y changes from 8 to 8.4. What will be the approximate change in z?

SOLUTION.

We first calculate $\dfrac{\partial z}{\partial x} = \frac{1}{2}x^{-1/2}y^{1/3}$ and $\dfrac{\partial z}{\partial y} = \frac{1}{3}x^{1/2}y^{-2/3}$. In this problem, $x_0 = 25$ and $y_0 = 8$. When these partial derivatives are evaluated at $(x, y) = (25, 8)$, we obtain

$$\frac{\partial z}{\partial x} = \frac{1}{2}(25)^{-1/2}(8)^{1/3} = \frac{1}{5}, \qquad \frac{\partial z}{\partial y} = \frac{1}{3}(25)^{1/2}(8)^{-2/3} = \frac{5}{12}.$$

Therefore Equation 3 becomes

(5) $\qquad \Delta z \approx \dfrac{1}{5}\Delta x + \dfrac{5}{12}\Delta y.$

In this problem, $\Delta x = 25.2 - 25 = 0.2$ and $\Delta y = 8.4 - 8 = 0.4$. Hence, from Equation 5,

(6) $\qquad \Delta z \approx \dfrac{1}{5}(0.2) + \dfrac{5}{12}(0.4) = \dfrac{1}{5}\cdot\dfrac{2}{10} + \dfrac{5}{12}\cdot\dfrac{4}{10} = \dfrac{31}{150},$

which is the desired answer.

(In this problem, we can use a hand calculator to compute the exact change in z. Let us do so, and then compare it with our approximate answer, $\frac{31}{150}$. The original value of z, call it z_0, is

$$z_0 = 25^{1/2}81^{1/3} = 10.$$

The new value of z, call it z_1, is

$$z_1 = (25.2)^{1/2}(8.4)^{1/3} = 10.2045.$$

The exact change in z is therefore

$$\Delta z = z_1 - z_0 = (25.2)^{1/2}(8.4)^{1/3} - 10,$$

which is 0.2045 to four significant figures; this is in close agreement with the approximate answer of $31/150 = 0.2067$.)

Example 2. Consider the production function

$$f(L, C) = 3L^2 - 5C.$$

At present, the values of L and C are $L = 10$ and $C = 8$.

a. Give the approximate formula for the change in production Δf when L changes by ΔL and C changes by ΔC.

b. Use this formula to predict the change in f when L increases to 11 and C decreases to 7.

c. Use the formula to predict the change in f when L decreases to 9 and C increases to 9.

SOLUTION.

a. The partial derivatives are

$$f_L = 6L \quad \text{and} \quad f_C = -5.$$

We evaluate these when $(L, C) = (10, 8)$ and obtain

$$f_L(10, 8) = 60 \quad \text{and} \quad f_C(10, 8) = -5.$$

Therefore

(6) $\Delta f \approx 60\, \Delta L - 5\, \Delta C.$

b. We substitute $\Delta L = 1$ and $\Delta C = -1$ in Equation 6 to obtain $\Delta f \approx 60(1) - 5(-1) = 65$. In words, f increases by about 65 when L increases by 1 and C decreases by 1.

c. We substitute $\Delta L = -1$ and $\Delta C = 1$ into Equation 6 and obtain $\Delta f \approx 60(-1) - 5(1) = -65$. In words, f decreases by about 65 when L decreases by 1 and C increases by 1.

Example 3. Poiseuille's law (see footnote 6) for laminar flow in an artery is of the form[10] $v = (R^2 - r^2)P$, where v is the velocity of a blood cell at a distance r from the central axis of the artery, R is the radius of the artery, and P is the pressure difference. Suppose $R = 3, r = 1$, and $P = 6$.

a. Give an approximate formula for the change in v produced by changing these values of R, r, and P by amounts $\Delta R, \Delta r$, and ΔP.

b. Use this formula to predict the approximate value of v when $R = 3.1$, $r = 1.1$, and $P = 5.9$.

SOLUTION.

a. We have $\dfrac{\partial v}{\partial R} = 2RP, \dfrac{\partial v}{\partial r} = -2rP$, and $\dfrac{\partial v}{\partial P} = R^2 - r^2$. When $R = 3$, $r = 1$, and $P = 6$, these partial derivatives have the values

$$\frac{\partial v}{\partial R} = 36, \qquad \frac{\partial v}{\partial r} = -12, \qquad \text{and} \qquad \frac{\partial v}{\partial P} = 8.$$

The general formula $\Delta v \approx \dfrac{\partial v}{\partial R}\Delta R + \dfrac{\partial v}{\partial r}\Delta r + \dfrac{\partial v}{\partial P}\Delta P$ (which corresponds to Equation 4) becomes, in this case,

(7) $\Delta v \approx 36\,\Delta R - 12\,\Delta r + 8\,\Delta P.$

b. We use Equation 7 to find Δv for the given changes in R, r, and P. This change in v, when added to the original value of v, will then give the predicted new value of v. The changes in R, r, and P are $\Delta R = 3 - 3.1 = -0.1, \Delta r = 1 - 1.1 = -0.1$, and $\Delta P = 6 - 5.9 = 0.1$, and so Equation 7 yields

$$\Delta v \approx 36(-0.1) - 12(-0.1) + 8(0.1) = -1.6.$$

The original value of v was

$$v = (R^2 - r^2)P = (3^2 - 1^2)6 = 48.$$

Therefore the predicted new value of v is $48 + (-1.6)$ or 46.4.

[10]E. Batschelet, *Introduction to Mathematics for Life Scientists* (Berlin: Springer-Verlag, 1975).

Problems

1. If $z = 4x^{10} - 6y^4$, find the approximate change in z that occurs when x changes from 1 to 0.9 and y changes from 2 to 2.3. Does z increase or decrease?

2. If $T = 3\sqrt{u} + 6\sqrt{v}$, find the approximate change in T that occurs when (u, v) changes from $(36, 4)$ to $(37, 3)$.

3. If $f(u, v, w) = ue^{2v} + w^2$, find the approximate change in f when (u, v, w) changes from $(100, 0, 5)$ to $(90, \frac{1}{2}, 6)$.

4. If $g(x, y) = xy^2$, find the approximate change in g when x changes from 3 to 3.2 and y changes from 10 to 10.3.

5. If $Q = 2x\sqrt{y}$, find the approximate change in Q when x changes from 6 to 6.5 and y changes from 4 to 3.9.

6. If $f(r, s, t) = 3r^2 + 5s^3 - 2t^5$, find the approximate change in f when r changes from 0.5 to 0.3, s changes from 2 to 2.02, and t changes from 1 to 1.01.

7. If $z = x^2y^3$, find the approximate change in z when (x, y) changes from $(5, 2)$ to $(4.7, 2)$.

8. In the study of pollution due to sewage disposal in rivers,[11] the equation $K = DU^{1/2}H^{-3/2}$ occurs, where K is the reaeration coefficient, D is the diffusion coefficient of oxygen in water, H is the average depth of the river, and U is the average flow of the river. Find the approximate change in K produced by changing D from 6 to 8, changing U from 9 to 9.7, and changing H from 1 to 1.1.

9. The formula $F = 24\,QD/tEh$ occurs in the study of heating efficiency in houses.[12] Suppose $Q = 3, D = 5, t = 12, E = 10$, and $h = 1$. If Q is increased to 4 and E is decreased to 8, what is the approximate change produced in F?

10. The total production output Q for New South Wales depended on the labor expenditure L and capital expenditure C according to the formula[13]

[11] *The McGraw-Hill Encyclopedia of Environmental Science* (1974), p. 517.

[12] *The McGraw-Hill Encyclopedia of Environmental Science* (1974), p. 253.

[13] P. H. Douglas and G. T. Gunn, "The Production Function for Australian Manufacturing," *Quarterly Journal of Economics,* LVI (1941), 108–129.

$Q = 1.14\ L^{4/5}\ C^{1/5}$. Suppose during a certain year, $L = 32$ and $C = 243$. If, during the next year, these values are changed to $L = 31$ and $C = 258$, will Q increase or decrease? By what amount, approximately?

Section 4 □ Higher-Order Partial Derivatives

Let us find partial derivatives of the function

(1) $z = 3x^5y + y^2$.

We obtain

(2) $\dfrac{\partial z}{\partial x} = 15x^4y,$

(3) $\dfrac{\partial z}{\partial y} = 3x^5 + 2y.$

It is sometimes useful (see, for example, the second derivative test on page 296) to continue this process of taking partial derivatives. The partial derivative of the right-hand side of Equation 2 with respect to x is $60x^3y$. This quantity, the partial derivative with respect to x of $\dfrac{\partial z}{\partial x}$, is denoted by $\dfrac{\partial^2 z}{\partial x^2}$ (the notation is an abbreviated form of $\dfrac{\partial}{\partial x} \cdot \dfrac{\partial z}{\partial x}$). Thus

(4) $\dfrac{\partial^2 z}{\partial x^2} = 60x^3y.$

If we take the partial derivative of Equation 2 with respect to y, the result would be denoted

(5) $\dfrac{\partial^2 z}{\partial y \partial x} = 15x^4$

(note how $\dfrac{\partial}{\partial y} \cdot \dfrac{\partial z}{\partial x}$ is abbreviated as $\dfrac{\partial^2 z}{\partial y \partial x}$). Similarly, the partial derivatives of Equation 3 would be written

(6) $\dfrac{\partial^2 z}{\partial x \partial y} = 15x^4,$

(7) $\dfrac{\partial^2 z}{\partial y^2} = 2.$

The quantities $\dfrac{\partial^2 z}{\partial x^2}, \dfrac{\partial^2 z}{\partial y \partial x}, \dfrac{\partial^2 z}{\partial x \partial y}$, and $\dfrac{\partial^2 z}{\partial y^2}$ are called the **second-order partial derivatives** of z. This concept is extended, in the natural way, to functions of more than two variables (see Example 3).

Example 1. Find the second-order partial derivatives of $z = 3x^2 + 4x^3y^2 - 8y^4$.

SOLUTION.
We must first calculate the partial derivatives of z.

(8) $\qquad \dfrac{\partial z}{\partial x} = 6x + 12x^2y^2,$

(9) $\qquad \dfrac{\partial z}{\partial y} = 8x^3y - 32y^3.$

To calculate the second-order partial derivatives, we take the partial derivatives of Equations 8 and 9. From Equation 8, we obtain

(10) $\qquad \dfrac{\partial^2 z}{\partial x^2} = 6 + 24xy^2$

(11) $\qquad \dfrac{\partial^2 z}{\partial y \partial x} = 24x^2y.$

From Equation 9, we obtain

(12) $\qquad \dfrac{\partial^2 z}{\partial x \partial y} = 24x^2y,$

(13) $\qquad \dfrac{\partial^2 z}{\partial y^2} = 8x^3 - 96y^2.$

The results in Equations 10, 11, 12, and 13 are the desired answers.

Example 2. Let $w = 5u^3v^4$.

a. Find $\dfrac{\partial^2 w}{\partial u \partial v}$ and $\dfrac{\partial^2 w}{\partial v^2}$.

b. Evaluate $\dfrac{\partial^2 w}{\partial u \partial v}$ and $\dfrac{\partial^2 w}{\partial v^2}$ when $u = 2$ and $v = 1$.

SOLUTION.

a. From $\dfrac{\partial w}{\partial v} = 20u^3v^3$, we obtain

(14) $\quad \dfrac{\partial^2 w}{\partial u \partial v} = 60u^2v^3 \quad$ and $\quad \dfrac{\partial^2 w}{\partial v^2} = 60u^3v^2.$

b. We evaluate Equation 14 when $(u, v) = (2, 1)$. The result is

$$\dfrac{\partial^2 w}{\partial u \partial v} = 240 \quad \text{when} \quad (u, v) = (2, 1)$$

$$\dfrac{\partial^2 w}{\partial v^2} = 480 \quad \text{when} \quad (u, v) = (2, 1).$$

Example 3. Let $P = 6r^3s\sqrt{t}$. Find $\dfrac{\partial^2 P}{\partial r \partial t}$ and $\dfrac{\partial^2 P}{\partial t^2}$.

SOLUTION.
From $P = 6r^3st^{1/2}$, we obtain

(15) $\quad \dfrac{\partial P}{\partial t} = 3r^3st^{-1/2}.$

From Equation 15, we obtain the desired second-order partial derivatives.

$$\dfrac{\partial^2 P}{\partial r \partial t} = 9r^2st^{-1/2},$$

$$\dfrac{\partial^2 P}{\partial t^2} = -\dfrac{3}{2}r^3st^{-3/2}.$$

Functional Notation. Suppose z is a function of x and y. If we use the functional notation

$$z = f(x, y),$$

then the various partial derivatives can also be denoted as follows.

$$f_x \text{ for } \dfrac{\partial z}{\partial x} \qquad f_y \text{ for } \dfrac{\partial z}{\partial y} \qquad f_{xx} \text{ for } \dfrac{\partial^2 z}{\partial x^2}$$

$$f_{yx} \text{ for } \dfrac{\partial^2 z}{\partial x \partial y} \qquad f_{xy} \text{ for } \dfrac{\partial^2 z}{\partial y \partial x} \qquad f_{yy} \text{ for } \dfrac{\partial^2 z}{\partial y^2}$$

(This notation comes from abbreviating $(f_x)_x$ by f_{xx}, $(f_x)_y$ by f_{xy}, etc.)

This functional notation is very convenient for designating the values at which a partial derivative is to be evaluated. For example,

$$\frac{\partial^2 z}{\partial x^2} \quad \text{when} \quad x = 3 \quad \text{and} \quad y = 7$$

can be denoted by

$$f_{xx}(3, 7).$$

Example 4. Let $f(x, y) = x^4 - 2x^3 y^3$. Find $f_x, f_x(5, 2), f_{xx}, f_{xx}(5, 2)$, and $f_{xy}(-3, 1)$.

SOLUTION.
From

(16) $f_x = 4x^3 - 6x^2 y^3$,

we obtain

(17) $f_x(5, 2) = 4(5)^3 - 6(5)^2(2)^3 = -700.$

From Equation 16, we also obtain

(18) $f_{xx} = 12x^2 - 12xy^3$.

Thus

$$f_{xx}(5, 2) = 12(5)^2 - 12(5)(2)^3 = -180.$$

From Equation 16, we find

(20) $f_{xy} = -18x^2 y^2$,

so

$$f_{xy}(-3, 1) = -18(-3)^2(1)^2 = -162.$$

(*Note:* In order to find $f_{xx}(5, 2)$, for example, one must first find f_{xx}. It would not work to first substitute $x = 5$ and $y = 2$ in the expression for f or f_x and then take partial derivatives.)

Equality of Mixed Partial Derivatives.   In Example 1, you may have noticed (compare Equations 11 and 12) that $\dfrac{\partial^2 z}{\partial x \partial y} = \dfrac{\partial^2 z}{\partial y \partial x}$. In Example 2, we found that for $w = 5u^3 v^4$, we have

(21) $$\frac{\partial^2 w}{\partial u \partial v} = 60u^2 v^3.$$

Let us see if $\dfrac{\partial^2 w}{\partial u \partial v} = \dfrac{\partial^2 w}{\partial v \partial u}$ also holds in this case. To do this, we first find

$$\frac{\partial w}{\partial u} = 15u^2 v^4,$$

and from this obtain

(22) $$\frac{\partial^2 w}{\partial v \partial u} = 60u^2 v^3.$$

By comparing Equations 21 and 22, we see that $\dfrac{\partial^2 w}{\partial u \partial v} = \dfrac{\partial^2 w}{\partial v \partial u}$. This raises the interesting question of whether these "mixed" partial derivatives are always the same, independently of the order. In functional notation $f(x, y)$, the question would be if f_{xy} is the same as f_{yx}.

The exact answer to this question is somewhat technical, and is beyond the scope of this course to state precisely. We can say that for all functions encountered in this course (and for nearly all functions encountered in actual applications), the answer is yes.

Example 5. Let $g(r, s, t) = r^2 t^3 - rs^4 + 2s^3 t^2$. Verify that $g_{rt} = g_{tr}$.

SOLUTION.
We shall calculate the mixed partial derivatives g_{rt} and g_{tr} and compare them. From

$$g_r = 2rt^3 - s^4 \qquad \text{and} \qquad g_t = 3r^2 t^2 + 4s^3 t,$$

we obtain

$$g_{rt} = 3 \cdot 2rt^{3-1} \qquad \text{and} \qquad g_{tr} = 2 \cdot 3r^{2-1} t^2.$$

Thus $g_{rt} = g_{tr}$.

Problems

---

1. Let $z = 4x^5 y^2$. Find all four second-order partial derivatives.

2. Let $z = 4x^5 y^2 + xy$. Find all four second-order partial derivatives.

3. Let $w = 3r^2 e^{5s}$. Find $\dfrac{\partial^2 w}{\partial r \partial s}$, $\dfrac{\partial^2 w}{\partial r^2}$, $\dfrac{\partial^2 w}{\partial s^2}$.

4. Let $f(x, y) = x^2 y - xy^3$. Find $f_{xx}, f_{xy}, f_{yy}, f_{xx}(2, 1), f_{xy}(2, 1)$, and $f_{yy}(2, 1)$.

5. Let $g(x, y, z) = x^2 + 2y^3 - 3z^2$. Find $g_{xy}, g_{yy}, g_{xz}, g_{xy}(2, -1, 3)$, $g_{yy}(2, -1, 3)$, and $g_{xz}(2, -1, 3)$.

6. Let $r = s^3 t + 4u^5$. Find $\dfrac{\partial^2 r}{\partial s^2}$, $\dfrac{\partial^2 r}{\partial s \partial t}$, $\dfrac{\partial^2 r}{\partial t^2}$, and $\dfrac{\partial^2 r}{\partial t \partial u}$. Find $\dfrac{\partial^2 r}{\partial s^2}$ when $s = 5$, $t = 2$, and $u = -3$.

7. Let $P(x, y) = 4x^{1.5} y^3$. Find P_{xx}, P_{xy}, P_{yy}, and $P_{xy}(4, 3)$.

8. Let $z = 5x^3 - 2y^4$. Find $\dfrac{\partial^2 z}{\partial x^2}$ when $(x, y) = (3, -1)$. Find $\dfrac{\partial^2 z}{\partial x \partial y}$ when $(x, y) = (3, 2)$.

9. Let $Q = 2e^x + x^2 y^3$. Find $\dfrac{\partial^2 Q}{\partial x \partial y}$. Find $\dfrac{\partial^2 Q}{\partial y^2}$ when $x = 1$ and $y = 5$.

10. Let $T = (2u + 5v)^4$. Find $\dfrac{\partial^2 T}{\partial u^2}$ when $(u, v) = (3, 1)$.

11. Let $E = (2x - 3y)^2 + (x - y)^3$. Find $\dfrac{\partial^2 E}{\partial x \partial y}$ when $(x, y) = (1, 0)$.

12. Let $f(p, q, r) = 2p^{0.5} - 4q^{4.5} + 3r^2$. Find $f_{pp}, f_{qq}, f_{pr}, f_{qr}, f_{pq}, f_{pp}(4, 1, 1)$, and $f_{pq}(4, 1, 1)$.

13. Let $w = 3x^2 - 5y^3 + 2xz^3$. Find $\dfrac{\partial^2 w}{\partial x^2}$, $\dfrac{\partial^2 w}{\partial x \partial y}$, $\dfrac{\partial^2 w}{\partial x \partial z}$, and $\dfrac{\partial^2 w}{\partial y \partial z}$. Find $\dfrac{\partial^2 w}{\partial x \partial z}$ when $(x, y, z) = (3, 2, 4)$.

14. Let $P = VT^5$. Find $\dfrac{\partial^2 P}{\partial V^2}$, $\dfrac{\partial^2 P}{\partial V \partial T}$, and $\dfrac{\partial^2 P}{\partial T^2}$. Find $\dfrac{\partial^2 P}{\partial T^2}$ when $V = 4$ and $T = 2$.

15. Let $s = e^{3x - 2y}$. Find $\dfrac{\partial^2 s}{\partial x^2}$, $\dfrac{\partial^2 s}{\partial x \partial y}$, and $\dfrac{\partial^2 s}{\partial y^2}$. Find $\dfrac{\partial^2 s}{\partial x^2}$ when $(x, y) = (1, 1)$. Find $\dfrac{\partial^2 s}{\partial x \partial y}$ when $(x, y) = (4, 6)$. Find $\dfrac{\partial^2 s}{\partial y^2}$ when $(x, y) = (0, 0)$.

16. Let $F(u, v) = u^3 v - uv^4$.
    a. Find F_{uu}, F_{uv}, and F_{vv}.
    b. Find $F_{uu}(3, 2), F_{uv}(1, 2)$, and $F_{vv}(-7, 3)$.

17. Let $Q(x, y) = e^x \ln y$. Find $Q_{xx}, Q_{xy}, Q_{yy}, Q_{xx}(2, e), Q_{xy}(3, 5)$, and $Q_{yy}(1, 2)$.

18. If $z = x^2yt^3 - x^3t + y^3$, find $\dfrac{\partial^2 z}{\partial x^2}, \dfrac{\partial^2 z}{\partial y^2}, \dfrac{\partial^2 z}{\partial t^2}, \dfrac{\partial^2 z}{\partial x \partial y}, \dfrac{\partial^2 z}{\partial x \partial t}$, and $\dfrac{\partial^2 z}{\partial y \partial t}$. Find $\dfrac{\partial^2 z}{\partial t^2}$ when $x = 1, y = 1$, and $t = 1$.

19. Verify that $f(x, t) = t^{-1/2}e^{-x^2/t}$ satisfies the *heat equation* $f_t = \frac{1}{4}f_{xx}$.

20. Verify that $z = x^3 - 3xy^2 + 2$ satisfies *Laplace's equation* $\dfrac{\partial^2 z}{\partial x^2} + \dfrac{\partial^2 z}{\partial y^2} = 0$.

21. Verify the equality of mixed partial derivatives for the function in Problem 18. That is, show that

$$\frac{\partial^2 z}{\partial x \partial y} = \frac{\partial^2 z}{\partial y \partial x}, \qquad \frac{\partial^2 z}{\partial x \partial t} = \frac{\partial^2 z}{\partial t \partial x}, \qquad \text{and} \qquad \frac{\partial^2 z}{\partial y \partial t} = \frac{\partial^2 z}{\partial t \partial y}.$$

Section 5 □ Maxima and Minima

In many applications, it is important to determine the values of x and y which make the value of z, a function of x and y, as large as possible.

Example 1. Consider a manufacturer who produces ale and beer. Let

(1) $p = 12 - x$

be the demand law for beer; that is, p is the highest price the manufacturer can charge in order to sell the entire output of x units of beer. Let

(2) $q = 30 - 2y$

be the demand law for ale; thus q is the highest price the manufacturer can charge in order to sell the entire output of y units of ale. Suppose his cost C to produce x units of beer and y units of ale is

(3) $C = 2x + 3y + 4.$

Let z be the manufacturer's profit when he produces x units of beer and y units of ale. We can calculate this function z; the income from beer is the price per unit times number of units sold, or $(12 - x)x$, and the income from ale is $(30 - 2y)y$. The profit z will be this total income less the cost:

(4) $z = (12 - x)x + (30 - 2y)y - (2x + 3y + 4).$

Obviously the manufacturer would like to know how much beer to produce (the value of x) and how much ale to produce (the value of y) so that the profit (the value of z) will be as large as possible.

Let z be a function of x and y, say $z = f(x, y)$. We say that z (or f) has a **relative maximum** when $x = x_0$ and $y = y_0$ if $f(x_0, y_0)$ is greater than or equal to $f(x, y)$ for all values of x near x_0 and all values of y near y_0; that is,

(5) $\quad f(x_0, y_0) \geqq f(x, y) \qquad$ for all x near x_0, y near y_0.

If Condition 5 holds for *all* x, y, then we say z (or f) has an **absolute maximum** (or simply **maximum**) when $x = x_0$ and $y = y_0$. The terms **relative minimum** and **absolute minimum** are defined similarly—just replace $\geqq$ by $\leqq$.

Figure 6 shows the graph of a function $z = f(x, y)$ which has a relative maximum at (x_0, y_0). Figure 7 shows the graph of a function which has a relative minimum at (x_0, y_0).

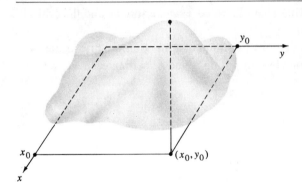

Figure 6

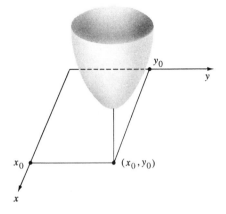

Figure 7

If $f(x,y)$ has a relative maximum (or minimum) at x_0, then the function of one variable $g(x) = f(x, y_0)$ has a relative maximum at $x = x_0$. Hence $g'(x_0) = 0$. Note that $g'(x_0)$ is the same as $f_x(x_0, y_0)$ $\left(\text{or } \dfrac{\partial z}{\partial x} \text{ at } (x, y) = (x_0, y_0)\right)$, so we conclude that $f_x(x_0, y_0) = 0$. Similarly, $f_y(x_0, y_0) = 0$.

Let us call (x_0, y_0) a **critical point** of $f(x, y)$ if $f_x(x_0, y_0) = 0$ and $f_y(x_0, y_0) = 0$. The reasoning in the preceding paragraph leads to the following.

Necessary Condition for Relative Maxima and Minima

If $f(x, y)$ has a relative maximum or minimum at (x_0, y_0), then (x_0, y_0) is a critical point of f; that is,

$$f_x(x_0, y_0) = 0 \quad \text{and} \quad f_y(x_0, y_0) = 0.$$

Figure 8 shows the graph of a function $z = f(x, y)$ which has a critical point at (x_0, y_0), yet (x_0, y_0) is neither a relative maximum nor a relative minimum of f. A surface of this shape is called a **saddle surface** and the critical point is called a **saddle point.**

To find the relative maxima or minima of a function, we first locate the critical points. This is done by computing f_x and f_y, setting each equal to zero, and solving for x and y.

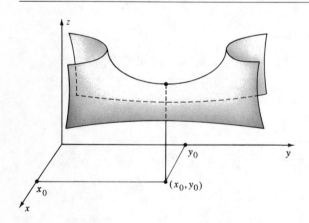

Figure 8

Example 2. Find the critical points of the function

(7) $\quad f(x, y) = 2x^2 + 15y^2 - 10xy - 10y + 3.$

SOLUTION.

We compute f_x and f_y, set each equal to zero, and then solve for x and y. The computations for f_x and f_y give

(8) $f_x = 4x - 10y$ and $f_y = 30y - 10x - 10.$

We set each of these equal to zero and obtain

(9) $\begin{cases} 4x - 10y = 0 \\ -10x + 30y = 10. \end{cases}$

Finally, we must solve this system for x and y. One way to do this is to solve the first equation of System 9 for y,

$$y = \frac{4x}{10} \quad \text{or} \quad y = \frac{2x}{5},$$

and then substitute this solution into the second equation:

(10) $-10x + 30\left(\frac{2x}{5}\right) = 10.$

We can find x from Equation 10:

$$-10x + 12x = 10, \qquad 2x = 10, \qquad x = 5.$$

Having found x, we can then find y by putting the value of x into the first equation of System 9:

$$4(5) - 10y = 0, \qquad -10y = -20, \qquad y = 2.$$

Thus $x = 5, y = 2$ is the only solution of System 9. Hence the only critical point of f in Equation 7 is $(x, y) = (5, 2)$. (Check the answer by using Equations 8: we see that $f_x(5, 2) = 0$ and $f_y(5, 2) = 0$, so $(5, 2)$ is a critical point.)

Second-derivative Test.   Consider the function of two variables

$$z = f(x, y).$$

Let (x_0, y_0) be a critical point of f. There is a test which is often useful for determining whether, at the point (x_0, y_0), f has a relative maximum, relative minimum, or neither. We shall state this test without proof.

Second-Derivative Test in Two Variables

Let (x_0, y_0) be a critical point of $f(x, y)$. Form the number

$$(11) \quad D = f_{xx}(x_0, y_0) f_{yy}(x_0, y_0) - [f_{xy}(x_0, y_0)]^2.$$

1. If $D > 0$, then f has a relative maximum or minimum at (x_0, y_0).

2. If $D < 0$, then f has neither a relative maximum nor a relative minimum at (x_0, y_0).

Suppose $D > 0$ (Case 1). Then

$$(12) \quad f_{xx}(x_0, y_0) > 0 \quad \text{implies a relative minimum,}$$

$$(13) \quad f_{xx}(x_0, y_0) < 0 \quad \text{implies a relative maximum.}$$

In the second-derivative test, it is possible that $D = 0$. In such a case, the test yields no conclusive result—(x_0, y_0) can be a relative maximum, relative minimum, or neither. If you are familiar with determinants, you will note that D can be expressed as

$$D = \begin{vmatrix} f_{xx}(x_0,y_0) & f_{xy}(x_0,y_0) \\ f_{xy}(x_0,y_0) & f_{yy}(x_0,y_0) \end{vmatrix}.$$

Example 3. Does the function

$$f(x,y) = 2x^2 + 15y^2 - 10xy - 10y + 3$$

from Example 2 have a relative maximum or relative minimum? If so, find it.

SOLUTION.

The first step is to find all critical points of f. This step was accomplished in Example 2. We shall use that result: the only critical point of f is $(x,y) = (5,2)$.

The next step is to use the second-derivative test to determine if $(5,2)$ gives a relative maximum or minimum. This means we must compute D, as in Equation 11. We find that

$$f_x = 4x - 10y \quad f_y = 30y - 10x - 10$$
$$f_{xx} = 4 \quad f_{xy} = -10 \quad f_{yy} = 30$$

and so $D = (4)(30) - (-10)^2$ or $D = 20$. Since $D > 0$, Case 1 of the test applies and $(5,2)$ is either a relative maximum or a relative minimum. We can use Equations 12 and 13 to determine which it is. Since $f_{xx}(5,2) = 4 > 0$, Equation 12 shows that we have a relative minimum. Thus f has only one extreme point (that is, a relative maximum or relative minimum); it is at $(x,y) = (5,2)$ and f attains a relative minimum there.

Example 4. In Example 1, we saw that the function (see Equation 4)

$$(14) \quad z = (12 - x)x + (30 - 2y)y - (2x + 3y + 4)$$

expresses the profit z earned by a manufacturer who produces x units of beer and y units of ale. What values of x and y yield maximum profit?

SOLUTION.
To solve this problem, we first find the critical points of Equation 14. We then test these points by means of the second-derivative test.

To find the critical points, simplify Equation 14.

$$z = f(x,y) = 10x + 27y - x^2 - 2y^2 - 4.$$

Calculate f_x and f_y, and set each equal to zero.

$$(15) \quad f_x = 10 - 2x \quad \text{and} \quad f_y = 27 - 4y,$$

$$(16) \quad \begin{cases} 10 - 2x = 0 \\ 27 - 4y = 0. \end{cases}$$

We solve System 16 for x and y, obtaining $x = 5$ and $y = \frac{27}{4}$. Thus $(5,\frac{27}{4})$ is the only critical point of the function.

We now test the critical point $(5,\frac{27}{4})$ with the second-derivative test. We have (from Equation 15)

$$f_{xx} = -2, \quad f_{xy} = 0, \quad \text{and} \quad f_{yy} = -4.$$

Therefore the quantity D in Equation 11 is

$$D = (-2)(-4) - (0)^2 = 8.$$

Since $D > 0$, we use Case 1. We now try to apply Equations 12 or 13; since $f_{xx}(5,\frac{27}{4}) = -2$ is negative, Equation 13 tells us that f has a relative maximum at $x = 5$ and $y = \frac{27}{4}$. It is actually true that this is an absolute maximum, and so the manufacturer can maximize profit by producing 5 units of beer and $\frac{27}{4}$ units of ale.

Problems

In Problems 1–6, find all the critical points of the given functions.

1. $z = x^2 - 8x + 2y^2 - 12y - 3.$

2. $z = 3x^2 - 24x + y^2 + 6y + 10$.

3. $f(x,y) = 7x - x^2 - xy - 2y^2 + 21y - 3$.

4. $g(u,v) = 3u + 5v - u^2 - \frac{1}{2}v^2 + uv$.

5. $Q = x^2 - y^2 - 6x + 2y + 8$.

6. $T = r^3 - 9r^2 + 3s^2 - 18s + 7$.

In Problems 7–12, use the second-derivative test to find all relative maxima and relative minima.

7. The function in Problem 1.

8. The function in Problem 2.

9. The function in Problem 3.

10. The function in Problem 4.

11. The function in Problem 5.

12. The function in Problem 6.

13. A farmer uses his land to grow lemons and oranges. If he harvests r units of lemons, he can sell them all at a unit price P, where $P = 50 - 5r$. If he harvests s units of oranges, he can sell them all at a unit price Q, where $Q = 200 - s$. His cost to harvest r units of lemons and s units of oranges is $C = 2r + s + 500$.

    a. Express R, his total profit, as a function of r and s.

    b. Find the values of r and s which produce the maximum profit.

14. A rectangular box has length l, width w, and height h (in inches). The post office will not accept a package for mailing unless the length plus the girth does not exceed 84 inches. (The *girth* of the box in Figure 9 is the length of the string around the middle.)

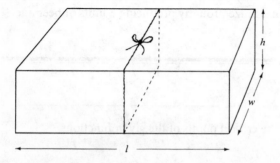

Figure 9

a. Find a formula for V, the volume of the box, as a function of w and h if the length plus the girth is exactly 84 inches.

b. What dimensions should the box have in order to have maximum volume and still be acceptable for mailing?

Section 6 □ Maxima and Minima with Constraints

In the preceding section, we tried to find the value of (x,y) which makes a function $z = f(x,y)$ attain its maximum (or minimum). In some applications, an additional complication may arise: the values of x and y under consideration are not arbitrary, but must satisfy a side condition called the **constraint.** The side conditions we consider are those where x and y satisfy an equation, say $g(x,y) = 0$. For example, if the sum of x and y must always equal 100 then $g(x,y)$ is $x + y - 100$.

Constrained Maximum-minimum Problem

Among all values of (x,y) which satisfy the side condition (constraint equation)

(1) $g(x,y) = 0,$

find the one that makes

(2) $z = f(x,y)$

a maximum or minimum.

Examples of constrained maximum-minimum problems can be found in many areas. In business, for example, $z = f(x,y)$ might be the profit a brewery realizes by producing x units of beer and y units of ale. Since the personnel and equipment at the brewery are of fixed size, not all possible values of x and y can be achieved. Perhaps only 100 total units can be produced. Then x and y must satisfy the side condition $x + y = 100$, so Constraint Equation 1 would apply with $g(x,y) = x + y - 100$.

The area called *utility theory* in economics treats problems of the following sort. The amount of satisfaction (or "utility") that a person derives by owning a house worth x dollars and a car worth y dollars is measured in some way and denoted by z. This is the function $z = f(x,y)$ in Equation 2. Suppose a person has $10,000 to spend for down payments on a house and a car, and suppose houses require a 20% down payment and cars require only a 10% down payment. Thus a house costing x requires $\frac{1}{5}x$ down payment, and a car costing y requires $\frac{1}{10}y$ down payment. To predict how this person will spend his \$10,000, we must maximize $z = f(x,y)$ subject to the constraint equation $\frac{1}{5}x + \frac{1}{10}y = 10,000$. In this case, $g(x,y)$ in Equation 1 is $g(x,y) = \frac{1}{5}x + \frac{1}{10}y - 10,000$. Obviously, variations of this kind of problem might occur in decision problems of sociology and psychology.

Let us return to the abstract formulation of the problem (see Equations 1 and 2). The straightforward method of solution is to solve Equation 1 for y and substitute this solution into Equation 2. Then z is expressed in terms of x alone. Since z is now a function of one variable, we can use the one-variable techniques of Chapter 3 to find the value x_0 of x which makes z a maximum or minimum. The corresponding value y_0 of y can then be found from Equation 1 by giving x the value x_0 and solving for y.

The following example illustrates this straightforward method for solving constrained maximum-minimum problems.

Example 1. Minimize the function

(3) $f(x,y) = 3x^2 + y^2$

subject to the constraint equation

(4) $3x + y = 8.$

SOLUTION.
From Equation 4, we have $y = 8 - 3x$. The quantity $z = f(x,y)$ in Equation 3 can therefore be written as a function of x alone.

(5) $z = 3x^2 + (8 - 3x)^2.$

We find the minimum of z in Equation 5 by computing $\dfrac{dz}{dx}$ and setting it equal to zero. After simplifying we find

$$\frac{dz}{dx} = 24x - 48, \qquad 0 = 24x - 48, \qquad x = 2.$$

The second derivative is $\dfrac{d^2z}{dx^2} = 24$; since this is positive, $x = 2$ makes z have a relative minimum.

When $x = 2$, the constraint equation gives $y = 8 - 3(2) = 2$. Thus $(x,y) = (2,2)$ gives the desired answer (we shall omit a discussion of why this relative minimum is actually an absolute minimum): $f(x,y) = 3x^2 + y^2$ attains a minimum, subject to the constraint $3x + y = 8$, when $(x,y) = (2,2)$. The value of the constrained minimum is $f(2,2) = 3(2)^2 + (2)^2 = 16.$

Lagrange Multipliers. The straightforward method of solving a constrained maximum-minimum problem as illustrated in Example 1 may not work in certain cases. For example, it may not be possible to solve the constraint equation $g(x,y) = 0$ for y in terms of x. This would occur, for example, when $g(x,y) = y^5 + xy + x^5$. In other cases, the method may fail because the one-variable problem it leads to is too complicated to solve. An alternative method was discovered by the mathematician Joseph Lagrange, called the *method of Lagrange multipliers.*[14] This method is easy to describe but difficult to justify. We shall first show how to use it, and then conclude with a short discussion about its justification. The method introduces a new variable, traditionally denoted by λ, the Greek letter lambda, and called the **Lagrange multiplier.**

Method of Lagrange Multipliers

Solve the following system of three equations for the three unknowns x, y, and λ.

(6) $\quad f_x(x,y) = \lambda g_x(x,y)$

(7) $\quad f_y(x,y) = \lambda g_y(x,y)$

(8) $\quad g(x,y) = 0.$

If (x_0, y_0) is a relative maximum or minimum of $f(x,y)$, subject to the constraint $g(x,y) = 0$, then (x_0, y_0, λ_0) occurs as a solution of the system consisting of Equations 6, 7, and 8.

In the problems treated in this course, there will be only one solution (x_0, y_0, λ_0) to the system of Equations 6–8. Assume that the constrained maximum-minimum problem has a solution. The method then allows us to conclude that (x_0, y_0) is the desired solution to the problem.

Example 2. If x dollars are spent on newspaper advertising and y dollars are spent on television advertising, the product can be expected to have sales totaling z dollars. Assume z depends on x and y according to the formula $z = 36x^{1/4}y^{3/4}$. The budget for advertising is $28,000. How much should be spent on television and how much with newspapers? Use the method of Lagrange multipliers.

[14]Joseph Louis Lagrange (1736–1813) was born in Italy. His mathematical discoveries won him worldwide recognition; he held professorships at universities in Turin, Berlin, and Paris.

SOLUTION.

The problem amounts to maximizing $f(x,y) = 36x^{1/4}y^{3/4}$, subject to the constraint $g(x,y) = 0$ where $g(x,y) = x + y - 28{,}000$. In this case the Lagrange equations (Equations 6–8) become

(9) $\qquad 9x^{-3/4}y^{3/4} = \lambda \cdot 1$

(10) $\qquad 27x^{1/4}y^{-1/4} = \lambda \cdot 1$

(11) $\quad x + y - 28{,}000 = 0.$

Equate the expressions for λ given by the first two equations. The result is

$$9x^{-3/4}y^{3/4} = 27x^{1/4}y^{-1/4};$$

multiplying each side by $x^{3/4}y^{1/4}$ gives

(12) $\quad 9y = 27x,\qquad$ or $\qquad y = 3x.$

Put $y = 3x$ in Equation 11 and obtain $4x = 28{,}000$, or $x = 7000$. From Equation 12, $y = 21{,}000$. Thus the sales will be maximized if \$7,000 is spent on newspaper advertising and \$21,000 is spent on television advertising.

Example 3. Assume that a certain person's satisfaction z from owning a house worth x dollars and a car worth y dollars is $z = f(x,y)$, where $f(x,y) = x^4y$. This person plans to use \$10,000 to provide the down payments for a house and a car. House purchases require a 20% down payment, and car purchases require a 10% down payment. How expensive a house will this person purchase? Use Lagrange multipliers.

SOLUTION.

The problem amounts to maximizing $f(x,y) = x^4y$ subject to the constraint $\frac{1}{5}x + \frac{1}{10}y = 10{,}000$ (see pages 299–300). The Lagrange equations (Equations 6–8) become, in this case,

(13) $\qquad 4x^3y = \dfrac{1}{5}\lambda$

(14) $\qquad x^4 = \dfrac{1}{10}\lambda$

(15) $\quad \dfrac{1}{5}x + \dfrac{1}{10}y = 10{,}000.$

Equate the expressions for λ in Equations 13 and 14; the result is

(16) $\quad 20x^3y = 10x^4,\qquad$ or $\qquad y = \dfrac{1}{2}x.$

We put this into Equation 15 and solve for x.

$$\frac{1}{5}x + \frac{1}{10}\left(\frac{1}{2}x\right) = 10{,}000, \quad \text{or} \quad x = 40{,}000.$$

This person will therefore purchase a $40,000 house (and a car worth $y = \frac{1}{2}x$ = $20,000!). (In transforming the first equation to the second in Equation 16, we divided by x, and therefore we should check that x was not zero. Since $x = 0$ would make $z = 0$, the value of x that maximizes z is clearly not zero. That is, $x \neq 0$ in Equation 16.)

The constrained maximum-minimum problem of Equations 1 and 2 has a geometric interpretation. The function $z = f(x,y)$ represents a surface S spread over the xy-plane (see Figure 10). The constraint equation $g(x,y) = 0$ represents a curve C in the xy-plane. The values of $z = f(x,y)$ that occur when (x,y) satisfies the constraint equation are represented by the heights of the surface S above the curve C. Thus the maximum of z subject to the constraint is the height of the point P_0 in Figure 10.

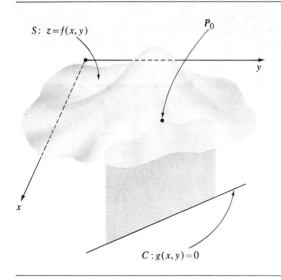

$S: z = f(x,y)$

P_0

y

x

$C: g(x,y) = 0$

Figure 10

The justification of the Lagrange multiplier method is not simple, but some illumination can be provided by total differentials. One has $\Delta f \approx f_x \Delta x + f_y \Delta y$ and this approximation remains valid even when divided by Δx. Thus

(17) $\quad \dfrac{\Delta f}{\Delta x} \approx f_x + f_y\dfrac{\Delta y}{\Delta x}, \quad$ and $\quad \dfrac{\Delta g}{\Delta x} \approx g_x + g_y\dfrac{\Delta y}{\Delta x},$

where the second equation was obtained similarly.

Now imagine that y is the function of x obtained by solving the constraint equation for y. Then $g(x,y)$ and $f(x,y)$ become functions of x alone; when $\Delta x \to 0$ in Equations 17, those approximations yield the equalities

(18) $\quad \dfrac{df}{dx} = f_x + f_y\dfrac{dy}{dx} \quad$ and $\quad \dfrac{dg}{dx} = g_x + g_y\dfrac{dy}{dx}.$

Furthermore, as functions of x alone, $g(x,y)$ is always zero and $\dfrac{df}{dx}$ is zero if evaluated at a value of x which makes $f(x,y)$ a relative maximum or minimum. At such a value of x, then, Equations 18 become

$$0 = f_x + f_y\dfrac{dy}{dx} \quad \text{and} \quad 0 = g_x + g_y\dfrac{dy}{dx}.$$

These two equations yield

$$\dfrac{-f_x}{f_y} = \dfrac{-g_x}{g_y}$$

so $f_x = \lambda g_x$ and $f_y = \lambda g_y$ for some λ. In this way, the first two Lagrange equations (Equations 6 and 7) arise. We know, of course, Equation 8, the constraint equation, is valid.

Problems

Solve Problems 1–6 by the straightforward method of substitution as well as by the method of Lagrange multipliers.

1. Find the value (x,y) which makes $f(x,y) = x^2 + y^2$ a minimum subject to the constraint $y + 3x = 30$.

2. Maximize $f(x,y) = y^2 - 3x^2$ subject to the constraint $y - x = 2$.

3. Maximize $f(x,y) = 2x^2 - 4y^2 + xy - 9x + 26y$ subject to the constraint $y - x = 1$.

4. Minimize $f(x,y) = x^2 + 3xy + 2y^2 - y - 28x$ subject to the constraint $x - 1 - y = 0$.

5. What value of (x,y) makes $f(x,y) = x^2 - 8y^2 + \frac{1}{3} - x$ a maximum subject to the constraint $x + y = \frac{1}{2}$?

6. What value of (x,y) makes $f(x,y) = y^2 - x^2 + 2xy - 70x$ a minimum subject to the constraint $2x = y$?

7. The total production output Q for New South Wales depended on labor expenditure L and capital expenditure C according to the formula[15] $Q = 1.14L^{4/5}C^{1/5}$. What values of L and C make Q a maximum if $L + C = 300$?

8. Patients in a mental hospital were rewarded with tokens for certain socially desirable behaviors. They could spend the tokens to purchase group therapy sessions or movie admission tickets. Assume that a patient's utility function is $z = x^3y^2$ where z denotes the patient's satisfaction from attending x hours of therapy and y hours of movie viewing. If therapy costs 2 tokens per hour and movies cost 1 token per hour, how will this patient spend a reward of 20 tokens?

Section 7 □ Least Squares Curve Fitting

Throughout this text, we have come across functions of practical importance that are described by mathematical formulas. Some of these formulas were originally derived by theoretical reasoning, some by empirical or statistical analysis, and some by a combination of these two methods (see "From where do the applied equations come?" on pages 100–101). The formulas for flow within a blood vessel (Problem 21 on page 280) were originally discovered by a physician, Jean Louis Poiseuille, by examining measurements from numerous experiments (the empirical method); later they were derived by physicists from the general principles of fluid dynamics (the theoretical method).

The combination method is perhaps the most common. Consider, for example, the problem in economics of finding a formula for a particular production function; that is, a formula for finding the total production output P of a commodity in terms of the amount spent on labor costs L and on capital costs C. In this case, an economist would use theoretical considerations to decide if the production function should have the form

$$P = aL^bC^{1-b}, \qquad a \text{ and } b \text{ are constants,}$$

[15]P. H. Douglas and G. T. Gunn, "The Production Function for Australian Manufacturing," *Quarterly Journal of Economics*, LVI (1941), pp. 108–129.

(which is called a *Cobb-Douglas production function*) or whether it should have the form

$$P = (aL^b + cC^b)^{1/b}, \qquad a, b, \text{ and } c \text{ are constants,}$$

(which is called a *homohyphallic production function*). Once that decision is made, the constants in the appropriate formula are adjusted so that the formula fits the data from past experience as well as possible.

The topic of this section is how to adjust those constants in order to obtain the best fitting formula. This topic is called **curve fitting,** and the constants to be adjusted are called **parameters.**

We shall discuss curve fitting for functions of one variable. Consider a quantity y which depends on x. Suppose that it is found, by experimentation, that when x takes the values $x_1, x_2, \ldots, x_n$, then y will have the values $y_1, y_2, \ldots, y_n$ respectively. We call the ordered pairs $(x_1, y_1), (x_2, y_2), \ldots, (x_n, y_n)$ the **data points.** If they are plotted on an xy-coordinate system, the graph obtained is called a **scatter diagram.** Let $y = f(x)$ be a formula which is supposed to approximate the true dependence of y upon x. Suppose that when x takes the value $x_1, x_2, \ldots, x_n$, this formula gives the values $\hat{y}_1, \hat{y}_2, \ldots, \hat{y}_n$ for y; that is,

$$\hat{y}_1 = f(x_1), \qquad \hat{y}_2 = f(x_2), \qquad \ldots, \qquad \hat{y}_n = f(x_n).$$

It is common to refer to $\hat{y}_1, \hat{y}_2, \ldots, \hat{y}_n$ as the **predicted** values of y; this is in contrast to the **actual** values $y_1, y_2, \ldots, y_n$ of y.

Example 1. The curve $y = \frac{1}{2}x^2$ is fitted to the data points $(0,1)$, $(1,3)$, $(2,3)$, $(3,6)$, $(6,15)$. What are the actual values y_1, y_2, y_3, y_4, and y_5 and the predicted values $\hat{y}_1, \hat{y}_2, \hat{y}_3, \hat{y}_4$, and $\hat{y}_5$ in this case? Draw the scatter diagram. Draw the graph of $y = \frac{1}{2}x^2$ in the scatter diagram and interpret the quantities $|y_1 - \hat{y}_1|, |y_2 - \hat{y}_2|, \ldots, |y_n - \hat{y}_n|$ as distances in the diagram.

SOLUTION.
If we let x_1, x_2, x_3, x_4, and x_5 be the x values in the order

$$x_1 = 0, \qquad x_2 = 1, \qquad x_3 = 2, \qquad x_4 = 3, \qquad \text{and} \qquad x_5 = 6,$$

then

$$y_1 = 1, \qquad y_2 = 3, \qquad y_3 = 3, \qquad y_4 = 6, \qquad \text{and} \qquad y_5 = 15$$

and

$$\hat{y}_1 = \frac{1}{2} \cdot x_1^2 = \frac{1}{2} \cdot 0^2 = 0,$$

$$\hat{y}_2 = \frac{1}{2} \cdot x_2^2 = \frac{1}{2} \cdot 1^2 = \frac{1}{2},$$

$$\hat{y}_3 = \frac{1}{2} \cdot x_3^2 = \frac{1}{2} \cdot 2^2 = 2,$$

$$\hat{y}_4 = \frac{1}{2} \cdot x_4^2 = \frac{1}{2} \cdot 3^2 = \frac{9}{2},$$

$$\hat{y}_5 = \frac{1}{2} \cdot x_5^2 = \frac{1}{2} \cdot 6^2 = 18.$$

The scatter diagram is shown in Figure 11. Figure 12 shows the curve $y = \frac{1}{2}x^2$ drawn on the scatter diagram; the value $|y_i - \hat{y}_i|$ for $i = 1, 2, 3, 4, 5$, which is the deviation of the predicted value of y from the actual value when $x = x_i$, is equal to the vertical distance from the data point (x_i, y_i) to the curve.

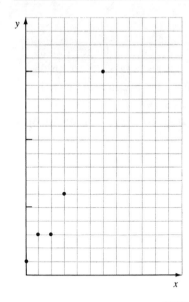

Figure 11

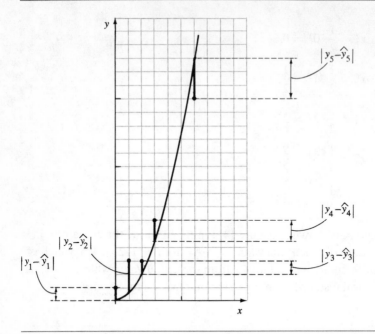

Figure 12

The Square-Sum Error. Let $(x_1, y_1), (x_2, y_2), \ldots, (x_n, y_n)$ be given data points. Let $y = f(x)$ be a formula that is used to fit this data. If this formula fits the data well, then the predicted values $\hat{y}_1, \hat{y}_2, \ldots, \hat{y}_n$ should be fairly close to the actual values $y_1, y_2, \ldots, y_n$. How does one describe the degree of goodness of the fit? The most common measure is by the size of the quantity

$$(1) \quad E = (y_1 - \hat{y}_1)^2 + (y_2 - \hat{y}_2)^2 + \cdots + (y_n - \hat{y}_n)^2,$$

which is the sum of the squares of the deviations of the actual values from the predicted values. The smaller the size of E, the better the fit is considered to be. Note that $E = 0$ means that there is a perfect fit; that is, all the predicted values agree exactly with the actual values.

The quantity E in Equation 1 may be called the **square-sum error.** It is often abbreviated by using the symbol Σ (the Greek letter sigma) as follows:

$$(2) \quad E = \sum_{i=1}^{n} (y_i - \hat{y}_i)^2;$$

we read "E is the sum of the quantities $(y_i - \hat{y}_i)^2$ as i runs from 1 to n." An even more abbreviated notation for Equation 1 is

$$(3) \quad E = \sum (y_i - \hat{y}_i)^2.$$

Example 2. Let y be the yield of a cornfield in suitable units when x tons of chemical fertilizer were applied during the growing season. Data from the four previous years gave the following values for (x, y).

x	0	3	8	24
y	0.5	1.5	4	6

a. Calculate the square-sum error E (see Equation 1) when the function $y = \sqrt{x + 1}$ is used to fit the data.

b. Calculate the square-sum error E when the function $y = \frac{1}{5}x + 1$ is used to fit the data.

c. Which of these two functions yields the better fit (in the sense of producing a smaller square-sum error)?

SOLUTION.

a. Label the given data by

$$x_1 = 0, \qquad x_2 = 3, \qquad x_3 = 8, \qquad \text{and} \qquad x_4 = 24$$
$$y_1 = 0.5, \qquad y_2 = 1.5, \qquad y_3 = 4, \qquad \text{and} \qquad y_4 = 6.$$

The first function $y = \sqrt{x + 1}$ produces the following predicted values.

$$\hat{y}_1 = \sqrt{x_1 + 1} = \sqrt{0 + 1} = 1 \qquad \hat{y}_2 = \sqrt{x_2 + 1} = \sqrt{3 + 1} = 2$$
$$\hat{y}_3 = \sqrt{x_3 + 1} = \sqrt{8 + 1} = 3 \qquad \hat{y}_4 = \sqrt{x_4 + 1} = \sqrt{24 + 1} = 5$$

The square-sum error for this function is

$$
\begin{aligned}
(4) \quad E = \sum (y_i - \hat{y}_i)^2 &= (y_1 - \hat{y}_1)^2 + (y_2 - \hat{y}_2)^2 + (y_3 - \hat{y}_3)^2 + (y_4 - \hat{y}_4)^2 \\
&= (0.5 - 1)^2 + (1.5 - 2)^2 + (4 - 3)^2 + (6 - 5)^2 \\
&= 0.25 + 0.25 + 1 + 1 = 2.5.
\end{aligned}
$$

b. The second function $y = \frac{1}{5}x + 1$ produces the predicted values

$$\hat{y}_1 = \frac{1}{5}x_1 + 1 = 1 \qquad\qquad \hat{y}_2 = \frac{1}{5}x_2 + 1 = 1.6$$

$$\hat{y}_3 = \frac{1}{5}x_3 + 1 = 2.6 \qquad\qquad \hat{y}_4 = \frac{1}{5}x_4 + 1 = 5.8.$$

The square-sum error for this function is

$$
\begin{aligned}
(5) \quad E = \sum (y_i - \hat{y}_i)^2 &= (0.5 - 1)^2 + (1.5 - 1.6)^2 + (4 - 2.6)^2 + (6 - 5.8)^2 \\
&= 0.25 + 0.01 + 1.96 + 0.04 = 2.26.
\end{aligned}
$$

We see from Equations 4 and 5 that the square-sum error E is smaller for the function $y = \frac{1}{5}x + 1$. We therefore say that this function fits the given data better than the function $y = \sqrt{x + 1}$ does. Figure 13 shows both functions superimposed on the scatter diagram.

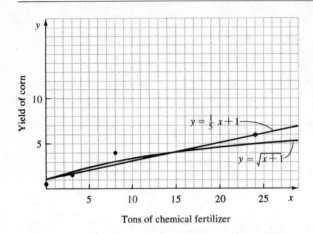

Tons of chemical fertilizer

Figure 13

The Least Squares Criterion. Consider again some given data points (x_1, y_1), (x_2, y_2), . . . , (x_n, y_n). Let $y = f(x)$ be an equation which contains some parameters, say a and b. (For example, $y = f(x)$ might be the equation $y = ax + b$; in that case any linear function can be obtained by an appropriate choice of the parameters. Or $y = f(x)$ might be the equation $y = ae^{bx}$, which yields various exponential functions for different choices of the parameters.) The curve-fitting problem in this situation is to determine the values of the parameters which yield the best fitting equation. The most widely accepted way to do this is to use the **least squares criterion,** which says that the parameters should be chosen so that the square-sum error E (see Equation 1) is as small as possible. The justifications for this criterion belong to the subject of statistics and will not be discussed here (see Problem 6 on page 315 for an enlightening consequence).

We now show how the use of the least squares criterion leads to a minimum problem for functions of several variables. We use the equation $y = f(x)$ to compute the predicted values $\hat{y}_1, \hat{y}_2$, . . . , $\hat{y}_n$. Since the equation $y = f(x)$ contains the parameters a and b, these predicted values $\hat{y}_1, \hat{y}_2$, . . . , $\hat{y}_n$ also depend on a and b, and therefore the square-sum error E depends on a and b. Thus E may be considered as a function of the two variables a and b.

To find its minimum, set the partial derivatives $\dfrac{\partial E}{\partial a}$ and $\dfrac{\partial E}{\partial b}$ equal to zero and solve for a and b.

Example 3. Find a function of the form $y = ax^2 + b$ which best fits the data given in the table below.

x	0	1	3
y	2	0	1

Use the least squares criterion.

SOLUTION.

$$x_1 = 0, \qquad x_2 = 1, \qquad x_3 = 3$$
$$y_1 = 2, \qquad y_2 = 0, \qquad y_3 = 1$$

and the predicted values calculated from the function $y = ax^2 + b$ are

$$\hat{y}_1 = a \cdot x_1^2 + b = a \cdot 0^2 + b = b$$
$$\hat{y}_2 = a \cdot x_2^2 + b = a \cdot 1^2 + b = a + b$$
$$\hat{y}_3 = a \cdot x_3^2 + b = a \cdot 3^2 + b = 9a + b.$$

The square-sum error or $E = \sum (y_i - \hat{y}_i)^2$ is

(6) $\qquad E = (2 - b)^2 + (0 - a - b)^2 + (1 - 9a - b)^2.$

To find the minimum of E, we calculate the partial derivatives $\dfrac{\partial E}{\partial a}$ and $\dfrac{\partial E}{\partial b}$, set them equal to zero, and solve for a and b. From Equation 6:

$$\frac{\partial E}{\partial a} = 0 - 2(-a - b) - 9 \cdot 2(1 - 9a - b) = 164a + 20b - 18,$$

$$\frac{\partial E}{\partial b} = -2(2 - b) - 2(-a - b) - 2(1 - 9a - b) = 20a + 6b - 6.$$

We set these partial derivatives equal to zero and obtain, after simplifying,

(7) $\qquad \begin{cases} 82a + 10b = 9 \\ 10a + 3b = 3. \end{cases}$

These simultaneous equations (7) can be solved by multiplying the first by 3, the second by 10, and subtracting. The result is $146a = -3$ or $a = -3/146$. This value of a can be substituted into either equation, which can then be

solved for b. The result is $b = 156/146 = 78/73$. Hence the best fitting
equation of the form $y = ax^2 + b$ is

$$y = \frac{78}{73} - \frac{3}{146}x^2.$$

The Least Squares Regression Line. We shall now derive a general formula
to solve the following problem.

> *Given data points* $(x_1, y_1), (x_2, y_2), \ldots, (x_n, y_n)$, *find the linear
> function* $y = mx + b$ *which best fits this data in the sense of the least
> squares criterion.*

The function $y = mx + b$ which results from solving this problem is called
the **least squares regression line.**

To solve the problem, we first note that the predicted values of y are
$\hat{y}_i = mx_i + b$ for $i = 1, 2, \ldots, n$. Hence the square-sum error is

(18) $\quad E = \sum (y_i - mx_i - b)^2$
$$= (y_1 - mx_1 - b)^2 + (y_2 - mx_2 - b)^2 + \cdots + (y_n - mx_n - b)^2.$$

To find the minimum of E, we calculate the partial derivatives $\dfrac{\partial E}{\partial m}$ and $\dfrac{\partial E}{\partial b}$,
set them equal to zero, and solve for m and b. From Equation 8, we find

$$\frac{\partial E}{\partial m} = -2x_1(y_1 - mx_1 - b) - 2x_2(y_2 - mx_2 - b) - \cdots$$
$$- 2x_n(y_n - mx_n - b),$$
$$\frac{\partial E}{\partial b} = -2(y_1 - mx_1 - b) - 2(y_2 - mx_2 - b) - \cdots$$
$$- 2(y_n - mx_n - b).$$

When these are set equal to zero, and after collecting like terms, we obtain

(9) $\begin{cases} (x_1^2 + x_2^2 + \cdots + x_n^2)m + (x_1 + x_2 + \cdots + x_n)b \\ \qquad\qquad\qquad\qquad\qquad = x_1y_1 + x_2y_2 + \cdots + x_ny_n \\ (x_1 + x_2 + \cdots + x_n)m + nb = y_1 + y_2 + \cdots + y_n. \end{cases}$

Let us write Equations 9 using the sigma abbreviation for sums.

(10) $\begin{cases} \left(\sum x_i^2\right)m + \left(\sum x_i\right)b = \sum x_iy_i \\ \left(\sum x_i\right)m + nb = \sum y_i \end{cases}$

We wish to solve Equations 10 for m and b. To this end, we multiply the first by $-n$ and the second by $\sum x_i$, and then add. The result is

$$\left(\sum x_i\right)^2 m - n\left(\sum x_i^2\right)m = \left(\sum x_i\right)\left(\sum y_i\right) - n\left(\sum x_i y_i\right).$$

Therefore

$$(11) \quad m = \frac{\left(\sum x_i\right)\left(\sum y_i\right) - n\left(\sum x_i y_i\right)}{\left(\sum x_i\right)^2 - n\left(\sum x_i^2\right)}.$$

Note that Equation 11 expresses m in terms of the given data. Having calculated this value of m, we can then find b from the second of Equations 10:

$$(12) \quad b = \frac{\left(\sum y_i\right) - \left(\sum x_i\right)m}{n}$$

Least Squares Regression Line

The linear function which fits the given data points $(x_1, y_1), (x_2, y_2), \ldots, (x_n, y_n)$ best according to the least squares criterion is $y = mx + b$ where

$$(13) \quad m = \frac{\left(\sum x_i\right)\left(\sum y_i\right) - n\left(\sum x_i y_i\right)}{\left(\sum x_i\right)^2 - n\left(\sum x_i^2\right)}$$

$$(14) \quad b = \frac{\left(\sum y_i\right) - \left(\sum x_i\right)m}{n}.$$

The least squares regression line is sometimes described as the straight line that best fits the scatter diagram. However, it should be remarked that the least squares regression line depends in a crucial way on which variable is chosen as the dependent variable. That is, if the x- and y- coordinates of the data points are interchanged, then it is *not* true that the new least squares regression line can be found by interchanging x and y in the equation for the original least squares regression line.

Example 4. The data of Example 2 is given again:

x	0	3	8	24
y	0.5	1.5	4	6

Here, y was the yield from a cornfield which received x tons of a chemical fertilizer during the growing season. Find the least squares regression line for this data, and then use it to predict the yield when 15 tons of chemical fertilizer are used.

SOLUTION.

The following table provides a convenient form for organizing the calculation:

x_i	y_i	x_i^2	$x_i y_i$
0	0.5	0	0
3	1.5	9	4.5
8	4	64	32
24	6	576	144
Sum 35	12	649	180.5

The last line of the table gives $\sum x_i = 35, \sum y_i = 12, \sum x_i^2 = 649$, and $\sum x_i y_i = 180.5$. In this case $n = 4$ (there are four data points). We substitute these values into Equation 13 and obtain

$$m = \frac{(35)(12) - 4(180.5)}{(35)^2 - 4(649)} = \frac{-302}{-1371} = 0.22.$$

We use this value in Equation 14 to obtain

$$b = \frac{12 - (35)(0.22)}{4} = \frac{4.3}{4} = 1.08.$$

Hence the best fitting linear function, in the least-squares sense, is

$$y = 0.22x + 1.08.$$

When $x = 15$, the predicted value of y is

$$(0.22)(15) + 1.08 = 4.38.$$

Problems

1. The curve $y = e^{x/2}$ is fitted to the data points

$$x_1 = 1, \quad x_2 = 2, \quad x_3 = 3.2, \quad x_4 = 5$$
$$y_1 = 3, \quad y_2 = 3.5, \quad y_3 = 4, \quad y_4 = 11.$$

Find the predicted values $\hat{y}_1, \hat{y}_2, \hat{y}_3$, and $\hat{y}_4$ and the square-sum error E defined in Equation 1 (use a hand calculator or Tables 1 and 2 in the appendix). Plot the scatter diagram, draw the curve $y = e^{x/2}$, and indicate the distances that correspond to the deviations $|y_1 - \hat{y}_1|, \ldots, |y_4 - \hat{y}_4|$.

2. An executive compares the total sales y of his company during a year with the amount x of money that was spent on advertising that year. The figures

for each of the five preceding years are all given in the table (x and y in millions of dollars).

x	.5	1	3	4.5
y	3	8	16.5	30

The executive decides to fit this data with the line $y = 6x + 1$. Find the square-sum error E defined in Equation 1. According to this executive's mathematical model, how much should be spent on advertising to achieve an annual sales total of 37 million dollars?

3. Find the least squares regression line for the data of Problem 1 and use it to predict the value of y when $x = 6$.

4. Find the least squares regression line for the data in Problem 2. Use this line to predict the total annual sales if 6 million dollars is spent on advertising during a year.

5. A biologist tests the potency of various concentrations of penicillin by placing a drop of penicillin solution on a Petri dish containing a homogeneous bacteria culture. Let x be the logarithm of the concentration of the penicillin solution. Let y be the diameter of the circular region of bacteria that are destroyed by the penicillin after 12 hours. When this experiment was performed for five different penicillin concentrations, the following results were obtained.[16]

x	0	1	2	3	4	5
y	15.8	17.8	19.5	21.3	23.1	24.8

Find the least squares regression line and use it to predict the value of y when $x = 7$.

6. A measurement experiment is repeated n times. Due to experimental error, the numerical outcomes $y_1, y_2, \ldots, y_n$ are not necessarily identical. Using a least squares criterion, we might estimate the true measurement by the value of y that minimizes the square-sum error $\sum (y_i - y)^2$.
   a. Prove that this value of y is precisely the average value of $y_1, y_2, \ldots, y_n$.

[16]O. L. Davies and P. L. Goldsmith, *Statistical Methods in Research and Production* (Edinburgh: Oliver and Boyd, 1972), p. 207.

b. Now suppose the true measurement is estimated by a value of y that minimizes the ordinary sum of the deviations $\sum |y_i - y|$. Show that this value of y is "unrealistic" by demonstrating that it would be 2 in the special case $y_1 = 1$, $y_2 = 2$, and $y_3 = 6$.

Summary

Suppose z is a function of more than one variable; say, for example, that $z = f(x, y)$. We can speak of the rate of change of z with respect to x by holding y constant; this is the *partial derivative* $\dfrac{\partial z}{\partial x}$ (also denoted by f_x).

Similarly, the partial derivative $\dfrac{\partial z}{\partial y}$ (or f_y) expresses the rate of change of z with respect to y when x is held constant. If x and y undergo the small changes Δx and Δy, then the resulting change in z, denoted by Δz, can be estimated from the approximation (*total differential*)

$$\Delta z \approx \frac{\partial z}{\partial x}\Delta x + \frac{\partial z}{\partial y}\Delta y.$$

To find the relative maxima and minima of a function such as $z = f(x, y)$, first find the *critical points*. The *second derivative test* is often useful for deciding if a critical point corresponds to a relative maximum, a relative minimum, or neither.

Constrained maximum-minimum problems were presented and the method of solution by *Lagrange multipliers* was explained.

The *least squares criterion* for curve fitting is extremely important for mathematical applications. The special case of the *least squares regression line* provided an interesting exercise in maximum-minimum problems for functions of several variables.

Miscellaneous Problems

1. Let production Q be a function of labor expenditure L and capital expenditure C. To be a realistic production function, the formula $Q = f(L, C)$ should be independent of the monetary units used to express Q, L, and C. Mathematically, this requirement can be stated as $f(kL, kC) = kf(L, C)$ for every constant k. Show that a Cobb-Douglas production

function $Q = aL^bC^{1-b}$ and a homohyphallic production function $Q = (a_1L^b + a_2C^b)^{1/b}$ each have this property of being independent of monetary units.

2. If $P = 3x^2y - e^x - ye^x$, find $\dfrac{\partial P}{\partial x}, \dfrac{\partial P}{\partial y}, \dfrac{\partial^2 P}{\partial x^2}, \dfrac{\partial^2 P}{\partial y^2}$, and $\dfrac{\partial^2 P}{\partial x \partial y}$.

3. If $f(u, v) = u(1 + v^2)^5$, find f_u, f_v, f_{uu}, f_{vv}, and f_{uv}. Also find $f_v(2, 0)$ and $f_{vv}(2, 0)$.

4. The production function for Australia in the 1930s was $Q = L^{0.64}C^{0.36}$.

   Find $\dfrac{\partial Q}{\partial L}$ (the marginal productivity of labor) when $L = 50$ and $C = 50$.

   Find $\dfrac{\partial Q}{\partial C}$ (the marginal productivity of capital) when $L = 30$ and $C = 30$.

5. Suppose that during one year in Australia, the total expenditure for labor was $L = 50$ and for capital was $C = 50$ in suitable units. According to the production function in Problem 4, would total production increase or decrease if, during the next year, L was changed to 48 and C was changed to 53? By what amount? Use the total differential to obtain approximate answers—exact answers are not required.

In Problems 6–9, find the critical points of the given function and then use the second derivative test in two variables to find if they correspond to a maximum or minimum.

6. $z = 4r + 4s - r^2 - 2s^2 - 6$.

7. $f(x, y) = y^4 + x^2 + 1 - 4xy$.

8. $g(x, y) = x^3 + 2y^2 - 27x + 3$.

9. $w = 8u - u^2 - 3v^2 + 6v - 1$.

10. Find the maximum and minimum of the function $f(x, y) = 3x - 5y + 1$ subject to the constraint $3x^2 + y^2 = 112$. Use Lagrange multipliers.

11. The production Q of goods in the United States was approximately $Q = L^{3/4}C^{1/4}$, where L is the expenditure for labor and C is the expenditure for capital.[17] In order to maximize production Q, how should a total expenditure of 100 units be apportioned between labor and capital (that is, to maximize Q subject to the constraint $L + C = 100$)? Use Lagrange multipliers.

12. Find the least squares regression line for the data points $(1, 11)$, $(6, 3)$, $(8, 1)$.

[17]P. H. Douglas and G. T. Gunn, "The Production Function for American Manufacturing in 1919," *American Economic Review*, XXXI (1941), pp. 67–80.

13. For theoretical reasons, a population of 100 bacteria is expected to increase according to a law of the form $p = mt + b$, where p is the natural logarithm of the population after t hours, m is a constant that depends on the particular environment, and b is a constant related to the population size at time $t = 0$. A biologist measures the population of a bacteria culture at times $t = 0$, $t = 3$, and $t = 6$ and obtains the approximate results $p = 4$, $p = 5$, and $p = 7$ respectively. Use the least squares regression line to find the values of m and b that best fit this data. Use these values to express the population P at time t by a law of the form $P = Ae^{kt}$, and then predict the population size when $t = 20$.

14. Find the least squares regression line for the following data.

x	2	3	5	6
y	4	15	60	50

15. Let D be the amount of gasoline sold during a month when the selling price of gasoline was P. Sales figures showed that (in suitable units) $D = 5.7$ when $P = 2$, $D = 4.2$ when $P = 3$, and $D = 3.3$ when $P = 4$. Find the least squares regression line (consider D the dependent variable and P the independent variable) and use it to predict the demand D if the price rises to $P = 6$.

eight

Differential Equations

The area of calculus which studies differential equations is a vast and important one. It is also an area in which current mathematical research is very active; indeed, many significant problems in this area are unsolved even today.

One important use of differential equations was mentioned in our earlier discussion about the origins of applied functions (see page 101). We saw there how differential equations are the sources for many of the mathematical functions used in applications of calculus. In recent years, it has become necessary for readers of professional journals in social science, life science, physical science, and business to have knowledge of the fundamentals of the theory of differential equations.

Section 1 □ Verifying Solutions.   General solutions, side conditions.

Section 2 □ Separation of Variables.

Section 3 □ Applications of Differential Equations.   Psychophysical laws, compound interest, diffusion, air resistance, logistic equation, chemical kinetics.

Section 4 □ Direction Fields.   Graphical solutions, numerical methods.

Miscellaneous Problems.

Section 1 □ Verifying Solutions

A differential equation is an equation involving a function and its derivatives. For example, let y denote some function of x; then

(1) $\quad \dfrac{dy}{dx} = 5y$

and

(2) $\quad \dfrac{d^2y}{dx^2} - \dfrac{1}{x}\dfrac{dy}{dx} + \dfrac{2}{x} = 0$

are differential equations. A **solution** of a differential equation is a function which makes the equation true for all values of the variable.

Let us show that the function $y = 2e^{5x}$ is a solution to the differential equation, Equation 1. For this function y, the left-hand side of Equation 1 is $(2)(5)e^{5x}$ or $10e^{5x}$. For this function y, the right-hand side of Equation 1 is $5y$, or $5(2e^{5x}) = 10e^{5x}$. Thus the left- and right-hand sides are equal for this function y; in other words, the function $y = 2e^{5x}$ is a solution of Equation 1.

You may wish to skim through Section 3 at this time to preview some of the uses of differential equations.

Example 1. Show that the function $y = 3x^2 + 2x$ is a solution to the differential equation

(3) $\quad \dfrac{d^2y}{dx^2} - \dfrac{1}{x}\dfrac{dy}{dx} + \dfrac{2}{x} = 0.$

SOLUTION.
For the function

(4) $\quad y = 3x^2 + 2x,$

we have

(5) $\quad \dfrac{dy}{dx} = 6x + 2$

and

(6) $\quad \dfrac{d^2y}{dx^2} = 6.$

We substitute these into Equation 3 and see if the result is true.

$$\frac{d^2y}{dx^2} - \frac{1}{x}\frac{dy}{dx} + \frac{2}{x} \overset{?}{=} 0$$

$$6 - \frac{1}{x}(6x + 2) + \frac{2}{x} \overset{?}{=} 0$$

$$6 - 6 - \frac{2}{x} + \frac{2}{x} \overset{?}{=} 0$$

$$0 \overset{?}{=} 0.$$

The equation is indeed true; this means that the function $y = 3x^2 + 2x$ is a solution to the given differential equation.

Example 2. Is the function $r = Ae^{-2t} + Be^{3t}$, where A and B are constants, a solution to the differential equation

(7) $$\frac{d^2r}{dt^2} - \frac{dr}{dt} = 6r$$

SOLUTION.
For the function

(8) $$r = Ae^{-2t} + Be^{3t},$$

we have

(9) $$\frac{dr}{dt} = -2Ae^{-2t} + 3Be^{3t}$$

and

(10) $$\frac{d^2r}{dt^2} = 4Ae^{-2t} + 9Be^{3t}.$$

We substitute these expressions into the given equation and see if the result is true.

$$\frac{d^2r}{dt^2} - \frac{dr}{dt} = 6r$$

$$(4Ae^{-2t} + 9Be^{3t}) - (-2Ae^{-2t} + 3Be^{3t}) \overset{?}{=} 6(Ae^{-2t} + Be^{3t})$$

$$4Ae^{-2t} + 9Be^{3t} + 2Ae^{-2t} - 3Be^{3t} \overset{?}{=} 6Ae^{-2t} + 6Be^{3t}$$

$$6Ae^{-2t} + 6Be^{3t} \overset{?}{=} 6Ae^{-2t} + 6Be^{3t}.$$

This equation is true; this means that the function given by Equation 8 is indeed a solution of Equation 7.

Example 3. Is the function $y = x^3 + c$, c a constant, a solution of the differential equation

(11) $$\frac{x}{2}\frac{d^2y}{dx^2} + \frac{dy}{dx} = 0$$

SOLUTION.

For the function

(12) $y = x^3 + c,$

we have

$$\frac{dy}{dx} = 3x^2$$

and

$$\frac{d^2y}{dx^2} = 6x.$$

We substitute these into Equation 11 and see if the result is true for all values of x.

$$\frac{x}{2}\frac{d^2y}{dx^2} + \frac{dy}{dx} = 0$$

$$\frac{x}{2}(6x) + (3x^2) \overset{?}{=} 0$$

$$3x^2 + 3x^2 \overset{?}{=} 0$$

$$6x^2 \overset{?}{=} 0.$$

This is not a true statement (except when $x = 0$); therefore Function 12 is not a solution of Equation 11.

Example 4 (Solutions in implicit form). Let y be a function of x defined implicitly by

(13) $y^3 + xy - 2x = 1.$

Show that this function y is a solution to the differential equation

(14) $$\frac{dy}{dx} = \frac{2 - y}{3y^2 + x}.$$

SOLUTION.

Differentiate each side of Equation 13 using the techniques of implicit differentiation (see Section 4 of Chapter 4).

$$3y^2\frac{dy}{dx} + x\frac{dy}{dx} + y \cdot 1 - 2 = 0.$$

Solve this equation for $\frac{dy}{dx}$.

$$\frac{dy}{dx}(3y^2 + x) = 2 - y,$$

$$\frac{dy}{dx} = \frac{2 - y}{3y^2 + x}.$$

This is the given differential equation in Equation 14. Thus the function y defined by Equation 13 is a solution of Equation 14.

General Solutions and Solutions Satisfying Side Conditions. In Example 2, it was shown that the function

(15) $r = Ae^{-2t} + Be^{3t}$,   where A and B are constants,

is a solution of the differential equation

(16) $\dfrac{d^2r}{dt^2} - \dfrac{dr}{dt} = 6r.$

As a special case of this result, we see immediately that

(17) $r = 7e^{-2t}$

is a solution of Equation 16 because it is obtained from Equation 15 by choosing $A = 7$ and $B = 0$. Similarly,

(18) $r = -5e^{3t}$

is also a solution of Equation 16 since it is obtained from Equation 15 by choosing $A = 0$ and $B = -5$. It is a fact that every solution of Equation 16 can be obtained from Equation 15 by choosing the constants A and B suitably. For that reason, we call Equation 15 the **general solution** of Equation 16. Functions in Equations 17 and 18 are called **particular solutions** of Equation 16 because their expressions do not contain arbitrary constants. These notions of *general solution* and *particular solution* apply to any differential equation although the methods of obtaining all solutions from the general solution may be more complicated than the one described above.

Example 5. The general solution of the differential equation

(19) $\quad x\dfrac{dy}{dx} - 2y = 2x$

is

(20) $\quad y = Cx^2 - 2x,$

where C is an arbitrary constant. Find the particular solution of Equation 19 which satisfies the side condition

(21) $\quad y = 10 \quad$ when $\quad x = 1.$

SOLUTION.
We must choose a suitable value of C in Equation 20 so that the resulting particular solution satisfies the condition $y = 10$ when $x = 1$. Thus we set $x = 1$ and $y = 10$ in Equation 20 and solve for C.

$$10 = C \cdot 1^2 - 2 \cdot 1$$
$$10 = C - 2$$
$$C = 12.$$

The answer is therefore

$$y = 12x^2 - 2x.$$

Example 6. Verify that $y = Ce^x + e^{-x}$ is a solution of the differential equation

(22) $\quad \dfrac{dy}{dx} + 2e^{-x} = y.$

Find the particular solution of Equation 22 which satisfies the side condition

(23) $\quad y = 0 \quad$ when $\quad x = 0.$

SOLUTION.
For the function

(24) $\quad y = Ce^x + e^{-x},$

we have

(25) $\quad \dfrac{dy}{dx} = Ce^x - e^{-x}.$

We substitute Equations 24 and 25 into Equation 22 and see if the result is a true statement.

$$\frac{dy}{dx} + 2e^{-x} \stackrel{?}{=} y$$

$$(Ce^x - e^{-x}) + 2e^{-x} \stackrel{?}{=} Ce^x + e^{-x}$$

$$Ce^x + e^{-x} \stackrel{?}{=} Ce^x + e^{-x}.$$

Since the resulting equation is true, $y = Ce^x + e^{-x}$ is indeed a solution of Equation 22.

We now determine the value of C so that the side condition $y = 0$ when $x = 0$ is satisfied. Thus we put $x = 0$ and $y = 0$ in Equation 24 and solve for C.

$$0 = Ce^0 + e^{-x},$$
$$0 = C + 1,$$
$$C = -1.$$

The answer is therefore

$$y = -e^x + e^{-x}.$$

Problems

In each of the following problems, (a) verify that the given function is a solution to the given differential equation; then (b) find the particular solution that satisfies the given side conditions.

1. Function: $y = Cx^3 + 1$.

   Differential equation: $\dfrac{x}{3}\dfrac{dy}{dx} = y - 1$.

   Side condition: $y = 0$ when $x = 1$.

2. Function: $y = Cxe^x$.

   Differential equation: $\dfrac{x}{x + 1}\dfrac{dy}{dx} = y$.

   Side condition: $y = 1$ when $x = 1$.

3. $y = Cx^{3/2} - 1$,

   $2x\dfrac{dy}{dx} = 3(y + 1)$,

   $y = 15$ when $x = 4$.

4. $y = x^3 - x + C$,

   $x\dfrac{d^2y}{dx^2} - 2\dfrac{dy}{dx} = 2$,

   $y = 1$ when $x = 0$.

5. $y = x + Ce^{3x}$,

$\dfrac{dy}{dx} = 1 + 3(y - x)$,

$y = 0$ when $x = 0$.

6. $s = \dfrac{1}{t + C}$,

$\dfrac{ds}{dt} + s^2 = 0$,

$s = \frac{1}{2}$ when $t = 0$.

7. $y^4 = 2x^2 + C$,

$y^3 \dfrac{dy}{dx} = x$,

$y = 0$ when $x = 0$.

8. $\ln(1 - 3q) = 3p + C$,

$\dfrac{dq}{dp} = 3q - 1$,

$q = 0$ when $p = \frac{1}{3}$.

9. $w = x \ln x + Cx$,

$x \dfrac{dw}{dx} = x + w$,

$w = 0$ when $x = e$.

10. $xy^2 - xy + C = 0$,

$\dfrac{dy}{dx} = \dfrac{y - y^2}{2xy - x}$,

$y = 2$ when $x = 1$.

11. $y^3 - xy = C$,

$\dfrac{dy}{dx} = \dfrac{y}{3y^2 - x}$,

$y = 2$ when $x = 0$.

12. $y + xe^y = C$,

$(x + e^{-y}) \dfrac{dy}{dx} + 1 = 0$,

$y = 0$ when $x = 0$.

13. $tP = C_1 + C_2 t$,

$\dfrac{d^2 P}{dt^2} + \dfrac{2}{t} \dfrac{dP}{dt} = 0$,

$P = 1$ and $\dfrac{dP}{dt} = -1$

when $t = 1$.

14. $y^2 = x^3 + C_1 x + C_2$,

$\dfrac{d^2 y}{dx^2} + \dfrac{1}{y}\left(\dfrac{dy}{dx}\right)^2 = \dfrac{3x}{y}$,

$y = 1$ when $x = 1$,   and

$y = 3$ when $x = 2$.

Section 2 □ Separation of Variables

In this section, we shall learn how to find the general solution to differential equations of a certain type.

A differential equation of the form

(1) $\qquad \dfrac{dy}{dx} = f(x,y)$

is called a *first-order differential equation*. The term "first-order" refers to the fact that only the first derivative of the unknown function y appears in the equation. In some cases, it may be possible to factor the right hand side of Equation 1 into a product of two expressions as follows.

(2) $\qquad \dfrac{dy}{dx} = p(x)q(y)$.

Here $p(x)$ is an expression involving only x, and $q(y)$ is an expression involving only y. Equations of this form are called **differential equations with variables separable;** these are the types we shall solve in this section. Equations of the form

(3) $\qquad \dfrac{dy}{dx} = p(x) \qquad$ or $\qquad \dfrac{dy}{dx} = q(y)$

are also considered to be of "variables separable" form.

Example 1. Which of the following differential equations can be put into the variables separable form?

a. $\dfrac{dy}{dx} = x + \dfrac{x}{y}$ $\qquad\qquad$ **b.** $e^{-x}\dfrac{dy}{dx} = 2 + \ln y$

c. $\dfrac{ds}{dr} = r^2$ $\qquad\qquad$ **d.** $\dfrac{dy}{dt} = e^{2v} + 1$

e. $\dfrac{dP}{dr} = P^2 r + P$ $\qquad\qquad$ **f.** $\dfrac{dy}{dx} = x + y$

SOLUTION.
Equation a can be written as

$$\dfrac{dy}{dx} = x\left(1 + \dfrac{1}{y}\right),$$

which is in the variables separable form; that is, it is of Form 2 with $p(x) = x$ and $q(y) = 1 + (1/y)$.

Equation b can be written as

$$\dfrac{dy}{dx} = e^x(2 + \ln y),$$

which is in the variables separable form.

Equations c and d are already in variables separable form (see Equation 3).

Equation e is not in variables separable form and it cannot be rewritten so that it takes that form. Of course, it can be rewritten as

$$\dfrac{dP}{dr} = P(Pr + 1),$$

but here the right-hand side is not the product of an expression in P alone by an expression in r alone.

Equation f also cannot be rewritten in variables separable form.

To solve a variables separable differential equation

(4) $$\frac{dy}{dx} = p(x)\,q(y),$$

we divide both sides by $q(y)$

$$\frac{1}{q(y)}\frac{dy}{dx} = p(x).$$

Then we integrate (antidifferentiate) with respect to x.

(5) $$\int \frac{1}{q(y)}\frac{dy}{dx}\,dx = \int p(x)\,dx.$$

The left side of Equation 5 can be rewritten: make the substitution $u = y$ and $du = (dy/dx)\,dx$ and we have $\int \frac{1}{q(u)}\,du$, which is essentially the same as $\int \frac{1}{q(y)}\,dy$. Thus Equation 5 is the same as

(6) $$\int \frac{1}{q(y)}\,dy = \int p(x)\,dx.$$

When the integrals in Equation 6 are evaluated, we obtain an equation in x and y, with essentially one arbitrary constant. This is the general solution in *implicit form.*

The development from Equation 4 to Equation 6 took several steps in the above discussion. In practice, it is customary to omit the intermediate steps and to go directly from

$$\frac{dy}{dx} = p(x)\,q(y)$$

to

$$\int \frac{1}{q(y)}\,dy = \int p(x)\,dx$$

by "multiplying by dx, dividing by $q(y)$, and integrating."

Example 2.   Find the general solution of the differential equation

(7) $$\frac{dy}{dx} = \frac{x}{y}.$$

SOLUTION.

We "multiply by dx and by y and integrate" to obtain

$$\int y\, dy = \int x\, dx.$$

Recall that $\int y\, dy = (y^2/2) + C_1$ and $\int x\, dx = (x^2/2) + C_2$. Thus

$$\frac{y^2}{2} + C_1 = \frac{x^2}{2} + C_2$$

or

(8) $y^2 - x^2 = 2C_2 - 2C_1.$

The term $2C_2 - 2C_1$ is an arbitrary constant. If we denote it by C, we can write Equation 8 in the simpler form

(9) $y^2 - x^2 = C.$

Equation 9 is the general solution of Equation 7, but it is in *implicit form*. It can be put into *explicit form* by solving for y.

(10) $y = \pm\sqrt{C - x^2}.$

Example 3.

a.   Find the general solution of

$$\frac{dy}{dx} = \frac{\sqrt{2x + 1}}{y^2 + 1}.$$

b.   Find the general solution of

$$\frac{dP}{dt} = P^3\, e^{2t}.$$

SOLUTION.

a.   Separate variables and integrate as follows.

$$(y^2 + 1)\, dy = \sqrt{2x + 1}\, dx$$

$$\int (y^2 + 1)\, dy = \int (2x + 1)^{1/2}\, dx$$

(11) $\dfrac{y^3}{3} + y = \dfrac{1}{3}(2x + 1)^{3/2} + C$

Equation 11 is the general solution in implicit form.

b.   Separate variables and integrate.

$$\frac{dP}{P^3} = e^{2t}\,dt$$

$$\int P^{-3}\,dP = \int e^{2t}\,dt$$

$$\frac{P^{-3+1}}{-3+1} = \frac{1}{2}e^{2t} + C_1$$

(12) $$P^{-2} = -e^{2t} - 2C_1$$

An equivalent general solution is

$$e^{2t} + P^{-2} = C,$$

where $-2C_1$ has been replaced by the simpler constant C.

Separation of variables will sometimes lead to integrals of the form

$$\int \frac{du}{u} \quad \text{or} \quad \int \frac{du}{au+b}$$

The following formula is used for evaluating these integrals.

---

(13) $$\int \frac{du}{au+b} = \begin{cases} \dfrac{1}{a}\ln(au+b) + C & \text{if } au+b>0 \\[2ex] \dfrac{1}{a}\ln[-(au+b)] + C & \text{if } au+b<0 \end{cases}$$

Example 4.

a. Find the general solution of

(14) $$\frac{dP}{dt} = 2P - 7$$

if the values of P satisfy $2P - 7 > 0$.

b. Find the general solution if $2P - 7 < 0$.

c. In Parts a and b, express the general solution *explicitly* for P as a function of t.

d. Find the particular solution of Equation 14 which satisfies the side conditions $P = 10$ when $t = 0$.

SOLUTION.

a. Separate variables and integrate as follows.

$$\frac{dP}{2P - 7} = dt$$

$$\int \frac{dP}{2P - 7} = \int dt$$

Since $2P - 7 > 0$, we use the first of Formulas 13 to continue.

$$\frac{1}{2}[\ln(2P - 7) + C_1] = t + C_2$$

(15) $$\ln(2P - 7) = 2t + C_3$$

(*Note:* $2C_2 - 2C_1$ has been replaced by C_3.)

b. We proceed as in Part a, except we use the second of Formulas 13 to evaluate $\int dP/(2P - 7)$.

$$\frac{dP}{2P - 7} = dt$$

$$\int \frac{dP}{2P - 7} = \int dt$$

$$\frac{1}{2}\{\ln[-(2P - 7)] + C_1\} = t + C_2$$

or

(16) $$\ln(7 - 2P) = 2t + C_3$$

c. To solve Equation 15 for P, exponentiate each side.

$$e^{\ln(2P - 7)} = e^{2t + C_3}$$

$$2P - 7 = e^{2t}e^{C_3}$$

or (with $C = \frac{1}{2}e^{C_3}$)

(17) $$P = Ce^{2t} + \frac{7}{2}.$$

To solve Equation 16 for P, proceed the same way.

$$e^{\ln(7 - 2P)} = e^{2t + C_3}$$

$$7 - 2P = e^{2t}e^{C_3}$$

or (with $C = -\frac{1}{2}e^{C_3}$)

(18) $P = Ce^{2t} + \dfrac{7}{2}.$

Note that Equations 17 and 18 have the same form in both Parts a and b.

d. Use the side condition to evaluate C in Equation 17.

$$10 = Ce^{2 \cdot 0} + \frac{7}{2}, \qquad 10 = C \cdot 1 + \frac{7}{2}, \qquad C = \frac{13}{2}.$$

With this value for C, Equation 17 gives the desired particular solution.

$$P = \frac{13}{2}e^{2t} + \frac{7}{2}.$$

Problems

In Problems 1–12, find the general solution in implicit form.

1. $\dfrac{dy}{dx} = \dfrac{x}{y^3}.$

2. $\dfrac{dy}{dx} = \dfrac{4x + 1}{y^2}.$

3. $\dfrac{ds}{dt} = t^3 s, \quad (s > 0).$

4. $\dfrac{dy}{dx} = 2y, \quad (y < 0).$

5. $\dfrac{dq}{dp} = q^2(2p + 3).$

6. $\dfrac{dP}{dx} = P^2 x + 2P^2.$

7. $\dfrac{dy}{dt} - \dfrac{t}{y} = \dfrac{e^{-2t}}{y}.$

8. $x\dfrac{dy}{dx} + 2y = xy, \quad (x > 0, y > 0).$

9. $\dfrac{dU}{ds} = 3U - 1, \quad (3U - 1 < 0).$

10. $\dfrac{dy}{dt} = yt + y + t + 1, (y + 1 > 0).$

11. $\dfrac{dy}{dx} = y \ln x, \quad (x > 0, y > 0).$

12. $\dfrac{1}{t}\dfrac{dP}{dt} = \dfrac{e^t}{P}.$

In Problems 13–24, find particular solutions to the above differential equations which satisfy the given side conditions. Leave your solutions in implicit form.

13. Solve Problem 1 if $y = 0$ when $x = 0.$

14. Solve Problem 2 if $y = 1$ when $x = 0.$

15. Solve Problem 3 if $s = 1$ when $t = 0.$

16. Solve Problem 4 if $y = -3$ when $x = 0$.

17. Solve Problem 5 if $q = 1$ when $p = 1$.

18. Solve Problem 6 if $P = 3$ when $x = 0$.

19. Solve Problem 7 if $y = 0$ when $t = 0$.

20. Solve Problem 8 if $y = 1$ when $x = 1$.

21. Solve Problem 9 if $U = 0$ when $s = \frac{1}{3}$.

22. Solve Problem 10 if $y = e - 1$ when $t = 0$.

23. Solve Problem 11 if $y = 1$ when $x = 1$.

24. Solve Problem 12 if $P = 0$ when $t = 0$.

25. In Problem 3, express the general solution s explicitly as a function of t.

26. In Problem 4, express the general solution y explicitly as a function of x.

27. In Problem 9, express the general solution U explicitly as a function of s.

28. In Problem 16, express the particular solution y explicitly as a function of x.

29. In Problem 23, express the particular solution y explicitly as a function of x.

Section 3 □ Applications of Differential Equations

In this section, we shall see how differential equations arise in practical applications. Six examples will be presented; they cover areas in business, psychology, life science, and physical science.[1]

Example 1 (Psychophysical laws). The intensity r of a person's perception of a stimulus is quite different from the actual intensity s of the stimulus. For example, a person will report a large increase in brightness when a 40-watt bulb is replaced by a 60-watt bulb, but he or she will report only a small increase in brightness when a 200-watt bulb is replaced by a 220-watt bulb. A similar situation occurs for other classes of stimuli: the loudness of a tone, the saltiness of a saline solution, the pressure felt when an object is pressed against the skin, and so on.

[1]It is not necessary to cover all of these examples.

Values of the stimulus intensity s can be measured directly (watts/m^2 for brightness or loudness, NaCl concentration for salinity, g/mm^2 for pressure). Since values of the perceived intensity r cannot be measured directly, we treat r as a function of s, for a given class of stimuli, and try to determine this function indirectly.

Consider a small change Δs in stimulus intensity s and the corresponding change Δr in perceived intensity r. It is reasonable to assume that Δr is approximately proportional to Δs but inversely proportional to s:

(1) $\quad \Delta r = k\dfrac{\Delta s}{s}, \quad k$ a constant.

We write this as $\Delta r/\Delta s = k/s$, let $\Delta s \to 0$, and arrive at the differential equation

(2) $\quad \dfrac{dr}{ds} = \dfrac{k}{s}, \quad s > 0.$

Solve Equation 2 by separation of variables; the result is

(3) $\quad r = k\ln s + C.$

(See Problem 1, page 342.) Equation 3 is often called the *Weber-Fechner psychophysical law.*

Equation 3 is useful for discussing Weber's law. Consider a fixed value of s. When s is increased very slightly, an observer will usually not report any change in r. The value of Δs which will produce a *just noticeable difference* (abbreviated *jnd*) is denoted by δs. Since, for example, a 5-watt increase is easier to notice in a low-wattage bulb than in a high-wattage bulb, we would expect δs to vary with s. It is reasonable to assume that small changes in r are proportional to jnd's; thus there is a constant K such that $\Delta r = K$ whenever r changes by 1 jnd. Then Equation 1 gives

(4) $\quad \dfrac{\delta s}{s} = \dfrac{K}{k} = $ constant.

This fact that $\delta s/s$ is a constant is called *Weber's law.*[2] It was on the basis of this law that Fechner[3] first derived the psychophysical Equation 3.

The value of the constant in Weber's law (that is, the value of K/k in Equation 4) has been experimentally determined for various classes of stimuli. Some typical values[4] are:

[2] Ernest Weber (1795–1878) was a German physiologist.

[3] Gustav Fechner (1801–1887) hoped to gain recognition as a philosopher by his mathematical analysis of psychological phenomena.

[4] E. G. Boring, H. S. Langfeld, and H. P. Weld, *Foundations of Psychology* (New York: Wiley Publishing Co., 1948).

Saline taste	0.2
Cutaneous pressure	0.136
Loudness	0.088
Visual brightness	0.016

Example 2 (Compound interest).   In Section 5 of Chapter 5, we showed that ordinary compound interest can be usefully approximated by continuously compounded interest. We shall discuss only continuously compounded interest here. Recall the formula for this case.

$$(5) \qquad P = P_0 e^{rt},$$

where P is the amount of money in dollars that results when P_0 dollars are left in an account for t years at the annual interest rate of $100r$ percent, (thus $r = 0.06$ if the interest rate is 6%). We shall first derive Equation 5 from the differential-equation point of view. Then we shall show how this viewpoint can be used to solve more complicated problems.

   To derive Equation 5, consider P as a function of t. At time t, there are P dollars in the account. When t increases by a small amount Δt, the compound interest earned by these P dollars is about the same as that earned by simple interest, namely $Pr\Delta t$. That is, the change ΔP in P is approximately $\Delta P = Pr\Delta t$. We divide by Δt and obtain

$$(6) \qquad \frac{\Delta P}{\Delta t} = Pr.$$

When we let $\Delta t \to 0$, this equation leads to the differential equation

$$(7) \qquad \frac{dP}{dt} = Pr.$$

Equation 7 can be solved by separation of variables (see Problem 2 on page 342); the solution is $P = P_0 e^{rt}$.

   The reasoning used above was not rigorous since we did not prove that the approximate Equation 6 leads to the exact equation in Equation 7. Nevertheless, such reasoning is very useful and therefore occurs quite frequently in applications of calculus. Let us use it to solve this problem:

   Each day 100 dollars are deposited in a bank account that earns 100r percent annual interest. How much is in the account after t years?

   Let P be the amount in the account after t years. Consider a value of t and the corresponding value of P. Let t increase by a small amount Δt. Then the bank balance increases in two ways. First, the amount P earns $100r$ percent interest for Δt years, which results in an increase of about $Pr\,\Delta t$ dollars.

Secondly, an amount has been added to the account by the daily deposits of $100. We assume that these deposits of $36,500 per year are made continuously, so that during the interval Δt in years, an amount of $36,500\Delta t$ dollars has been added to the account. Thus we have, approximately,

$$\Delta P = Pr\,\Delta t + 36,500\,\Delta t.$$

Divide this equation by Δt and then let $\Delta t \to 0$. This leads to the differential equation

(8) $$\frac{dP}{dt} = Pr + 36,500.$$

The solution of Equation 8 (see Problem 3 on page 342) is

(9) $$P = \frac{36,500}{r}(e^{rt} - 1);$$

this solves the problem.

Example 3 (Simple diffusion).   Consider a living cell surrounded by a homogeneous solution of large volume. The solution contains a dissolved chemical called the **solute.** Molecules of solute can pass through the cell membrane into or from the liquid cell interior; this is called **diffusion** (see Figure 1).

Let A be the concentration of solute in the solution outside of the cell. Let y be the concentration of solute inside the cell. Because of diffusion, y will be a function of time t. Since the solution surrounding the cell is very large compared to the cell, diffusion will hardly effect its concentration A; therefore we treat A as a constant. Suppose that at time $t = 0$, A is much larger than y.

Let m be the total mass of solute inside the cell at time t. It is known that m will increase rapidly when A is much larger than y, and will increase slowly when A is only slightly larger than y. This fact is stated precisely by Fisk's law, which says

(10) $$\frac{dm}{dt} = k(A - y)$$

where k is a positive constant that depends on the area and the permeability of the cell membrane.[5] The concentration is, by definition,

$$y = \frac{m}{V},$$

where V is the volume of the cell. Therefore

[5] Adolf Fisk (1829–1901) was a German physiologist.

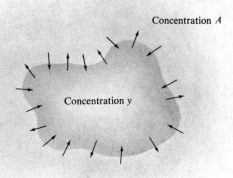

Concentration A

Concentration y

Figure 1

(11) $\quad \dfrac{dy}{dt} = \dfrac{1}{V}\dfrac{dm}{dt}$

since V is a constant. From Equations 10 and 11, we obtain

(12) $\quad \dfrac{dy}{dt} = \dfrac{k}{V}(A - y).$

The solution of the above differential equation is

(13) $\quad y = A - (A - y_0)e^{-kt/V}, \qquad k, V \text{ positive constants,}$

where y_0 is the concentration of solute inside the cell at time $t = 0$ (see Problem 4 on page 342).

Example 4 (Falling bodies and air resistance). We shall illustrate how physicists use differential equations. Consider an object that is released from rest at some height above the earth's surface and then falls freely to earth. After t seconds, it has a speed v in feet per second and it has fallen s feet (see Figure 2). The problem is to derive equations for v and for s as functions of t.

We shall use Newton's second law of motion

(14) $\quad F = ma,$

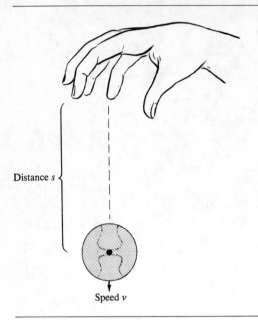

Distance s

Speed v

Figure 2

where F is the net force on an object, m is its mass, and a is the acceleration which the force produces. Recall that acceleration is the rate of change of velocity; that is, $a = dv/dt$. Thus Equation 14 will become a differential equation for the function v if we can find a suitable expression for F on the left-hand side.

One force which acts on the falling object is the force of gravitation; denote it by F_G. Another significant force is that of air resistance, which we shall denote by F_R. There are other forces (buoyancy, viscosity, etc.) but they are small; we shall neglect them. We now need suitable expressions for F_G and F_R.

The force of gravity can be found from Newton's law of universal gravitation

(15) $\quad F_G = G\dfrac{mM}{r^2},$

where m is the mass of the object, M is the mass of the earth, and r is the distance between the center of the object and the center of the earth. As the object falls, the value of r will actually decrease. However, this change in r is extremely small in comparison to r. The error in F_G that occurs if we treat r as constant is negligible, and so we shall use the simpler expression

(16) $\quad F_G = gm, \qquad g$ a constant,

for the gravitation force. (In Equation 16, g arose by combining the "constants" GM/r^2.)

Next we seek an expression for F_R. Anyone who has held an umbrella in a strong wind knows that the air resistance F_R increases with the speed of the air flowing past it. We shall assume that this increase is a simple proportionality,

(17) $F_R = kv$,

where k is a constant and v is the speed of the object through the air.

We mentioned that Newton's second law (Equation 14) can be written as

(18) $F = m\dfrac{dv}{dt}.$

Now $F = F_G - F_R$; therefore we can use Equations 16 and 17 to rewrite Equation 18 as

(19) $gm - kv = m\dfrac{dv}{dt}.$

Equation 19 can be solved by separation of variables (recall that g, m, and k are constants[6]). The particular solution to Equation 19 which satisfies $v = 0$ when $t = 0$ is

(20) $v = \dfrac{gm}{k}(1 - e^{-kt/m}),$ g, m, k positive constants

(see Problem 5, page 342).

Equation 20 is the desired formula expressing v in terms of t. From it one can find s, the distance the object has fallen by time t. Recall that $v = ds/dt$ (speed is the rate of change of distance). Thus Equation 20 is itself a differential equation:

(21) $\dfrac{ds}{dt} = \dfrac{gm}{k}(1 - e^{-kt/m})$

The solution of Equation 21 can also be found by separation of variables. The details are left as an exercise (see Problem 6, page 343); the solution turns out to be

$$s = \frac{gm}{k}\left(t + \frac{m}{k}e^{-kt/m} - \frac{m}{k}\right).$$

[6]The constants g, m, k in Equation 19 have to be determined experimentally. The approximate value of g, which varies slightly in different locations on earth, is $g = 32$. The mass of the object is found from its weight by weight $= 32 \times$ mass. The value of k depends on the shape of the object and on air density; a typical value might be $k = 2$.

Example 5 (The logistic equation). The logistic equation occurs in a number of surprisingly different contexts. Here we discuss it in the context where it first arose, that of population growth.

Let y be the number of individuals in a certain population at time t (as is usual in such applications, we neglect the fact that y takes only integer values and treat y as if it were a continuous function of t). A reasonable assumption is that dy/dt is proportional to y. We saw in Section 5 of Chapter 5 that this leads to a differential equation whose solution is an exponential function $y = Ae^{kt}$, $k > 0$. In that case $y \to +\infty$ as $t \to +\infty$.

A more likely hypothesis is that the population size y can never surpass some number M, and approaches M asymptotically as $t \to +\infty$. In this case, as y approaches M, the population will increase at a slower and slower rate; that is, dy/dt may be approximately proportional to the difference $M - y$.

Verhulst suggested a way to combine these assumptions[7]; he supposed that dy/dt is proportional to both y and $M - y$. This led him to the differential equation

(22) $$\frac{dy}{dt} = ky(M - y), \qquad k \text{ a positive constant,}$$

which he called the **logistic equation.**

Equation 22 can be solved by separation of variables. The general solution to Equation 22 is

[7]P. F. Verhulst (1804–1849) was a Belgian mathematician.

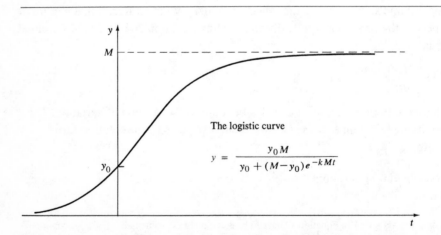

The logistic curve

$$y = \frac{y_0 M}{y_0 + (M - y_0)e^{-kMt}}$$

Figure 3

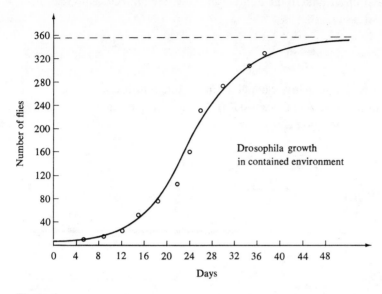

Figure 4

$$(23) \quad y = \frac{M}{1 + Ce^{-kMt}}$$

(see Problems 7 and 8 on page 343). Let y_0 be the population at time $t = 0$; then C can be evaluated; the result is

$$(24) \quad y = \frac{y_0 M}{y_0 + (M - y_0)e^{-kMt}}.$$

The function in Equations 23 or 24 is called the logistic function. Its graph is shown in Figure 3. Figure 4 shows the results of an experiment in raising a population of *Drosophila* (fruit flies) in a confined environment.[8]

Example 6 (Chemical kinetics).   A solution originally consists of two chemicals A and B. When an A molecule collides with a B molecule, they combine to form an AB molecule. Let a and b be the number of A and B molecules in the solution at time $t = 0$. Let y be the number of AB molecules at time t. Then y is a function of t, $y = 0$ when $t = 0$, and at time t there are $a - y$ and $b - y$ free molecules of A and B in solution. The rate of change dy/dt is called the **reaction rate.** Since the chance collisions between A and B

[8]A. J. Lotka, *Elements of Mathematical Biology* (New York: Dover, 1956), p. 69.

molecules are proportional to the number of these free molecules present, it is reasonable to assume that

$$(25) \quad \frac{dy}{dt} = k(a - y)(b - y),$$

where k is a constant (actually, k depends on the temperature so we must assume the temperature remains constant). The solution to Equation 25 that satisfies the initial condition is

$$(26) \quad y = a\left(1 + \frac{b - a}{a - be^{k(b-a)t}}\right)$$

(see Problem 9 on page 343).

Problems

1. (See Example 1). Solve the Weber-Fechner psychophysical differential equation $\frac{dr}{ds} = \frac{k}{s}$, $(s > 0)$, by separation of variables.

2. (See Example 2). Let P be a function of t.
   a. Use separation of variables to find the general solution of the compound interest differential equation $dP/dt = Pr$, $P > 0$, r a constant.
   b. Suppose P_0 is the value of P when $t = 0$; express this particular solution P explicitly as a function of t.

3. (See Example 2). Let P be a function of t.
   a. Use separation of variables to find the general solution of the differential equation $\frac{dP}{dt} = Pr + 36{,}500$, $Pr + 36{,}500 > 0$, r a constant.
   b. Suppose $P = 0$ when $t = 0$; express this particular solution P explicitly as a function of t.

4. (See Example 3). Let y be a function of t.
   a. Use separation of variables to find the general solution of the diffusion equation $\frac{dy}{dt} = \frac{k}{V}(A - y)$, k, A, V constants, where $A - y > 0$.
   b. Suppose $y = y_0$ when $t = 0$; express this particular solution y explicitly as a function of t.

5. (See Example 4). Let v be a function of t.
   a. Use separation of variables to solve the differential equation for free

fall with air resistance, $gm - kv = m\dfrac{dv}{dt}$, g, k, m constants, if

$gm - kv > 0$ (that is, v is increasing) and $v > 0$.

  b.   Suppose $v = 0$ when $t = 0$; express this particular solution v explicitly as a function of t.

6. (See Example 4). Let s be a function of t.
    a.   Use separation of variables to find the general solution of

$$\frac{ds}{dt} = \frac{gm}{k}(1 - e^{-kt/m}), \; g, k, m \text{ constants.}$$

    b.   Suppose $s = 0$ when $t = 0$; find an explicit expression for this particular solution.

7. (See Example 5).
    a.   Verify directly that the function in Equation 23 is a solution to the logistic differential Equation 22.
    b.   Suppose the function y in Equation 23 satisfies $y = y_0$ when $t = 0$; show that $C = (M - y_0)/y_0$, and thereby derive the formula in Equation 24.

8. (See Example 5). Let y be a function of t. Use separation of variables to find the general solution to the logistic differential equation

$$\frac{dy}{dt} = ky(M - y), \; k, M \text{ constants,}$$

if $M - y > 0$ and $y > 0$. (*Hint:* In order to evaluate $\int dy/y(M - y)$, see Formula 8 in the Table of Integers on page 389, or else use the

identity $\dfrac{1}{y(M - y)} = \dfrac{1}{M}\left(\dfrac{1}{y} + \dfrac{1}{M - y}\right).\Big)$

9. (See Example 6). Let y be a function of t.
    a.   Use separation of variables to solve the differential equation of

chemical kinetics $\dfrac{dy}{dt} = k(a - y)(b - y)$, $a, b,$ and k constants, if

$a - y > 0$ and $b - y > 0$. (*Hint:* To evaluate $\int dy/(a - y)(b - y)$,

use the identity $\dfrac{1}{(a - y)(b - y)} = \dfrac{1}{b - a}\left(\dfrac{1}{a - y} - \dfrac{1}{b - y}\right)$ or use

Formula 8 in the Table of Integrals on page 389 together with an appropriate substitution.)

    b.   Use the initial condition, $y = 0$ when $t = 0$, to put your answer to Part a in the form of Equation 26.

10. (Random equipment failures).   Certain equipment fails at a random rate, independent of its time in operation. If there are n pieces of equipment operating, they will fail at a constant rate of approximately $0.03n$ per week. Suppose there are 10,000 pieces of such equipment present at time $t = 0$. Let $f(t)$ be the number of failures that occurred up to time t in weeks. Find a differential equation for the function $y = f(t)$ and solve it.

11. (Compound interest annuity).   A man plans to withdraw money continuously from his bank account at the constant rate of $600 per year. The account pays 5% annual interest compounded continuously. At time $t = 0$, the account has a balance of $10,000.
   a.   Find a formula for the balance P after t years.
   b.   When will the balance reach zero?
   c.   When the balance reaches zero, how much money has been withdrawn?

12. (Elasticity of demand).   Let $q = f(p)$ be a demand function for some demand function for some commodity; that is, q is the quantity that will be in demand if the selling price is p. The *elasticity of demand* is defined to be $\dfrac{p}{q}\dfrac{dq}{dp}$.

   A situation of economic importance occurs when the elasticity of demand is a constant k, independent of the price. Find the general demand function $q = f(p)$, which has constant elasticity of demand.

13. Let y be the concentration function in Equation 13 of Example 3. Find $\lim\limits_{t \to +\infty} y$. Interpret your answer.

14. Let v be the velocity function in Equation 20 of Example 4. Find $\lim\limits_{t \to +\infty} v$. Interpret your answer.

15. Let y be the population function in Equation 24 of Example 5. Find $\lim\limits_{t \to +\infty} y$. Interpret your answer.

16. Let y be the function in Equation 26 of Example 6 when a, b, and k have the values $a = 100$, $b = 300$, and $k = 1$. Find $\lim\limits_{t \to +\infty} y$. Interpret your answer.

17. Differentiate Equation 22 to show that the logistic curve has a point of inflection when $y = M/2$.

In this section, we shall discuss the geometric interpretation of first-order differential equations

(1) $\quad \dfrac{dy}{dx} = f(x,y).$

The geometric interpretation will lead to a better understanding of differential equations. It will also reveal the basic ideas behind computer techniques for solving them.[9] The geometric interpretation will allow us to obtain information about the solutions to a differential equation without actually solving it.

Direction Fields.   Consider a differential equation in the form of Equation 1; it is useful to think of $f(x,y)$ as a concrete expression involving x and y; for example,

$$f(x,y) = x - y \quad \text{or} \quad f(x,y) = \frac{y^2 + xy}{y^3 + x}.$$

At each point (x,y) in an xy-coordinate system, draw a line segment centered at (x,y) with slope $f(x,y)$. The resulting figure is called the **direction field** for the differential equation. Note that a curve in this figure will be the graph of a solution to Equation 1, provided that at each point P on the curve, the line in the direction field centered at P is tangent to the curve at P. In short, *solutions have slopes that match the direction field.*

It is, of course, impossible to construct a complete direction field since there are infinitely many line segments to draw. We can, however, draw partial direction fields as in Example 1; computers can draw very extensive ones as illustrated in Figure 6 on page 347.

Suppose we have drawn a (partial) direction field for a differential equation. Starting at any point in the field, we can sketch the graph of a solution by moving our pencil along the flow of the direction field.

[9]Computer solutions for differential equations have become very important. One reason is that it is usually not possible to solve Equation 1 unless $f(x,y)$ has a very special form such as "variables separable"; that is, even though there is a solution to Equation 1, it might be impossible to express it in closed form. Another reason is that $f(x,y)$ might be an empirically-derived function which consists only of a table of approximate values.

Example 1.

a. Draw the direction field for the differential equation

(2) $\dfrac{dy}{dx} = x - y;$

show only the lines of the direction field that are centered at the 36 points (x,y) where x and y are the integers from 0 through 5.

b. Sketch the graphs of the solutions to Equation 2 which satisfy each of the following side conditions: $y = 5$ when $x = 0$; $y = 3$ when $x = 0$; $y = 1$ when $x = 0$.

SOLUTION.

a. At each of the specified points (x,y), we center a short line segment of slope $x - y$. For example, at $(0,5)$ we draw a segment of slope -5; at $(1,5)$, a segment of slope -4; at $(2,5)$, a segment of slope -3; and so on. The result is shown in Figure 5. (Although this figure is not very complete, we can easily imagine the nature of the field at the remaining points; Figure 6 shows a more complete field drawn by computer.)

b. Place your pencil at $(0,5)$ and then let it follow the flow of the direction field. The result is the graph labeled as Curve 1 in Figure 7. The solution

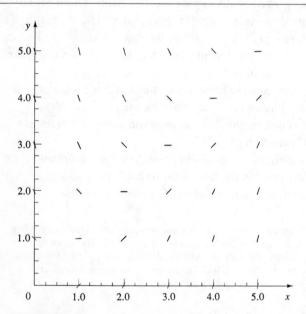

Figure 5

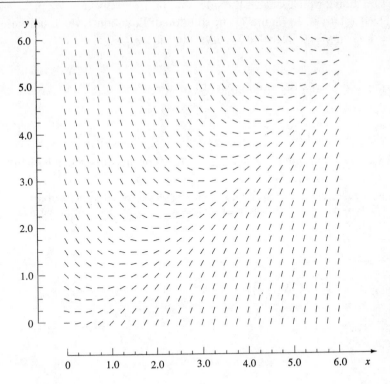

Figure 6

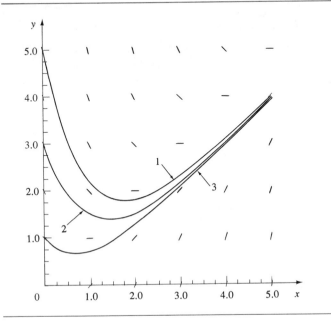

Figure 7

of Equation 2 which satisfies the side condition, $y = 3$ when $x = 0$, is labeled as curve 2 in Figure 7. The solution of Equation 2 which satisfies $y = 1$ when $x = 0$ is labeled as curve 3 in Figure 7.

Example 2. Consider the differential equation[10]

(3) $$\frac{dy}{dt} = \frac{1}{15} y \ln \frac{y}{160}.$$

Figure 8 shows a computer drawn direction field for this differential equation.

[10]Equation 3 is called a Gompertz equation, after the English actuary Benjamin Gompertz (1779–1865). The solutions of such equations have been used in population biology and in life insurance statistics.

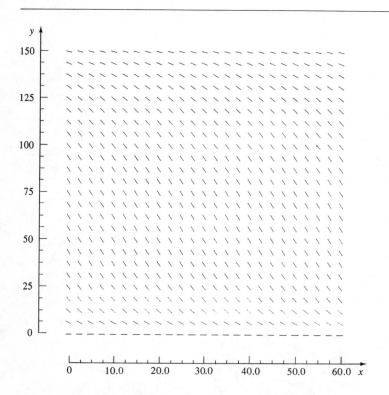

Figure 8 Direction field for Gompertz's differential equation

$$\frac{dy}{dx} = \frac{1}{15} y \ln \frac{y}{160}$$

Sketch in the graph of the solution to Equation 3 that satisfies the side conditions $y = 150$ when $t = 0$. (The solution is given in Figure 9.)

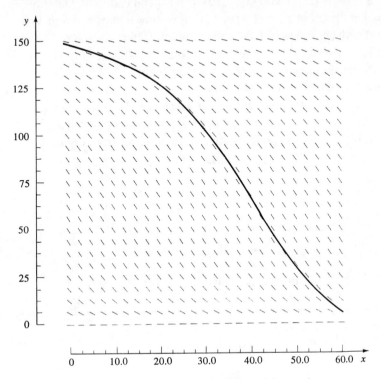

Figure 9 Direction field for Gompertz's differential equation

$$\frac{dy}{dx} = \frac{1}{15} y \ln \frac{y}{160}$$

Numerical Solutions by Euler's Method. We wish to find an approximate solution to a first-order differential equation with a side condition. Let us denote the differential equation by

(4) $\dfrac{dy}{dx} = f(x, y)$

and the initial conditions by

(5) $y = y_0$ when $x = x_0.$

The following procedure (Euler's method) is a very natural way to systematize, for computer use, the geometric method used in Examples 1 and 2.

Begin at the point (x_0, y_0) in an xy-coordinate system. Draw a short segment S_0 of slope $f(x_0, y_0)$ with (x_0, y_0) as its left-hand endpoint. Determine the right-hand endpoint of S_0; call it (x_1, y_1). Next draw a short segment S_1 of slope $f(x_1, y_1)$ with (x_1, y_1) as its left-hand endpoint. Determine the right hand endpoint of S_1; call it (x_2, y_2). Continue in this manner. The result is a curve made up of the line segments $S_0, S_1, S_2, \ldots$ (see Figure 10). This curve is an approximate solution to Equations 4 and 5.

Let us make this procedure even more systematic. We shall draw the segments $S_0, S_1, S_2, \ldots$ so that their horizontal projections (labeled $h_0, h_1, h_2, \ldots$ in Figure 10) all have the same length Δx. Then we see from Figure 11 that

$$\frac{y_{i+1} - y_i}{x_{i+1} - x_i} = f(x_i, y_i)$$

or, since $x_{i+1} - x_i = \Delta x$,

$$y_{i+1} = y_i + f(x_i, y_i)\Delta x.$$

We summarize this procedure, called Euler's Method, as follows.

Euler's Method
The graph of the solution to

(6) $\quad \dfrac{dy}{dx} = f(x, y), \qquad y = y_0 \qquad \text{when} \qquad x = x_0$

can be approximated by a polygonal curve with vertices

(7) $\quad (x_0, y_0), \quad (x_1, y_1), \quad (x_2, y_2), \ldots, (x_n, y_n)$

where

(8) $\quad x_1 = x_0 + \Delta x, \; x_2 = x_0 + 2\Delta x, \; x_3 = x_0 + 3\Delta x, \ldots, x_n = x_0 + n\Delta x$

and

(9) $\quad \begin{aligned} y_1 &= y_0 + f(x_0, y_0)\Delta x, \\ y_2 &= y_1 + f(x_1, y_1)\Delta x, \\ y_3 &= y_2 + f(x_2, y_2)\Delta x, \end{aligned}$

$$\vdots$$

$$y_n = y_{n-1} + f(x_{n-1}, y_{n-1})\Delta x.$$

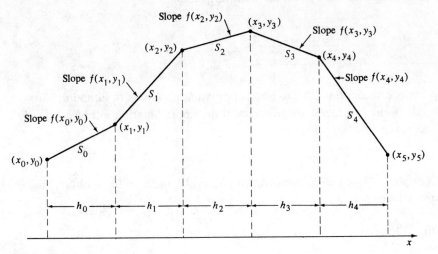

Figure 10

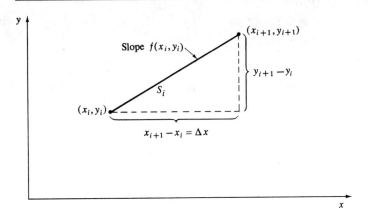

Figure 11

Euler's method yields the following table.

x	x_0	x_1	x_2	$\cdots$	x_n
y	y_0	y_1	y_2	$\cdots$	y_n

We may consider this table to be an approximate solution to Equation 6 in tabular form. In general, the accuracy of this approximation will be good only if Δx is very small.

Example 3. Use Euler's method, with $\Delta x = 0.1$ and $n = 10$, to obtain an approximate solution to

$$(10) \quad \frac{dy}{dx} = y, \quad y = 1 \quad \text{when} \quad x = 0.$$

SOLUTION.
We apply Equation 9 with $\Delta x = 0.1, f(x, y) = y$ and $y_0 = 1$, and obtain the following values.

$$y_1 = 1 + (1)(0.1) = 1.1$$
$$y_2 = 1.1 + (1.1)(0.1) = 1.21$$
$$y_3 = 1.21 + (1.21)(0.1) \approx 1.33$$
$$y_4 = 1.33 + (1.33)(0.1) \approx 1.46$$
$$y_5 = 1.46 + (1.46)(0.1) \approx 1.61$$
$$y_6 = 1.61 + (1.61)(0.1) \approx 1.77$$
$$y_7 = 1.77 + (1.77)(0.1) \approx 1.95$$
$$y_8 = 1.95 + (1.95)(0.1) \approx 2.15$$
$$y_9 = 2.15 + (2.15)(0.1) \approx 2.37$$
$$y_{10} = 2.37 + (2.37)(0.1) \approx 2.60$$

The approximate solution in tabular form is shown below.

x	0	0.1	0.2	0.3	0.4	0.5	0.6	0.7	0.8	0.9	0.10
y	1	1.1	1.21	1.33	1.46	1.61	1.77	1.95	2.15	2.37	2.60

Example 4. Use Euler's method with $\Delta x = 0.5$ and $n = 8$ to obtain an approximate solution to

(11) $\qquad \dfrac{dy}{dx} = x - y, \qquad y = 5 \qquad$ when $\qquad x = 0.$

(Compare this problem with Example 1.)

SOLUTION.
We use Equations 8 and 9 with

$$
\begin{array}{lll}
x_0 = 0 & x_1 = 0.5 & x_2 = 1 \\
x_3 = 1.5 & x_4 = 2 & x_5 = 2.5 \\
x_6 = 3 & x_7 = 3.5 & x_8 = 4
\end{array}
$$

and $y_0 = 5, f(x, y) = x - y$ to obtain

$$
\begin{aligned}
y_1 &= 5 + (0 - 5)(0.5) = 2.5 \\
y_2 &= 2.5 + (0.5 - 2.5)(0.5) = 1.5 \\
y_3 &= 1.5 + (1 - 1.5)(0.5) = 1.25 \\
y_4 &= 1.25 + (1.5 - 1.25)(0.5) \approx 1.38 \\
y_5 &= 1.38 + (2 - 1.38)(0.5) \approx 1.69 \\
y_6 &= 1.69 + (2.5 - 1.69)(0.5) \approx 2.1 \\
y_7 &= 2.1 + (3 - 2.1)(0.5) \approx 2.55 \\
y_8 &= 2.55 + (3.5 - 2.55)(0.5) \approx 3.03.
\end{aligned}
$$

The table of values for this approximate solution is shown below.

x	0	0.5	1.0	1.5	2.0	2.5	3.0	3.5	4.0
y	5	2.5	1.5	1.25	1.38	1.69	2.1	2.55	3.03

Problems

1. Figure 12 shows a computer-drawn direction field for the differential equation $\dfrac{dy}{dx} = y - x^2$. In this figure, sketch the particular solution of this differential equation which satisfies each of the following.

   a. The side condition $y = 3$ when $x = 0$.

   b. The side condition, $y = 1.5$ when $x = 0$.

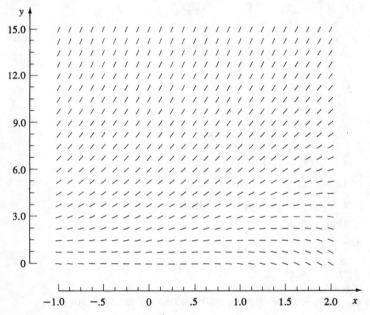

Figure 12   Direction field for $\dfrac{dy}{dx} = y - x^2$

2. Figure 13 shows a computer drawn direction field for the differential equation $\dfrac{dy}{dx} = x^2 + y^2$. In this figure sketch the particular solution which satisfies each of the following.

   a. The side condition $y = -2$ when $x = 0$.

   b. The side condition $y = 0$ when $x = 0$.

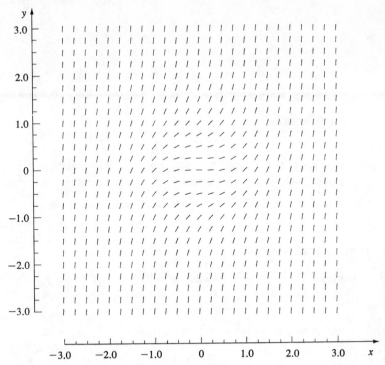

Figure 13 Direction field for $\dfrac{dy}{dx} = x^2 + y^2$

3. Use Euler's method with $\Delta x = 0.1$ and $n = 10$ to obtain an approximate solution to

$$\frac{dy}{dx} = x^2 + y^2, y = 1 \text{ when } x = 0.$$

4. Use Euler's method, with $\Delta x = 0.25$ and $n = 4$ to obtain an approximate solution to

$$\frac{dy}{dx} = 2x + y, y = 0 \text{ when } x = 0.$$

5. Use Euler's method, with $\Delta x = 0.5$ and $n = 5$ to obtain an approximate solution to

$$\frac{dy}{dx} = \frac{(y - xy)}{x}, y = 0.3 \text{ when } x = 0.5.$$

6. Use Euler's method with $\Delta x = 1$ and $n = 5$, to obtain an approximate solution to

$$\frac{dy}{dx} = \frac{1}{x}, y = 0 \text{ when } x = 1.$$

Miscellaneous Problems

In Problems 1–4, verify that the given function is a solution to the given differential equation. Then find the particular solution that satisfies the side conditions.

1. Function: $\quad\quad\quad\quad\quad y = Ce^{-x} - x.$

Differential equation: $\quad y + x + 1 + \dfrac{dy}{dx} = 0.$

Side condition: $\quad\quad\quad y = 10$ when $x = 0.$

2. Function: $\quad\quad\quad\quad\quad y = C_1 e^x + C_2 e^{6x}.$

Differential equation: $\quad \dfrac{d^2 y}{dx^2} - 7\dfrac{dy}{dx} + 6y = 0.$

Side condition: $\quad\quad\quad y = 0$ and $\dfrac{dy}{dx} = 5$ when $x = 0.$

3. Function (implicit): $\quad\quad x^2 + y^2 = cx.$

Differential equation: $\quad 2x\dfrac{dy}{dx} = \dfrac{y^2 - x^2}{y}.$

Side condition: $\quad\quad\quad y = 1$ when $x = 1.$

4. Function: $\quad\quad\quad\quad\quad y = 2x - \dfrac{1}{x} + Ce^{-1/x}.$

Differential equation: $\quad x^2\dfrac{dy}{dx} = y + \dfrac{1}{x} + 1 + 2x^2 - 2x.$

Side condition: $\quad\quad\quad y = 0$ when $x = 1.$

In Problems 5–12, use the method of separation of variables, if it applies, to find the general solution in implicit form.

5. $\dfrac{x}{y} + \dfrac{dy}{dx} = \dfrac{1}{y}.$

6. $\dfrac{dp}{dt} = tp + pe^{3t}, p > 0.$

7. $2\sqrt{t} + \dfrac{ds}{dt} = s\sqrt{t}, s - 2 > 0.$

8. $\dfrac{dy}{dx} = xy + x^2.$

9. $\dfrac{dP}{dq} = e^{-q}(2P + 1)^{-4}.$

10. $\dfrac{dR}{dt} = \dfrac{3R - 1}{e^{2t}},\ 3R - 1 < 0.$

11. $\dfrac{dY}{dt} = 1 + tY.$

12. $\dfrac{dy}{dx} = \dfrac{x}{y} \cdot \dfrac{e^x}{(1 + y^2)^{10}}.$

13. Solve

$$\frac{t}{1 + t}\frac{dp}{dt} = p - 2, \qquad p = 3 \text{ when } t = 1.$$

Express this particular solution p explicitly as a function of t.

14. A bank account pays 6% annual interest. The present balance is $100. If daily deposits of $2 are made for t years, find the final balance P. Assume that interest is compounded continuously, and that the deposits, which amount to $730 per year, are made continuously.

15. A man has $50,000 in a bank account that pays 8% annual interest. He wishes to quit work and cruise the south seas in his sailboat. He will need $5,000 a year to maintain himself and his boat. How long will his bank account support his cruising? (Assume that interest is compounded continuously, and that he withdraws the $5,000 per year at a continuous rate.)

16. A simple model of glucose metabolism is shown in Figure 14. The blood and glucose mixture is considered to be in a single pool and to be kept perfectly mixed. The mixture flows out of the pool at a rate r in cubic centimeters, is metabolized, and blood without glucose returns to the pool. The volume V in cubic centimeters of blood in the pool is constant. Suppose that an initial glucose injection at time $t = 0$ produces an initial

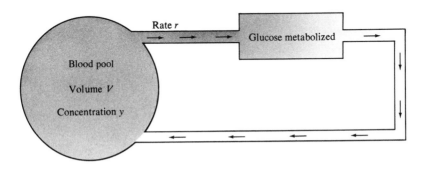

Figure 14

glucose concentration in the pool of y_0 in cubic centimeters of glucose per cubic centimeter of blood. Let y be the glucose concentration in the pool at time t in minutes. Derive the approximation $\Delta y = -ry\,\Delta t/V$ if Δt is small. Obtain a differential equation and solve it to find that $y = y_0 e^{-rt/V}$. (Models like this have been used to experimentally determine values such as r and V.)

17. Use Euler's method with $\Delta x = 0.2$ and $n = 5$ to fill in the following table of values for the solution of $\dfrac{dy}{dx} = x + y^2$, $y = 0$ when $x = 0$.

x	0	0.2	0.4	0.6	0.8	1.0
y						

appendix one

Algebra: Review and Self-Tests

Number Systems. The **natural numbers** are the counting numbers 1, 2, 3, . . .
The **integers** are the natural numbers, their negatives, and zero:

$$. . . , -3, -2, -1, 0, 1, 2, 3, . . .$$

The **rational numbers** are numbers that can be expressed as fractions with
integral numerators and denominators, but *zero is never allowed as a denominator.* Here are some examples of rational numbers:

$$\frac{1}{2}, \quad \frac{-2}{3}, \quad \frac{6}{-2}, \quad \frac{-125}{-671}, \quad \frac{790}{1}.$$

The **real numbers** are all numbers which can be expressed as decimals, either
terminating or nonterminating. Some examples of terminating decimals are
the real numbers

$$72.015, \quad -1.0015, \quad 0.526, \quad 12.7771777.$$

The real numbers

$$\frac{1}{3} = 0.3333 . . . \quad \text{and} \quad \pi = 3.14159265358979323846 2643 . . .$$

are examples of nonterminating decimals.

The real numbers contain the rational numbers, which contain the
integers, which contain the natural numbers. The word *number* will often be
used to mean *real number.*

Number Line, Inequalities. On a straight line, mark one point with 0 and call
it the **origin.** Choose any convenient distance (for example, an inch, a centimeter, a foot, etc.) and call it the **unit distance.** We imagine that each point is
labeled with the number that describes its distance, in terms of the unit
distance, from the origin; however, we use positive numbers for points to the
right of the origin and negative numbers for points to the left of the origin.
In this way, each point is labeled with a unique real number, and each real
number is the label of a unique point. The line, so labeled, is called a **number
line** or **number axis.** Figure A–1 shows a number line and the location of the
points labeled 0, 1, 2, 3, 4, −1, −2, −3, −4, −2.75, $\sqrt{2} = 1.414$ . . . , and
$\pi = 3.1415$ . . . It is customary to make no linguistic distinction between a
point on the number line and the *real number* that labels it; for example, we
speak of "the point $\frac{1}{2}$" rather than of "the point labeled $\frac{1}{2}$."

If a and b are numbers (that is, real numbers), then $a < b$ (or $b > a$)
means *a **is less than** b;* we can also interpret the meaning as *a is to the left of b.*

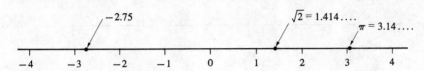

Figure A-1   A number line

For example,

$$-3 < -1, \quad -\frac{1}{2} < -\frac{1}{4}, \quad -\frac{1}{8} < 0, \quad 0 < 0.001, \quad \text{and} \quad 2 < \pi.$$

We also define the following symbols.

$$a \leqq b \quad \text{means} \quad a < b \quad \text{or} \quad a = b$$
$$a < x < b \quad \text{means} \quad a < x \quad \text{and} \quad x < b.$$

The symbol $b \geqq a$ means the same as $a \leqq b$.

Arithmetic Operations.   To perform subtraction with negative numbers, the following rules can be used.

$$a - (-b) = a + b$$
$$-a - (-b) = -a + b = b - a$$
$$-a - b = -(a + b)$$

Example 1.   Calculate each of the following.
(a) $5 - (-1)$.
(b) $-3 - (-8)$.
(c) $-7 - 12$.
(d) $-7 - (-4)$.
(e) $-9 - 6$.
(f) $5 - 13$.

SOLUTION.
(a) 6.   (b) 5.   (c) -19.   (d) -3.   (e) -15.   (f) -8.

When working with fractions, we use the following rules.

$$a = \frac{a}{1}$$

$$\frac{-a}{b} = \frac{a}{-b} = -\frac{a}{b}$$

$$\frac{a}{c} + \frac{b}{c} = \frac{a+b}{c}$$

$$\frac{a}{b} + \frac{c}{d} = \frac{ad + bc}{bd}$$

$$\frac{a}{b} \cdot \frac{c}{d} = \frac{ac}{bd}$$

Example 2.

$$5 \cdot \left(\frac{3}{7}\right) = \frac{5}{1} \cdot \frac{3}{7} = \frac{15}{7}$$

$$\frac{1}{-3} - \frac{-4}{9} = \frac{-3}{9} + \frac{4}{9} = \frac{-3+4}{9} = \frac{1}{9}$$

$$\frac{3}{8} \cdot \frac{5}{12} = \frac{3 \cdot 5}{8 \cdot 12} = \frac{1 \cdot 5}{8 \cdot 4} = \frac{5}{32}$$

$$\frac{2}{3} - \frac{4}{5} = \frac{10 - 12}{15} = \frac{-2}{15} = -\frac{2}{15}$$

$$\frac{\dfrac{1}{2}}{\dfrac{2}{3}} = \frac{\dfrac{3}{2} \cdot \dfrac{1}{2}}{\dfrac{3}{2} \cdot \dfrac{2}{3}} = \frac{\dfrac{3}{4}}{1} = \frac{3}{4}$$

The use of exponents is reviewed fully in Chapter 4 (see pages 143–144). You should first study those pages and then use the following material for review.

In the expression b^n, we call b the **base** and n the **exponent**. If n is a natural number, then

$$b^n = \underbrace{b \cdot b \cdot b \cdots b}_{n \text{ times}}$$

and

$$b^{1/n} = \sqrt[n]{b},$$

where $\sqrt[n]{b}$ is the principal nth root of b; that is, the number (if it exists) or the positive number, if there are two, whose nth power is b. The meaning of rational exponents follows from these definitions.

$$b^{m/n} = (b^m)^{1/n} = \sqrt[n]{b^m}$$
$$b^0 = 1 \qquad (b \neq 0)$$

The expression 0^0 is never used, we say it is *undefined.*

Exponents obey the rules (m and n are rational numbers here).

Rules of Exponents

Given rational numbers m and n:

$$b^{m+n} = b^m b^n$$
$$(b^m)^n = b^{mn}$$
$$(ab)^n = a^n b^n$$
$$\left(\frac{a}{b}\right)^n = \frac{a^n}{b^n}$$

Example 3.

$$25^{1/2} = \sqrt{25} = 5$$
$$-8^{2/3} = \sqrt[3]{(-8)^2} = \sqrt[3]{64} = 4$$
$$\frac{2^{-3}}{5^{-4} \cdot 3^2} = \frac{5^4}{2^3 \cdot 3^2} = \frac{5 \cdot 5 \cdot 5 \cdot 5}{2 \cdot 2 \cdot 2 \cdot 3 \cdot 3} = \frac{625}{72}$$
$$32^{0.8} = (2^5)^{8/10} = 2^{40/10} = 2^4 = 16$$
$$(a^{-3/2}b^{1/3})^6(a^8b^{-1}) = (a^{-9}b^2)(a^8b^{-1}) = a^{-9+8}b^{2-1} = a^{-1}b = \frac{b}{a}$$

Absolute Value. Let x be a real number other than 0. The **absolute value of x,** also called the **magnitude of x,** is denoted by $|x|$; it is defined to be the positive value of $\pm x$. The absolute value of 0 is defined to be 0. For example, $|2| = 2, |-3| = 3, |5| = 5, |0| = 0, |-\tfrac{1}{3}| = \tfrac{1}{3}, |\tfrac{2}{3}| = \tfrac{2}{3}$, and $|-3.14| = 3.14$.

Absolute value is useful in expressing the distance (in terms of the unit distance) between two numbers on the number line: **The distance between a and b is $|b - a|$.** Note that $|a - b|$ and $|b - a|$ are equal, so the order of subtraction does not matter.

Example 1. The distance between 2 and 6 is $|6 - 2| = |4| = 4$ (*Note:* The same answer results when reversing the order of the numbers; $|2 - 6| = |-4| = 4$).

The distance between -5 and -2 is $|-5 - (-2)| = |-5 + 2| = |-3| = 3$ or $|-2 - (-5)| = |-2 + 5| = |3| = 3$.

The distance between -1 and 6 is $|6 - (-1)| = |6 + 1| = |7| = 7$ or $|-1 - 6| = |-7| = 7$.

Self-Test I. This test is for use with Number Systems, Number Line, Inequalities, Arithmetic Operations, and Absolute Value sections. The answers are given at the end of the test.

1. Express 0.516 as a fraction with integral numerator and denominator.

2. Express 13.0102 as a fraction with integral numerator and denominator.

3. Draw a number line and indicate the points 0, ± 1, ± 2, ± 3, -2.5, -2.25, $\frac{1}{3}$, $\frac{2}{3}$ and 2.5.

In Problems 4–13, write $<$ or $>$ to make a correct inequality.

4. $-3 \ \square \ -1$.

5. $6 \ \square \ 2$.

6. $-1 \ \square \ 4$.

7. $3 \ \square \ 5$.

8. $2.413457 \ \square \ 2.414006$.

9. $\frac{6}{7} \ \square \ \frac{29}{35}$.

10. $-\frac{3}{4} \ \square \ -\frac{1}{5}$.

11. $\frac{1}{3} \ \square \ 0.33333$.

12. $-2.1 \ \square \ -1.3 \ \square \ 4$.

13. $0.1 \ \square \ 0.001 \ \square \ 0.0009$.

14. Express $\frac{30}{200}$ as a terminating decimal.

15. Express $\frac{2}{3}$ as a nonterminating decimal.

16. Answer true or false.
    (a) $0 < 0.0003$.
    (b) $-0.01 < 0.0003$.
    (c) $-3 < 5$.
    (d) $-3 < 2$.
    (e) $6 > -8$.
    (f) $2 \leqq 2.1$.
    (g) $2 < 2.01 < 2.1$.
    (h) $-5 < -1 < \frac{1}{2}$.

In Problems 17–32, complete each calculation and simplify.

17. $-1 - (-4)$. 18. $-2 - 8$. 19. $3 - (-2)$.

20. $-8 - (-6)$. 21. $-1 - (-1)$. 22. $4 - 6$.

23. $-7 - (-1)$. 24. $12 - (-4)$. 25. $6 - 4$.

26. $-4 - (-7)$. 27. $6 - (-1)$. 28. $-3 - 6$.

29. $(-5)(8)$. 30. $(-5)(-8)$. 31. $|-7|$.

32. $|-3 - 6|$.

33. Find each distance on a number line.
    (a) From -5 to 7.   (b) From 7 to 2.
    (c) From -1 to -3.   (d) From 6 to -6.

In Problems 34–43, complete each calculation and simplify.

34. $(\frac{2}{3})(7)$.

35. $(\frac{-1}{4})(\frac{3}{-5})(\frac{-10}{27})$.

36. $\frac{2}{3} - \frac{1}{7}$.

37. $\frac{-3}{4} + \frac{6}{-7}$.

38. $\frac{2}{5} + (-0.31)$.

39. $\dfrac{\frac{1}{2} - \frac{2}{3}}{\frac{1}{3} \quad \frac{1}{4}}$.

40. $64^{-1/2}$.

41. $8^{4/3}$.

42. $4^{1.5}$.

43. $(25^{1/3})^{3/2}$.

44. Express in the form b^n.

    (a) $3\sqrt[5]{3}$.

    (b) $\dfrac{\sqrt{3}}{\sqrt[3]{3}}$.

    (c) $\dfrac{2^6 \cdot \sqrt{8}}{4^3 \cdot 2^{-4}}$.

    (d) $\sqrt{2}\sqrt{5}$.

Answers to Self-Test I. **1.** $\frac{516}{1000}$. **2.** $\frac{130102}{1000000}$. **3.** See Figure A–2. **4.** $<$.
5. $>$. **6.** $<$. **7.** $<$. **8.** $<$. **9.** $>$. **10.** $<$. **11.** $>$.
12. $<, <$. **13.** $>, >$. **14.** 0.15. **15.** 0.666 . . . . **16.** All true.
17. 3. **18.** -10. **19.** 5. **20.** -2. **21.** 0. **22.** -2. **23.** -6.
24. 16. **25.** 2. **26.** 3. **27.** 7. **28.** -9. **29.** -40. **30.** 40.
31. 7. **32.** 9. **33.** (a) 12, (b) 5, (c) 2, (d) 12. **34.** $\frac{14}{3}$.
35. $\frac{1}{18}$. **36.** $\frac{11}{21}$. **37.** $-\frac{45}{28}$. **38.** 0.09. **39.** $-\frac{7}{6}$. **40.** $\frac{1}{8}$.
41. 2^4, or 16. **42.** 8. **43.** 5. **44.** (a) $3^{6/5}$, (b) $3^{3/10}$, (c) $2^{11/2}$,
(d) $10^{1/2}$.

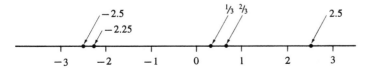

Figure A-2

Factoring.   Factoring is based on the **distributive property** of real numbers,

(1) $a(b + c) = ab + ac,$

and generalizations such as

(2) $a(b + c + d) = ab + ac + ad$

and

$$(a + e)(b + c) = a(b + c) + e(b + c)$$
$$= ab + ac + eb + ec.$$

Example 1.   Multiply.
(a) $2x(3x + 1).$ (b) $(3x + 2)(3x - 2).$
(c) $(2x + 1)^2.$ (d) $(x - 1)(x + 4).$

SOLUTION.

(a) $2x(3x + 1) = 6x^2 + 2x.$

(b) $(3x + 2)(3x - 2) = 3x(3x - 2) + 2(3x - 2)$
$$= 9x^2 - 6x + 6x - 4 = 9x^2 - 4.$$

(c) $(2x + 1)^2 = (2x + 1)(2x + 1) = 2x(2x + 1) + 1(2x + 1)$
$$= 4x^2 + 2x + 2x + 1 = 4x^2 + 4x + 1.$$

(d) $(x - 1)(x + 4) = x(x + 4) - 1(x + 4)$
$$= x^2 + 4x - x - 4 = x^2 + 3x - 4.$$

A **polynomial** in one variable x is a sum or difference of terms of the form ax^n, where a is a real number and n is a nonnegative integer. Thus

$$5x^3 - 3x^2 + 2x + 1, \qquad x^3, \qquad 3, \qquad \text{and} \qquad 10x^{11} - 2x + 1$$

are polynomials, but

$$3\sqrt{x} + 1, \qquad x^{2/3} + 5x + 1, \qquad \text{and} \qquad x^2 + \frac{3}{x} + 1$$

are not.

The process of factoring a polynomial into other polynomials amounts to reversing the multiplication procedure used in Example 1. The reverse of the procedure used in Part a is called *factoring out a common term.* For example, $5x^2 - 10x$ is factored as

$$5x^2 - 10x = 5x(x - 10)$$

and $x^6 - 2x^3 + x^2$ is factored as

$$x^6 - 2x^3 + x^2 = x^2(x^4 - 2x + 1).$$

The reverse of the procedure in Part (b) of Example 1 is called *factoring as a difference of squares*. Here are some illustrations.

$$x^2 - 1 = (x + 1)(x - 1)$$
$$x^2 - 25 = (x + 5)(x - 5)$$
$$x^2 - 3 = (x + \sqrt{3})(x - \sqrt{3})$$
$$2x^2 - 5 = 2\left(x^2 - \frac{5}{2}\right) = 2\left(x + \sqrt{\frac{5}{2}}\right)\left(x - \sqrt{\frac{5}{2}}\right)$$

or

$$2x^2 - 5 = (\sqrt{2}x + \sqrt{5})(\sqrt{2}x - \sqrt{5}).$$

Example 2. Factor.[1]
(a) $3x^2 - 4x$.   (b) $2x^3 - \frac{1}{2}x^2$.
(c) $x^2 - 16$.   (d) $3x^2 - 1$.

SOLUTION.

(a) $3x^2 - 4x = x(3x - 4)$.

(b) $2x^3 - \frac{1}{2}x^2 = x^2(2x - \frac{1}{2})$.

(c) $x^2 - 16 = (x + 4)(x - 4)$.

(d) $3x^2 - 1 = (\sqrt{3}x + 1)(\sqrt{3}x - 1)$, or
     $3x^2 - 1 = 3(x^2 - \frac{1}{3}) = 3(x + \sqrt{\frac{1}{3}})(x - \sqrt{\frac{1}{3}})$.

The reverse of the procedure in Part d of Example 1 is less automatic and requires more skill than the two factorization methods just discussed. The pattern on which it is based is this.

$$(x + a)(x + b) = x^2 + (a + b)x + ab.$$

If we want to factor $x^2 + x - 2$, for example, we can write

$$x^2 + x - 2 = (x + \square)(x + \square);$$

[1] To be precise, we should say, "Factor into a product of polynomials." Otherwise, Part a might be factored as $3x^2 - 4x = (\sqrt{3}x + 2\sqrt{x})(\sqrt{3}x - 2\sqrt{x})$.

we then try to replace the boxes with numbers which have a product of -2 and a sum of $+1$. Some trial and error leads to

$$x^2 + x - 2 = (x + 2)(x - 1).$$

Example 3. Factor.

(a) $x^2 + x - 2$. (b) $x^2 + 4x + 3$.
(c) $x^2 + 2x - 3$. (d) $x^2 - 4x + 3$.
(e) $x^2 + 10x + 25$. (f) $x^2 - 6x + 9$.

SOLUTION.

(a) $x^2 + x - 2 = (x - 1)(x + 2)$.

(b) $x^2 + 4x + 3 = (x + 3)(x + 1)$.

(c) $x^2 + 2x - 3 = (x + 3)(x - 1)$.

(d) $x^2 - 4x + 3 = (x - 3)(x - 1)$.

(e) $x^2 + 10x + 25 = (x + 5)(x + 5) = (x + 5)^2$.

(f) $x^2 - 6x + 9 = (x - 3)(x - 3) = (x - 3)^2$.

Solving Equations.   A **quadratic equation** is an equation of the form

(1) $ax^2 + bx + c = 0, \qquad a \neq 0,$

where a, b, and c are given numbers and x is an unknown. The **solutions** of Equation 1 are the values of x that make the equation true. There cannot be more than two solutions to a quadratic equation, and there may be only one or none.

The basis for solving quadratic equations is the following property of numbers.

(2)   If $ab = 0$, then $a = 0$ or $b = 0$.

Example 1.   Solve $x^2 + x - 2 = 0$.

SOLUTION.
First factor the left-hand side; the result (see Part a of Example 3) is

$$(x - 1)(x + 2) = 0.$$

Now use Property 2 to conclude

$$x - 1 = 0 \quad \text{or} \quad x + 2 = 0$$

and so

$$x = 1 \quad \text{or} \quad x = -2.$$

Example 2. Solve.

(a) $x^2 + 2x - 3 = 0$. (b) $x^2 - 10x + 25 = 0$.
(c) $3x^2 - 12x = 0$. (d) $2x^2 = 16x$.
(e) $2x^2 = 32$. (f) $50 - 2x^2 = 0$.

SOLUTION.

(a) $x^2 + 2x - 3 = 0; (x - 1)(x + 3) = 1; x - 1 = 0$ or $x + 3 = 0;$
     $x = 1$ or $x = -3$.

(b) $x^2 - 10x + 25 = 0; (x - 5)(x - 5) = 0; x - 5 = 0$ or $x - 5 = 0;$
     $x = 5$.

(c) $3x^2 - 12x = 0; 3x(x - 4) = 0; 3x = 0$ or $x - 4 = 0; x = 0$ or $x = 4$.

(d) $2x^2 = 16x; 2x^2 - 16x = 0; 2x(x - 8) = 0; x = 0$ or $x = 8$.

(e) $2x^2 = 32; x^2 = 16; x = \pm\sqrt{16}; x = 4$ or $x = -4$.

(f) $50 - 2x^2 = 0; -2x^2 = -50; x^2 = 25; x = \pm\sqrt{25}; x = \pm 5$.

If you wish to solve a quadratic equation in the form $ax^2 + bx + c = 0$, with $a \neq 0$, and cannot factor it, then the **quadratic formula** can be used (this formula will not be needed for the problems in the main body of this textbook).

Quadratic Formula

$$x = \frac{-b \pm \sqrt{b^2 - 4ac}}{2a}.$$

Example 3. Solve.

(a) $2x^2 + 6x + 3 = 0$. (b) $x^2 - 2x + 3 = 0$.

SOLUTION.

(a) $$x = \frac{-6 \pm \sqrt{6^2 - 4 \cdot 2 \cdot 3}}{2 \cdot 2} = \frac{-6 \pm \sqrt{12}}{4} = -\frac{3}{2} \pm \frac{\sqrt{3}}{2}$$

(b) $$x = \frac{2 \pm \sqrt{(-2)^2 - 4 \cdot 1 \cdot 3}}{2 \cdot 1} = \frac{2 \pm \sqrt{-8}}{2};$$

since $\sqrt{-8}$ is not a real number, we conclude that there are no real solutions.

Self-Test II. This test is for use with Factoring and Solving Equations sections. The answers are given at the end of the test.

Find each product.

1. $2x(x^5 - 4x^4 + 2x^3 - x + 1)$.   2. $(x + 1)(x + 3)$.

3. $(x + 4)(x - 2)$.   4. $(x - 1)(x + 5)$.

5. $(x - 3)(x - 2)$.

In Problems 6–33, factor the given expression into polynomials.

6. $3x^6 - x^3$.   7. $2x^3 + x$.

8. $5x^2 - 10x$.   9. $2x^2 + 4x$.

10. $x^5 - 2x^3 + 5x^2$.   11. $3x^4 + 12x^2 - 9x$.

12. $x^2 - 36$.   13. $x^2 - 100$.

14. $4x^2 - 49$.   15. $x^2 - 2$.

16. $x^2 - 12$.   17. $3x^2 - 27$.

18. $x^3 - 36x$.   19. $x^3 - 16x$.

20. $4x^4 - 4x^2$.   21. $x^3 - 2x$.

22. $x^2 - 8x + 12$.   23. $x^2 + 6x + 5$.

24. $x^2 - 8x + 15$.   25. $x^2 - 8x + 7$.

26. $x^2 + 7x + 12$.   27. $x^2 - 7x + 12$.

28. $x^2 - 4x - 12$.   29. $x^2 + 4x - 12$.

30. $x^2 + 2x - 15$.   31. $x^2 - 2x - 15$.

32. $6x^2 - 5x + 1$.   33. $5x^2 - 11x + 2$.

In Problems 34–41, solve for x.

34. $2x - 8 = 0$.   **35.** $x + 2 = 10 - 3x$.

36. $20 - 5x^2 = 0$.   **37.** $3x^2 = 12x$.

38. $x^2 + x = 0$.   **39.** $x^2 - 10x = 0$.

40. $x^2 - 8x + 12 = 0$.   **41.** $x^2 - 8x + 15 = 0$.

42. Solve for t: $t^2 - 4t - 12 = 0$.

43. Use the quadratic formula to solve the equation $x^2 + 5x + 2 = 0$.

Answers to Self-Test II.   **1.** $2x^6 - 8x^5 + 4x^4 - 2x^2 + 2x$.   **2.** $x^2 + 4x + 3$.
3. $x^2 + 2x - 8$.   **4.** $x^2 + 4x - 5$.   **5.** $x^2 - 5x + 6$.   **6.** $x^3(3x^3 - 1)$.
7. $x(2x^2 + 1)$.   **8.** $5x(x - 2)$.   **9.** $2x(x + 2)$.   **10.** $x^2(x^3 - 2x + 5)$.
11. $3x(x^3 + 4x - 3)$.   **12.** $(x + 6)(x - 6)$.   **13.** $(x + 10)(x - 10)$.
14. $(2x + 7)(2x - 7)$, or $4(x + \frac{7}{2})(x - \frac{7}{2})$.   **15.** $(x + \sqrt{2})(x - \sqrt{2})$.
16. $(x + \sqrt{12})(x - \sqrt{12})$, or $(x + 2\sqrt{3})(x - 2\sqrt{3})$.   **17.** $3(x + 3)(x - 3)$.
18. $x(x + 6)(x - 6)$.   **19.** $x(x + 4)(x - 4)$.   **20.** $4x^2(x + 1)(x - 1)$.
21. $x(x + \sqrt{2})(x - \sqrt{2})$.   **22.** $(x - 2)(x - 6)$.   **23.** $(x + 1)(x + 5)$.
24. $(x - 5)(x - 3)$.   **25.** $(x - 7)(x - 1)$.   **26.** $(x + 3)(x + 4)$.
27. $(x - 3)(x - 4)$.   **28.** $(x + 2)(x - 6)$.   **29.** $(x - 2)(x + 6)$.
30. $(x - 3)(x + 5)$.   **31.** $(x + 3)(x - 5)$.   **32.** $(2x - 1)(3x - 1)$.
33. $(5x - 1)(x - 2)$.   **34.** $x = 4$.   **35.** $x = 2$.   **36.** $x = \pm 2$.
37. $x = 0, 4$.   **38.** $x = 0, -1$.   **39.** $x = 0, 10$.   **40.** $x = 2, 6$.
41. $x = 3, 5$.   **42.** $t = -2, 6$.   **43.** $x = \dfrac{-5 + \sqrt{17}}{2}, \dfrac{-5 - \sqrt{17}}{2}$.

appendix two

Graphing: Curve Catalog

Slope and Distance. Let P_1 and P_2 be points in the plane with coordinates

$$P_1: (x_1, y_1),\ P_2: (x_2, y_2).$$

If the unit distances on the horizontal and vertical axes are equal (see Figure A-3), then the following hold.

---

slope of $\overline{P_1P_2}$ is $\dfrac{y_2 - y_1}{x_2 - x_1}$

length of $\overline{P_1P_2}$ is $\sqrt{(x_2 - x_1)^2 + (y_2 - y_1)^2}$

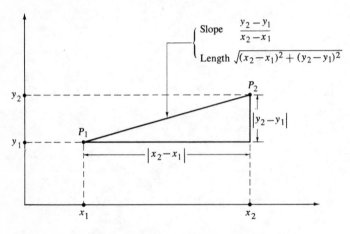

Figure A-3 The slope formula and the distance formula

The Linear Equation $Ax + By = C.$ If $A = 0$ and $B \neq 0$, this equation can be put into the form

$\quad$ $y = b$; a horizontal line.

If $B = 0$ and $A \neq 0$, it can be put into the form

$\quad$ $x = a$; a vertical line.

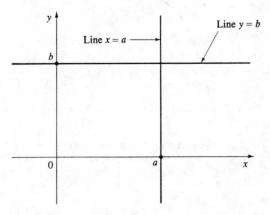

Figure A-4   Equations for horizontal and vertical lines

If $B \neq 0$, it can be put into the form

$\quad$ $y = mx + b$; a line with slope m and y-intercept b.

Finally, we define parallel and perpendicular lines.

The slopes of parallel lines and of perpendicular lines are related as follows.

If line L_1 has slope m_1 and line L_2 has slope m_2, then

$\quad$ $m_1 = m_2$ implies L_1 is parallel to L_2

$\quad$ $m_1 \cdot m_2 = -1$ implies L_1 is perpendicular to L_2

The Parabola $y = ax^2 + bx + c$.   The parabola may open up or down; it always intersects the y-axis, but may intersect the x-axis in two points, one point, or not at all (see Parts a and b of Figure A–5). The **vertex** is the highest or lowest point on the parabola.

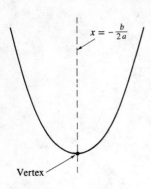

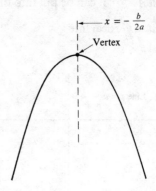

Figure A-5a   Graph of $y = ax^2 + c$
when $a > 0$

Figure A-5b   Graph of $y = ax^2 + bx + c$
when $a < 0$

Parabola $y = ax^2 + bx + c$

$a > 0$: opens upward ($y \to +\infty$ as $x \to \pm\infty$)

$a < 0$: opens downward ($y \to -\infty$ as $x \to \pm\infty$)

x-coordinate of vertex $= -\dfrac{b}{2a}$.

The Circle $x^2 + y^2 = r^2$. The graph of $x^2 + y^2 = r^2$, where r is a positive constant, is a circle centered at the origin with radius r. More generally, if h, k, and $r > 0$ are constants (see Figure A-6). then

$$(x - h)^2 + (y - k)^2 = r^2$$

is a circle with center (h, k) and radius r.

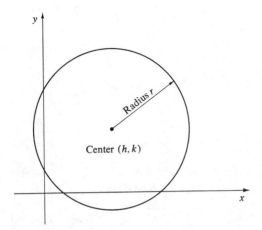

Figure A-6   Graph of $(x - h)^2 + (y - k)^2 = r^2$ $(r > 0)$

The Ellipse $\dfrac{x^2}{a^2} + \dfrac{y^2}{b^2} = 1.$ The center of this ellipse is at the origin and the intercepts are $x = \pm a$ (x-intercepts) and $y = \pm b$ (y-intercepts) (see Figure A–7).

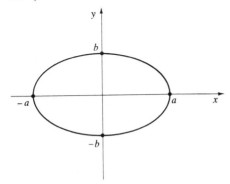

Figure A-7   Graph of $\dfrac{x^2}{a^2} + \dfrac{y^2}{b^2} = 1 (a > 0, b > 0)$

The Hyperbola $y = \dfrac{a}{x} + b.$ The vertical asymptote is $x = 0$ and the horizontal asymptote is $y = b$ (see Figure A–8).

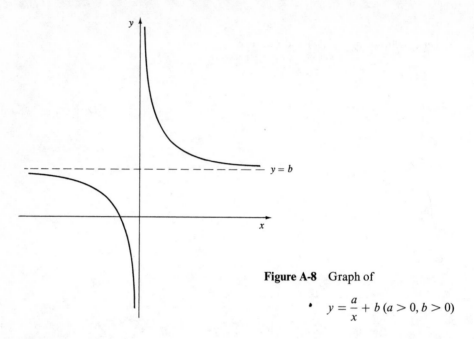

Figure A-8   Graph of

$$y = \frac{a}{x} + b \; (a > 0, b > 0)$$

The Exponential Functions $y = e^x$ *and* $y = e^{-x}.$ The graphs of these functions are shown in Parts a and b of Figure A–9. In $y = e^x$, $y \to 0$ as $x \to -\infty$ and $y \to +\infty$ as $x \to +\infty$. In $y = e^{-x}$, $y \to +\infty$ as $x \to -\infty$ and $y \to 0$ as $x \to +\infty$.

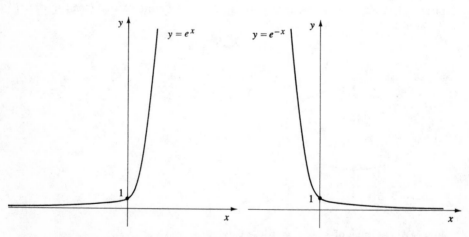

Figure A-9a   Graph of
$y = e^x$

Figure A-9b   Graph of
$y = e^{-x}$

The Natural Logarithm $y = \ln x$. The graph is shown in Figure A–10; $y \to -\infty$ as $x \to 0+$ and $y \to +\infty$ as $x \to +\infty$.

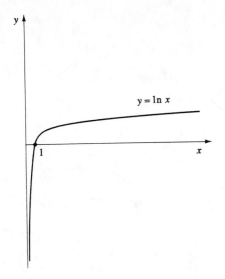

$y = \ln x$

Figure A-10 Graph of $y = \ln x$

The Logistic Function. Figure A–11 shows the graph of the logistic function

$$y = \frac{B}{1 + Ae^{-kx}}, \qquad A > 0, B > 0, \text{ and } k > 0.$$

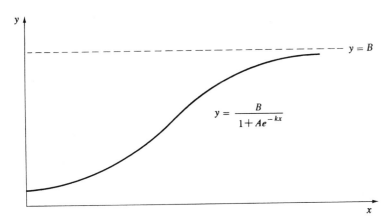

$y = B$

$y = \dfrac{B}{1 + Ae^{-kx}}$

Figure A-11 Graph of $y = \dfrac{B}{1 + Ae^{-kx}}$ $(B > 0, A > 0, k > 0)$

Miscellaneous Functions. Figures A–12 through A–15 show graphs for

$$y = |x|, \qquad y = x^{2/3}, \qquad y = e^{-x^2},$$

and the Gompertz function

$$y = ae^{-be^{kt}}, \qquad a > 0, b > 0, \text{ and } k > 0.$$

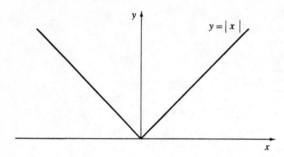

Figure A-12   Graph of
$y = |x|$

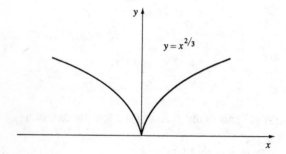

Figure A-13   Graph of
$y = x^{2/3}$

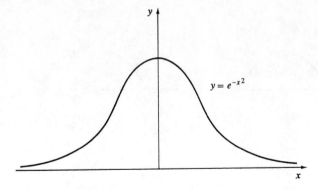

Figure A-14   Graph of $y = e^{-x^2}$

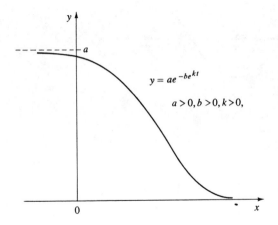

$y = ae^{-be^{kt}}$

$a > 0, b > 0, k > 0,$

Figure A-15   Graph of the Gompertz function

The functions

$$y = \sin x, \qquad y = \cos x, \qquad y = \tan x,$$
$$y = \arcsin x, \qquad y = \arccos x, \qquad y = \arctan x,$$

though not discussed in this book, are used in certain applications of calculus and are needed to evaluate some integrals (see the Table of Integrals on pages 388–392); the graphs of these functions are shown in Figures A–16 to A–21.

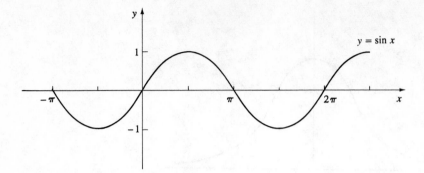

Figure A-16 Graph of $y = \sin x$

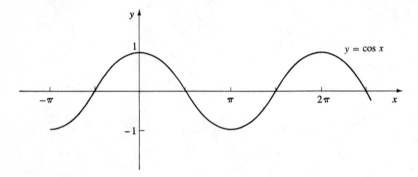

Figure A-17 Graph of $y = \cos x$

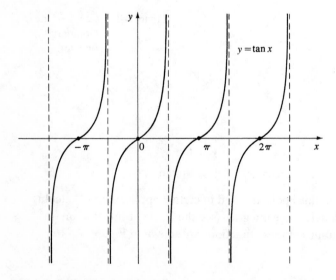

Figure A-18 Graph of $y = \tan x$

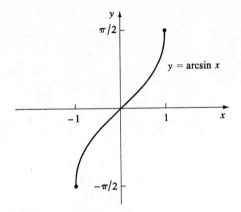

Figure A-19 Graph of $y = \arcsin x$

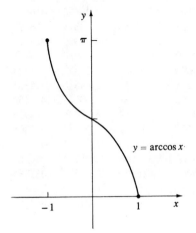

Figure A-20 Graph of $y = \arccos x$

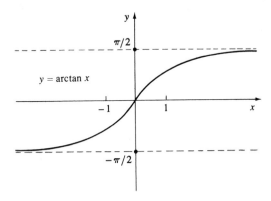

Figure A-21 Graph of $y = \arctan x$

Table 1 Squares, Square Roots, and Reciprocals

n	n^2	$\sqrt{n}$	$\sqrt{10n}$	$\dfrac{1}{n}$	n	n^2	$\sqrt{n}$	$\sqrt{10n}$	$\dfrac{1}{n}$
1	1	1.000000	3.162278	1.000000	51	2601	7.141428	22.58318	.019608
2	4	1.414214	4.472136	.500000	52	2704	7.211103	22.80351	.019231
3	9	1.732051	5.477226	.333333	53	2809	7.280110	23.02173	.018868
4	16	2.000000	6.324555	.250000	54	2916	7.348469	23.23790	.018519
5	25	2.236068	7.071068	.200000	55	3025	7.416198	23.45208	.018182
6	36	2.449490	7.745967	.166667	56	3136	7.483315	23.66432	.017857
7	49	2.645751	8.366600	.142857	57	3249	7.549834	23.87467	.017544
8	64	2.828427	8.944272	.125000	58	3364	7.615773	24.08319	.017241
9	81	3.000000	9.486833	.111111	59	3481	7.681146	24.28992	.016949
10	100	3.162278	10.00000	.100000	60	3600	7.745967	24.49490	.016667
11	121	3.316625	10.48809	.090909	61	3721	7.810250	24.69818	.016393
12	144	3.464102	10.95445	.083333	62	3844	7.874008	24.89980	.016129
13	169	3.605551	11.40175	.076923	63	3969	7.937254	25.09980	.015873
14	196	3.741657	11.83216	.071429	64	4096	8.000000	25.29822	.015625
15	225	3.872983	12.24745	.066667	65	4225	8.062258	25.49510	.015385
16	256	4.000000	12.64911	.062500	66	4356	8.124038	25.69047	.015152
17	289	4.123106	13.03840	.058824	67	4489	8.185353	25.88436	.014925
18	324	4.242641	13.41641	.055556	68	4624	8.246211	26.07681	.014706
19	361	4.358899	13.78405	.052632	69	4761	8.306624	26.26785	.014493
20	400	4.472136	14.14214	.050000	70	4900	8.366600	26.45751	.014286
21	441	4.582576	14.49138	.047619	71	5041	8.426150	26.64583	.014085
22	484	4.690416	14.83240	.045455	72	5184	8.485281	26.83282	.013889
23	529	4.795832	15.16575	.043478	73	5329	8.544004	27.01851	.013699
24	576	4.898979	15.49193	.041667	74	5476	8.602325	27.20294	.013514
25	625	5.000000	15.81139	.040000	75	5625	8.660254	27.38613	.013333
26	676	5.099020	16.12452	.038462	76	5776	8.717798	27.56810	.013158
27	729	5.196152	16.43168	.037037	77	5929	8.774964	27.74887	.012987
28	784	5.291503	16.73320	.035714	78	6084	8.831761	27.92848	.012821
29	841	5.385165	17.02939	.034483	79	6241	8.888194	28.10694	.012658
30	900	5.477226	17.32051	.033333	80	6400	8.944272	28.28427	.012500
31	961	5.567764	17.60682	.032258	81	6561	9.000000	28.46050	.012346
32	1024	5.656854	17.88854	.031250	82	6724	9.055385	28.63564	.012195
33	1089	5.744563	18.16590	.030303	83	6889	9.110434	28.80972	.012048
34	1156	5.830952	18.43909	.029412	84	7056	9.165151	28.98275	.011905
35	1225	5.916080	18.70829	.028571	85	7225	9.219544	29.15476	.011765
36	1296	6.000000	18.97367	.027778	86	7396	9.273618	29.32576	.011628
37	1369	6.082763	19.23538	.027027	87	7569	9.327379	29.49576	.011494
38	1444	6.164414	19.49359	.026316	88	7744	9.380832	29.66479	.011364
39	1521	6.244998	19.74842	.025641	89	7921	9.433981	29.83287	.011236
40	1600	6.324555	20.00000	.025000	90	8100	9.486833	30.00000	.011111
41	1681	6.403124	20.24846	.024390	91	8281	9.539392	30.16621	.010989
42	1764	6.480741	20.49390	.023810	92	8464	9.591663	30.33150	.010870
43	1849	6.557439	20.73644	.023256	93	8649	9.643651	30.49590	.010753

Table 1 Squares, Square Roots, and Reciprocals (*continued*)

n	n^2	$\sqrt{n}$	$\sqrt{10n}$	$\dfrac{1}{n}$	n	n^2	$\sqrt{n}$	$\sqrt{10n}$	$\dfrac{1}{n}$
44	1936	6.633250	20.97618	.022727	94	8836	9.695360	30.65942	.010638
45	2025	6.708204	21.21320	.022222	95	9025	9.746794	30.82207	.010526
46	2116	6.782330	21.44761	.021739	96	9216	9.797959	30.98387	.010417
47	2209	6.855655	21.67948	.021277	97	9409	9.848858	31.14482	.010309
48	2304	6.928203	21.90890	.020833	98	9604	9.899495	31.30495	.010204
49	2401	7.000000	22.13594	.020408	99	9801	9.949874	31.46427	.010101
50	2500	7.071068	22.36068	.020000	100	10000	10.00000	31.62278	.010000

Table 2 Exponential Function. Natural (Napierian) Logarithm Function

x	e^x	$\ln x$	x	e^x	$\ln x$
0	1.0000	$-\infty$	1.9	6.6859	.6419
.1	1.1052	-2.303	2.0	7.3891	.6931
.2	1.2214	-1.609	2.1	8.1662	.7419
.3	1.3499	-1.204	2.2	9.0250	.7885
.4	1.4918	$-.916$	2.3	9.9742	.8329
.5	1.6487	$-.693$	2.4	11.023	.8755
.6	1.8221	$-.511$	2.5	12.182	.9163
.7	2.0138	$-.357$	2.6	13.464	.9555
.8	2.2255	$-.223$	2.7	14.880	.9933
.9	2.4596	$-.105$	2.8	16.445	1.0296
1.0	2.7183	.0000	2.9	18.174	1.0647
1.1	3.0042	.0953	3	20.086	1.0986
1.2	3.3201	.1823	4	54.598	1.3863
1.3	3.6693	.2624	5	148.41	1.6094
1.4	4.0552	.3365	6	403.4	1.7918
1.5	4.4817	.4055	7	1096.6	1.9459
1.6	4.9530	.4700	8	2981.0	2.0794
1.7	5.4739	.5306	9	8103.1	2.1972
1.8	6.0496	.5878	10	22026.0	2.3026

Table 3   Common (Base 10) Logarithm Function, Log n

n	n	$n.1$	$n.2$	$n.3$	$n.4$	$n.5$	$n.6$	$n.7$	$n.8$	$n.9$
10	1.0000	1.0043	1.0086	1.0128	1.0170	1.0212	1.0253	1.0294	1.0334	1.0374
11	1.0414	1.0453	1.0492	1.0531	1.0569	1.0607	1.0645	1.0682	1.0719	1.0755
12	1.0792	1.0828	1.0864	1.0899	1.0934	1.0969	1.1004	1.1038	1.1072	1.1106
13	1.1139	1.1173	1.1206	1.1239	1.1271	1.1303	1.1335	1.1367	1.1399	1.1430
14	1.1461	1.1492	1.1523	1.1553	1.1584	1.1614	1.1644	1.1673	1.1703	1.1732
15	1.1761	1.1790	1.1818	1.1847	1.1875	1.1903	1.1931	1.1959	1.1987	1.2014
16	1.2041	1.2068	1.2095	1.2122	1.2148	1.2175	1.2201	1.2227	1.2253	1.2279
17	1.2304	1.2330	1.2355	1.2380	1.2405	1.2430	1.2455	1.2480	1.2504	1.2529
18	1.2553	1.2577	1.2601	1.2625	1.2648	1.2672	1.2695	1.2718	1.2742	1.2765
19	1.2788	1.2810	1.2833	1.2856	1.2878	1.2900	1.2923	1.2945	1.2967	1.2989
20	1.3010	1.3032	1.3054	1.3075	1.3096	1.3118	1.3139	1.3160	1.3181	1.3201
21	1.3222	1.3243	1.3263	1.3284	1.3304	1.3324	1.3345	1.3365	1.3385	1.3404
22	1.3424	1.3444	1.3464	1.3483	1.3502	1.3522	1.3541	1.3560	1.3579	1.3598
23	1.3617	1.3636	1.3655	1.3674	1.3692	1.3711	1.3729	1.3747	1.3766	1.3784
24	1.3802	1.3820	1.3838	1.3856	1.3874	1.3892	1.3909	1.3927	1.3945	1.3962
25	1.3979	1.3997	1.4014	1.4031	1.4048	1.4065	1.4082	1.4099	1.4116	1.4133
26	1.4150	1.4166	1.4183	1.4200	1.4216	1.4232	1.4249	1.4265	1.4281	1.4298
27	1.4314	1.4330	1.4346	1.4362	1.4378	1.4393	1.4409	1.4425	1.4440	1.4456
28	1.4472	1.4487	1.4502	1.4518	1.4533	1.4548	1.4564	1.4579	1.4594	1.4609
29	1.4624	1.4639	1.4654	1.4669	1.4683	1.4698	1.4713	1.4728	1.4742	1.4757
30	1.4771	1.4786	1.4800	1.4814	1.4829	1.4843	1.4857	1.4871	1.4886	1.4900
31	1.4914	1.4928	1.4942	1.4955	1.4969	1.4983	1.4997	1.5011	1.5024	1.5038
32	1.5051	1.5065	1.5079	1.5092	1.5105	1.5119	1.5132	1.5145	1.5159	1.5172
33	1.5185	1.5198	1.5211	1.5224	1.5237	1.5250	1.5263	1.5276	1.5289	1.5302
34	1.5315	1.5328	1.5340	1.5353	1.5366	1.5378	1.5391	1.5403	1.5416	1.5428
35	1.5441	1.5453	1.5465	1.5478	1.5490	1.5502	1.5514	1.5527	1.5539	1.5551
36	1.5563	1.5575	1.5587	1.5599	1.5611	1.5623	1.5635	1.5647	1.5658	1.5670
37	1.5682	1.5694	1.5705	1.5717	1.5729	1.5740	1.5752	1.5763	1.5775	1.5786
38	1.5798	1.5809	1.5821	1.5832	1.5843	1.5855	1.5866	1.5877	1.5888	1.5899
39	1.5911	1.5922	1.5933	1.5944	1.5955	1.5966	1.5977	1.5988	1.5999	1.6010
40	1.6021	1.6031	1.6042	1.6053	1.6064	1.6075	1.6085	1.6096	1.6107	1.6117
41	1.6128	1.6138	1.6149	1.6160	1.6170	1.6180	1.6191	1.6201	1.6212	1.6222
42	1.6232	1.6243	1.6253	1.6263	1.6274	1.6284	1.6294	1.6304	1.6314	1.6325
43	1.6335	1.6345	1.6355	1.6365	1.6375	1.6385	1.6395	1.6405	1.6415	1.6425
44	1.6435	1.6444	1.6454	1.6464	1.6474	1.6484	1.6493	1.6503	1.6513	1.6522
45	1.6532	1.6542	1.6551	1.6561	1.6571	1.6580	1.6590	1.6599	1.6609	1.6618
46	1.6628	1.6637	1.6646	1.6656	1.6665	1.6675	1.6684	1.6693	1.6702	1.6712
47	1.6721	1.6730	1.6739	1.6749	1.6758	1.6767	1.6776	1.6785	1.6794	1.6803
48	1.6812	1.6821	1.6830	1.6839	1.6848	1.6857	1.6866	1.6875	1.6884	1.6893
49	1.6902	1.6911	1.6920	1.6928	1.6937	1.6946	1.6955	1.6964	1.6972	1.6981
50	1.6990	1.6998	1.7007	1.7016	1.7024	1.7033	1.7042	1.7050	1.7059	1.7067
51	1.7076	1.7084	1.7093	1.7101	1.7110	1.7118	1.7126	1.7135	1.7143	1.7152
52	1.7160	1.7168	1.7177	1.7185	1.7193	1.7202	1.7210	1.7218	1.7226	1.7235
53	1.7243	1.7251	1.7259	1.7267	1.7275	1.7284	1.7292	1.7300	1.7308	1.7316
54	1.7324	1.7332	1.7340	1.7348	1.7356	1.7364	1.7372	1.7380	1.7388	1.7396

Table 3   Common (Base 10) Logarithm Function, Log *n* (*continued*)

n	n	n.1	n.2	n.3	n.4	n.5	n.6	n.7	n.8	n.9
						log				
55	1.7404	1.7412	1.7419	1.7427	1.7435	1.7443	1.7451	1.7459	1.7466	1.7474
56	1.7482	1.7490	1.7497	1.7505	1.7513	1.7520	1.7528	1.7536	1.7543	1.7551
57	1.7559	1.7566	1.7574	1.7582	1.7589	1.7597	1.7604	1.7612	1.7619	1.7627
58	1.7634	1.7642	1.7649	1.7657	1.7664	1.7672	1.7679	1.7686	1.7694	1.7701
59	1.7709	1.7716	1.7723	1.7731	1.7738	1.7745	1.7752	1.7760	1.7767	1.7774
60	1.7782	1.7789	1.7796	1.7803	1.7810	1.7818	1.7825	1.7832	1.7839	1.7846
61	1.7853	1.7860	1.7868	1.7875	1.7882	1.7889	1.7896	1.7903	1.7910	1.7917
62	1.7924	1.7931	1.7938	1.7945	1.7952	1.7959	1.7966	1.7973	1.7980	1.7987
63	1.7993	1.8000	1.8007	1.8014	1.8021	1.8028	1.8035	1.8041	1.8048	1.8055
64	1.8062	1.8069	1.8075	1.8082	1.8089	1.8096	1.8102	1.8109	1.8116	1.8122
65	1.8129	1.8136	1.8142	1.8149	1.8156	1.8162	1.8169	1.8176	1.8182	1.8189
66	1.8195	1.8202	1.8209	1.8215	1.8222	1.8228	1.8235	1.8241	1.8248	1.8254
67	1.8261	1.8267	1.8274	1.8280	1.8287	1.8293	1.8299	1.8306	1.8312	1.8319
68	1.8325	1.8331	1.8338	1.8344	1.8351	1.8357	1.8363	1.8370	1.8376	1.8382
69	1.8388	1.8395	1.8401	1.8407	1.8414	1.8420	1.8426	1.8432	1.8439	1.8445
70	1.8451	1.8457	1.8463	1.8470	1.8476	1.8482	1.8488	1.8494	1.8500	1.8506
71	1.8513	1.8519	1.8525	1.8531	1.8537	1.8543	1.8549	1.8555	1.8561	1.8567
72	1.8573	1.8579	1.8585	1.8591	1.8597	1.8603	1.8609	1.8615	1.8621	1.8627
73	1.8633	1.8639	1.8645	1.8651	1.8657	1.8663	1.8669	1.8675	1.8681	1.8686
74	1.8692	1.8698	1.8704	1.8710	1.8716	1.8722	1.8727	1.8733	1.8739	1.8745
75	1.8751	1.8756	1.8762	1.8768	1.8774	1.8779	1.8785	1.8791	1.8797	1.8802
76	1.8808	1.8814	1.8820	1.8825	1.8831	1.8837	1.8842	1.8848	1.8854	1.8859
77	1.8865	1.8871	1.8876	1.8882	1.8887	1.8893	1.8899	1.8904	1.8910	1.8915
78	1.8921	1.8927	1.8932	1.8938	1.8943	1.8949	1.8954	1.8960	1.8965	1.8971
79	1.8976	1.8982	1.8987	1.8993	1.8998	1.9004	1.9009	1.9015	1.9020	1.9025
80	1.9031	1.9036	1.9042	1.9047	1.9053	1.9058	1.9063	1.9069	1.9074	1.9079
81	1.9085	1.9090	1.9096	1.9101	1.9106	1.9112	1.9117	1.9122	1.9128	1.9133
82	1.9138	1.9143	1.9149	1.9154	1.9159	1.9165	1.9170	1.9175	1.9180	1.9186
83	1.9191	1.9196	1.9201	1.9206	1.9212	1.9217	1.9222	1.9227	1.9232	1.9238
84	1.9243	1.9248	1.9253	1.9258	1.9263	1.9269	1.9274	1.9279	1.9284	1.9289
85	1.9294	1.9299	1.9304	1.9309	1.9315	1.9320	1.9325	1.9330	1.9335	1.9340
86	1.9345	1.9350	1.9355	1.9360	1.9365	1.9370	1.9375	1.9380	1.9385	1.9390
87	1.9395	1.9400	1.9405	1.9410	1.9415	1.9420	1.9425	1.9430	1.9435	1.9440
88	1.9445	1.9450	1.9455	1.9460	1.9465	1.9469	1.9474	1.9479	1.9484	1.9489
89	1.9494	1.9499	1.9504	1.9509	1.9513	1.9518	1.9523	1.9528	1.9533	1.9538
90	1.9542	1.9547	1.9552	1.9557	1.9562	1.9566	1.9571	1.9576	1.9581	1.9586
91	1.9590	1.9595	1.9600	1.9605	1.9609	1.9614	1.9619	1.9624	1.9628	1.9633
92	1.9638	1.9643	1.9647	1.9652	1.9657	1.9661	1.9666	1.9671	1.9675	1.9680
93	1.9685	1.9689	1.9694	1.9699	1.9703	1.9708	1.9713	1.9717	1,9722	1.9727
94	1.9731	1.9736	1.9741	1.9745	1.9750	1.9754	1.9759	1.9763	1.9768	1.9773
95	1.9777	1.9782	1.9786	1.9791	1.9795	1.9800	1.9805	1.9809	1.9814	1.9818
96	1.9823	1.9827	1.9832	1.9836	1.9841	1.9845	1.9850	1.9854	1.9859	1.9863
97	1.9868	1.9872	1.9877	1.9881	1.9886	1.9890	1.9894	1.9899	1.9903	1.9908
98	1.9912	1.9917	1.9921	1.9926	1.9930	1.9934	1.9939	1.9943	1.9948	1.9952
99	1.9956	1.9961	1.9965	1.9969	1.9974	1.9978	1.9983	1.9987	1.9991	1.9996

Table 4 Table of Integrals

1. $\displaystyle\int x^n \, dx = \frac{x^{n+1}}{n+1} + C, n \neq -1.$

2. $\displaystyle\int \frac{dx}{x} = \ln x + C.$

3. $\displaystyle\int a^x \, dx = \frac{a^x}{\ln a} + C.$

4. $\displaystyle\int \frac{dx}{x^2 + a^2} = \frac{1}{a} \arctan \frac{x}{a} + C.$

5. $\displaystyle\int \frac{dx}{x^2 - a^2} = \frac{1}{2a} \ln \frac{x-a}{x+a} + C, \text{if } x^2 > a^2.$

$\displaystyle\int \frac{dx}{x^2 - a^2} = \frac{1}{2a} \ln \frac{a-x}{a+x} + C, \text{if } x^2 < a^2.$

6. $\displaystyle\int \frac{dx}{\sqrt{a^2 - x^2}} = \arcsin \frac{x}{a} + C.$

7. $\displaystyle\int \frac{dx}{\sqrt{x^2 \pm a^2}} = \ln\left(x + \sqrt{x^2 \pm a^2}\right) + C.$

8. $\displaystyle\int \frac{dx}{x(a + bx)} = \frac{1}{a} \ln \frac{x}{a+bx} + C.$

9. $\displaystyle\int \frac{dx}{x^2(a + bx)} = -\frac{1}{ax} + \frac{b}{a^2} \ln \frac{a+bx}{x} + C.$

10. $\displaystyle\int \frac{dx}{x(a + bx)^2} = \frac{1}{a(a+bx)} - \frac{1}{a^2} \ln \frac{a+bx}{x} + C.$

11. $\displaystyle\int x\sqrt{a + bx} \, dx = \frac{2(3bx - 2a)\sqrt{(a+bx)^3}}{15b^2} + C.$

12. $\displaystyle\int \frac{x \, dx}{\sqrt{a + bx}} = \frac{2(bx - 2a)\sqrt{a+bx}}{3b^2} + C.$

13. $\displaystyle\int x^2\sqrt{a + bx} \, dx = \frac{2(15b^2x^2 - 12abx + 8a^2)\sqrt{(a+bx)^3}}{105b^3} + C.$

14. $\displaystyle\int \frac{x^2 \, dx}{\sqrt{a + bx}} = \frac{2(3b^2x^2 - 4abx + 8a^2)\sqrt{a+bx}}{15b^3} + C.$

15. $\displaystyle\int \frac{dx}{x\sqrt{a + bx}} = \frac{1}{\sqrt{a}} \ln \frac{\sqrt{a+bx} - \sqrt{a}}{\sqrt{a+bx} + \sqrt{a}} + C, \text{if } a > 0.$

$\displaystyle\int \frac{dx}{x\sqrt{a + bx}} = \frac{2}{\sqrt{-a}} \arctan \sqrt{\frac{a+bx}{-a}} + C, \text{if } a < 0.$

16. $\displaystyle\int \frac{\sqrt{a + bx} \, dx}{x} = 2\sqrt{a+bx} + a \int \frac{dx}{x\sqrt{a+bx}} + C.$

17. $\displaystyle\int \frac{dx}{x^2\sqrt{a + bx}} = -\frac{\sqrt{a+bx}}{ax} - \frac{b}{2a} \int \frac{dx}{x\sqrt{a+bx}} + C.$

18. $\displaystyle\int \sqrt{a^2 - x^2} \, dx = \frac{1}{2}\left(x\sqrt{a^2 - x^2} + a^2 \arcsin \frac{x}{a}\right) + C.$

Table 4   Table of Integrals (*continued*)

19. $\displaystyle\int x\sqrt{a^2-x^2}\,dx = -\tfrac{1}{3}(a^2-x^2)^{3/2}+C.$

20. $\displaystyle\int x^2\sqrt{a^2-x^2}\,dx = \frac{x}{8}(2x^2-a^2)\sqrt{a^2-x^2}+\frac{a^4}{8}\arcsin\frac{x}{a}+C.$

21. $\displaystyle\int \frac{x\,dx}{\sqrt{a^2-x^2}} = -\sqrt{a^2-x^2}+C.$

22. $\displaystyle\int \frac{x^2\,dx}{\sqrt{a^2-x^2}} = -\frac{x}{2}\sqrt{a^2-x^2}+\frac{a^2}{2}\arcsin\frac{x}{a}+C.$

23. $\displaystyle\int (a^2-x^2)^{3/2}\,dx = \frac{x}{8}(5a^2-2x^2)\sqrt{a^2-x^2}+\frac{3a^4}{8}\arcsin\frac{x}{a}+C.$

24. $\displaystyle\int \frac{dx}{(a^2-x^2)^{3/2}} = \frac{x}{a^2\sqrt{a^2-x^2}}+C.$

25. $\displaystyle\int \frac{x\,dx}{(a^2-x^2)^{3/2}} = \frac{1}{\sqrt{a^2-x^2}}+C.$

26. $\displaystyle\int \frac{x^2\,dx}{(a^2-x^2)^{3/2}} = \frac{x}{\sqrt{a^2-x^2}}-\arcsin\frac{x}{a}+C.$

27. $\displaystyle\int \frac{dx}{x\sqrt{a^2-x^2}} = \frac{1}{a}\ln\frac{a-\sqrt{a^2-x^2}}{x}+C.$

28. $\displaystyle\int \frac{dx}{x^2\sqrt{a^2-x^2}} = -\frac{\sqrt{a^2-x^2}}{a^2x}+C.$

29. $\displaystyle\int \frac{dx}{x^3\sqrt{a^2-x^2}} = -\frac{\sqrt{a^2-x^2}}{2a^2x^2}+\frac{1}{2a^3}\ln\frac{a-\sqrt{a^2-x^2}}{x}+C.$

30. $\displaystyle\int \frac{\sqrt{a^2-x^2}}{x}\,dx = \sqrt{a^2-x^2}-a\ln\frac{a+\sqrt{a^2-x^2}}{x}+C.$

31. $\displaystyle\int \frac{\sqrt{a^2-x^2}}{x^2}\,dx = \frac{-\sqrt{a^2-x^2}}{x}-\arcsin\frac{x}{a}+C.$

32. $\displaystyle\int \sqrt{x^2\pm a^2}\,dx = \tfrac{1}{2}[x\sqrt{x^2\pm a^2}\pm a^2\ln(x+\sqrt{x^2\pm a^2})]+C.$

33. $\displaystyle\int x\sqrt{x^2\pm a^2}\,dx = \tfrac{1}{3}(x^2\pm a^2)^{3/2}+C.$

34. $\displaystyle\int x^2\sqrt{x^2\pm a^2}\,dx = \frac{x}{8}(2x^2\pm a^2)\sqrt{x^2\pm a^2}-\frac{a^4}{8}\ln(x+\sqrt{x^2\pm a^2})+C.$

35. $\displaystyle\int \frac{x\,dx}{\sqrt{x^2\pm a^2}} = \sqrt{x^2\pm a^2}+C.$

36. $\displaystyle\int \frac{x^2\,dx}{\sqrt{x^2\pm a^2}} = \frac{1}{2}[x\sqrt{x^2\pm a^2}\mp a^2\ln(x+\sqrt{x^2\pm a^2})]+C.$

37. $\displaystyle\int (x^2\pm a^2)^{3/2}\,dx = \frac{x}{8}(2x^2+5a^2)\sqrt{x^2\pm a^2}+\frac{3a^4}{8}\ln(x+\sqrt{x^2\pm a^2})+C.$

38. $\displaystyle\int \frac{dx}{(x^2\pm a^2)^{3/2}} = \frac{\pm x}{a^2\sqrt{x^2\pm a^2}}+C.$

Table 4 Table of Integrals (*continued*)

39. $\displaystyle\int \frac{x\,dx}{(x^2 \pm a^2)^{3/2}} = \frac{-1}{\sqrt{x^2 \pm a^2}} + C.$

40. $\displaystyle\int \frac{x^2\,dx}{(x^2 \pm a^2)^{3/2}} = -\frac{x}{\sqrt{x^2 \pm a^2}} + \ln\left(x + \sqrt{x^2 \pm a^2}\right) + C.$

41. $\displaystyle\int \frac{dx}{x^2\sqrt{x^2 \pm a^2}} = \mp\frac{\sqrt{x^2 \pm a^2}}{a^2 x} + C.$

42. $\displaystyle\int \frac{dx}{x^3\sqrt{x^2 - a^2}} = \frac{\sqrt{x^2 - a^2}}{2a^2 x^2} + \frac{1}{2a^3}\arccos\frac{a}{x} + C.$

43. $\displaystyle\int \frac{\sqrt{x^2 - a^2}}{x}\,dx = \sqrt{x^2 - a^2} - a\arccos\frac{a}{x} + C.$

44. $\displaystyle\int \frac{\sqrt{x^2 \pm a^2}}{x^2}\,dx = -\frac{\sqrt{x^2 \pm a^2}}{x} + \ln\left(x + \sqrt{x^2 \pm a^2}\right) + C.$

45. $\displaystyle\int \frac{dx}{x\sqrt{x^2 + a^2}} = \frac{1}{a}\ln\frac{x}{a + \sqrt{x^2 + a^2}} + C.$

46. $\displaystyle\int \frac{dx}{x\sqrt{x^2 - a^2}} = \frac{1}{a}\arccos\frac{a}{x} + C.$

47. $\displaystyle\int \frac{dx}{x^3\sqrt{x^2 + a^2}} = -\frac{\sqrt{x^2 + a^2}}{2a^2 x^2} + \frac{1}{2a^3}\ln\frac{a + \sqrt{x^2 + a^2}}{x} + C.$

48. $\displaystyle\int \frac{\sqrt{x^2 + a^2}}{x}\,dx = \sqrt{x^2 + a^2} - a\ln\frac{a + \sqrt{x^2 + a^2}}{x} + C.$

49. $\displaystyle\int \sqrt{2ax - x^2}\,dx = \frac{x - a}{2}\sqrt{2ax - x^2} + \frac{a^2}{2}\arcsin\frac{x - a}{a} + C.$

50. $\displaystyle\int \frac{dx}{\sqrt{2ax - x^2}} = 2\arcsin\sqrt{\frac{x}{2a}} + C = \arccos\left(1 - \frac{x}{a}\right) + C.$

51. $\displaystyle\int \frac{x^n\,dx}{\sqrt{2ax - x^2}} = \frac{-x^{n-1}\sqrt{2ax - x^2}}{n} + \frac{a(2n - 1)}{n}\int \frac{x^{n-1}\,dx}{\sqrt{2ax - x^2}} + C.$

52. $\displaystyle\int \frac{dx}{x^n\sqrt{2ax - x^2}} = \frac{\sqrt{2ax - x^2}}{a(1 - 2n)x^n} + \frac{n - 1}{(2n - 1)a}\int \frac{dx}{x^{n-1}\sqrt{2ax - x^2}} + C.$

53. $\displaystyle\int x^n\sqrt{2ax - x^2}\,dx = \frac{-x^{n-1}(2ax - x^2)^{3/2}}{n + 2}$
$$+ \frac{(2n + 1)a}{n + 2}\int x^{n-1}\sqrt{2ax - x^2}\,dx + C.$$

54. $\displaystyle\int \frac{\sqrt{2ax - x^2}}{x^n}\,dx = \frac{(2ax - x^2)^{3/2}}{(3 - 2n)ax^n} + \frac{n - 3}{(2n - 3)a}\int \frac{\sqrt{2ax - x^2}}{x^{n-1}}\,dx + C.$

55. $\displaystyle\int \frac{dx}{(2ax - x^2)^{3/2}} = \frac{x - a}{a^2\sqrt{2ax - x^2}} + C.$

56. $\displaystyle\int \frac{dx}{\sqrt{2ax - x^2}} = \ln\left(x + a + \sqrt{2ax + x^2}\right) + C.$

Table 4   Table of Integrals (*continued*)

57. $\displaystyle\int \frac{dx}{a + bx + cx^2} = \frac{2}{\sqrt{4ac - b^2}} \arctan \frac{2cx + b}{\sqrt{4ac - b^2}} + C.$

58. $\displaystyle\int \frac{dx}{a + bx - cx^2} = \frac{1}{\sqrt{b^2 + 4ac}} \ln \frac{\sqrt{b^2 + 4ac} - b + 2cx}{\sqrt{b^2 + 4ac} + b - 2cx} + C.$

59. $\displaystyle\int \frac{dx}{\sqrt{a + bx - cx^2}} = \frac{1}{\sqrt{c}} \arcsin \frac{2cx - b}{\sqrt{b^2 + 4ac}} + C.$

60. $\displaystyle\int \frac{dx}{\sqrt{a + bx + cx^2}} = \frac{1}{\sqrt{c}} \ln \left(2cx + b + 2\sqrt{c}\, \sqrt{a + bx + cx^2}\right) + C.$

61. $\displaystyle\int \sqrt{a + bx + cx^2}\, dx = \frac{2cx + b}{4c} \sqrt{a + bx + cx^2}$
$$- \frac{b^2 - 4ac}{8c^{3/2}} \ln \left(2cx + b + 2\sqrt{c}\, \sqrt{a + bx + cx^2}\right) + C.$$

62. $\displaystyle\int \sqrt{a + bx - cx^2}\, dx = \frac{2cx - b}{4c} \sqrt{a + bx - cx^2}$
$$+ \frac{b^2 + 4ac}{8c^{3/2}} \arcsin \frac{2cx - b}{\sqrt{b^2 + 4ac}} + C.$$

63. $\displaystyle\int \frac{x\, dx}{\sqrt{a + bx - cx^2}} = \frac{-\sqrt{a + bx - cx^2}}{c} + \frac{b}{2c^{3/2}} \arcsin \frac{2cx - b}{\sqrt{b^2 + 4ac}} + C.$

64. $\displaystyle\int \frac{x\, dx}{\sqrt{a + bx + cx^2}} = \frac{\sqrt{a + bx + cx^2}}{c}$
$$- \frac{b}{2c^{3/2}} \ln \left(2cx + b + 2\sqrt{c}\, \sqrt{a + bx + cx^2}\right) + C.$$

65. $\displaystyle\int x^n e^{ax}\, dx = \frac{x^n e^{ax}}{a} - \frac{n}{a} \int x^{n-1} e^{ax}\, dx + C.$

66. $\displaystyle\int x^n \ln x\, dx = x^{n+1} \left[\frac{\ln x}{n + 1} - \frac{1}{(n + 1)^2}\right] + C.$

67. $\displaystyle\int x^n \ln^m x\, dx = \frac{x^{n+1} \ln^m x}{n + 1} - \frac{m}{n + 1} \int x^n \ln^{m-1} x\, dx + C.$

68. $\displaystyle\int \frac{dx}{a + be^{nx}} = \frac{1}{an}[nx - \ln (a + be^{nx})] + C.$

69. $\displaystyle\int \frac{dx}{ae^{nx} + be^{-nx}} = \frac{1}{n\sqrt{ab}} \arctan \left(e^{nx} \sqrt{\frac{a}{b}}\right) + C.$

70. $\displaystyle\int \frac{xe^x\, dx}{(1 + x)^2} = \frac{e^x}{1 + x} + C.$

Table 4   Table of Integrals (*continued*)

71. $\displaystyle \int x^m (a + bx^n)^p \, dx = \frac{x^{m-n+1}(a + bx^n)^{p+1}}{b(np + m + 1)} - \frac{(m - n + 1)a}{b(np + m + 1)}$

$$\int x^{m-n}(a + bx^n)^p \, dx + C,$$

$$= \frac{x^{m+1}(a + bx^n)^p}{np + m + 1} + \frac{npa}{np + m + 1}$$

$$\int x^m (a + bx^n)^{p-1} \, dx + C,$$

$$= \frac{x^{m+1}(a + bx^n)^{p+1}}{a(m + 1)} - \frac{b(np + n + m + 1)}{a(m + 1)}$$

$$\int x^{m+n}(a + bx^n)^p \, dx + C,$$

$$= \frac{-x^{m+1}(a + bx^n)^{p+1}}{na(p + 1)} + \frac{np + n + m + 1}{na(p + 1)}$$

$$\int x^m (a + bx^n)^{p+1} \, dx + C.$$

Solutions to Odd-Numbered Problems

Chapter 1, Section 1, pages 7–10

1. (a) See Figure 1. (b) About 0.6 hr. (36 min.). (c) See Figure 2.
(d) About 1.5 hours, according to Figure 2 (the answer depends on how the curve is drawn; any answer from about 1.5 to 2.5 hours would be reasonable).
3. (a) About 5 million bu. per mo.. (b) $q = 5$ when $p = 1.5$; $q = 11$ when $p = 2.5$; q increased by about 6.
5. See Figure 3. As x changes from 0 to 0.5, y increases by about 0.7; as x changes from 0.5 to 1.0, y increases by about 0.3.
7. See Figure 4. Jan. 15 to Feb. 15: T increased by about 7°; Nov. 15 to Dec. 15: T decreased by about 8°.
9.

C	F
0	32
50	122
100	212

$F = 77$ when $C = 25$; $F = 167$ when $C = 75$.

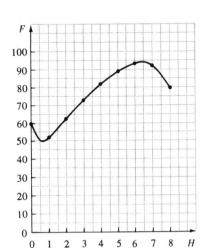

Figure 1

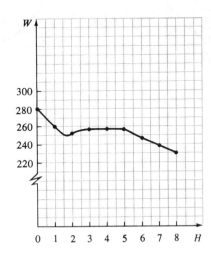

Figure 2

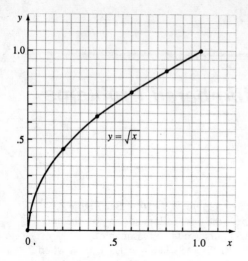

Figure 3

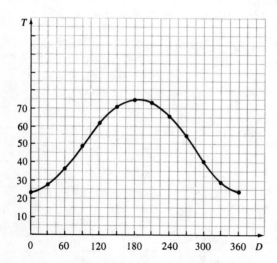

Figure 4

Chapter 1, Section 2, pages 12–13

1. See Figure 5.   **3.** (a) $v = -13$.   (b) $u = 9$.
5. Legs: 3 and 4; Hypotenuse: 5.   **7.** 6.

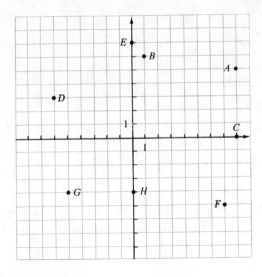

Figure 5

Chapter 1, Section 3, pages 19–21

1. a and d (In the other figures, we can find vertical lines which cross the graphs in more than one point, so they are not the graphs of functions.)

3. Dependent; $s = 2$ when $t = 6$; $s = 4$ when $t = 3$; $s = 1.5$ when $t = 7$.

5. y increases by 3 (from $y = 1$ to $y = 4$); maximum of y is $y = 4$; this maximum occurs when $x = 7$.

7. $y = 12$; it occurs when $x = 10$.   **9.** $R = 170$; it occurs when $p = 10$.

11. $E = 46$; it occurs when $t = 10$.   **13.** $L = 60$; it occurs when $x = 5$.

15. $y = 25$; it occurs when $x = 5$.   **17.** $y = 16$; it occurs when $x = 4$.

19. $y = 25$; it occurs when $x = 7$.   **21.** $s = 4$; it occurs when $t = 12$.

23. $w = 8$; it occurs when $u = 0$.   **25.** $y \approx 24.2$; it occurs when $x \approx 3.2$.

Chapter 1, Section 4, pages 25–26

1. $h(3) = 14$; $h(100) = 305$; $h(-2) = -1$.   **3.** $f(3) = \frac{8}{3}$; $f(1) = 2$; $f(0) = \frac{5}{2}$.

5. $f(3) - g(2) = 10 - 6 = 4$.   **7.** $(2, 5)$ and $(4, 7)$.

9. $f(x) = 2x + 1$ is perhaps the simplest answer.

11. See Figure 6; $f(x)$ is minimum when $x = 1$. The value of the minimum is 1.

13. (a) $f(4) = 9$.   (b) $f(1) = 3$.   (c) $f(4) - f(1) = 9 - 3 = 6$.

(d) y increases by 6.   (e) True.

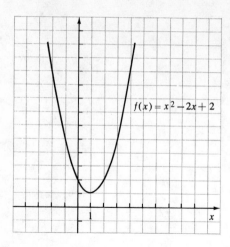

$$f(x) = x^2 - 2x + 2$$

Figure 6

Chapter 1, Section 5, pages 31–32

1. $\dfrac{\Delta y}{\Delta x} = \dfrac{60}{2} = 30.$ **3.** $\dfrac{\Delta Q}{\Delta P} = \dfrac{-93 - (-3)}{4 - 1} = \dfrac{-90}{3} = -30.$

5. $\dfrac{f(5) - f(4)}{5 - 4} = \dfrac{29 - 23}{1} = 6.$ **7.** $\dfrac{\Delta s}{\Delta t} = \dfrac{64 - 36}{0.5} = 56.$

9. $\dfrac{\Delta y}{\Delta x} = \dfrac{11 - 7}{18 - 12} = \dfrac{2}{3}$ **11.** $\dfrac{\Delta D}{\Delta x} = \dfrac{129 - 17}{4 - 2} = 56.$

13. $\dfrac{\Delta c}{\Delta P} = \dfrac{43.5 - 20}{0.5} = 47$ mB/DOE's per trillion dollars.

15. (a) $\dfrac{\Delta h}{\Delta t} = \dfrac{23 - 15}{9 - 4} = \dfrac{8}{5}$, or 1.6 in. per da.

  (b) $\dfrac{\Delta h}{\Delta t} = \dfrac{79 - 71}{100 - 81} = \dfrac{8}{19}$, or about 0.42 in. per da.

17. $\dfrac{\Delta N}{\Delta c} = \dfrac{10 - 16.67}{20,000 - 10,000} = 0.000667$ bacterium per dollar.

19. $\dfrac{\Delta p}{\Delta t} = \dfrac{53 - 15}{4 - 2} = \dfrac{38}{2} = 19$, or \$19 per year.

Chapter 1, Section 6, pages 36–37

1. L_1: $-\frac{1}{4}$; L_2: $\frac{7}{4}$; L_3: $\frac{3}{8}$; L_4: $\frac{5}{7}$; L_5: 0; L_6: -2.

3. (a) $\dfrac{18-6}{7-3}=3$. (b) $\dfrac{1-7}{2-5}=-2$. (c) $\dfrac{13-10}{6-4}=\dfrac{3}{2}$.

(d) $\dfrac{3-8}{1-8}=\dfrac{5}{7}$. (e) $\dfrac{1-37}{20-2}=-2$. (f) $\dfrac{12-9}{2-7}=-\dfrac{3}{5}$. (g) 0.

(h) undefined; the line is vertical. (i) $\dfrac{-5-(-7)}{8-6}=1$.

(j) $\dfrac{1-(-5)}{2-5}=-2$.

5. $\dfrac{y-7}{3-2}=5$, so $y=12$. **7.** Slope $=\dfrac{\sqrt{25}-\sqrt{4}}{25-4}=\dfrac{1}{7}$.

9. Slope $=\dfrac{3^2-1^2}{3-1}=4$. **11.** Slope $=\dfrac{(1.1)^2-1^2}{1.1-1}=2.1$.

13. Slope $=\dfrac{(1.001)^2-1^2}{1.001-1}=2.001$. **15.** Slope $=\dfrac{(1.00001)^2-1^2}{1.00001-1}=2.00001$.

Chapter 1, Section 7, pages 40–42

1. (a) $\frac{2}{15}$ (or about 0.13) million people per year. (b) 1 million people per year.
(c) 0.4 million people per year. (d) -1 million people per year.

3. $\dfrac{\Delta u}{\Delta t}=40$ billion dollars per year; $\dfrac{\Delta v}{\Delta t}=12.5$ billion dollars per year.

5. Analytic method; when $q=1$, $c=4$ and when $q=3$, $c=22$.
$\dfrac{\Delta c}{\Delta q}=\dfrac{c_2-c_1}{q_2-q_1}=\dfrac{22-4}{3-1}=9$.

Chapter 1, Section 8, pages 47–49

1. 9; $(0,-2)$. **3.** 3; $(0,4)$. **5.** -2; $(0,7)$. **7.** 1; $(0,0)$.
9. 0; $(0,2)$. **11.** $-\frac{1}{3}$; $(0,0)$.
13. (a) $y=6x+3$. (b) $y=-\frac{1}{2}x+2$. (c) $y=x+5$.
(d) $y=-x+9$. (e) $y=3x+1$. (f) $y=7$. (g) $y=-\frac{1}{2}x+8$.
(h) $y=2x+3$.
15. $y=-4000x+35{,}000$.
17. There is no single correct answer because of the subjectivity of "best fitting."
A standard statistical method ("least squares") yields $y=4.13-0.23x$.

Chapter 1, Miscellaneous Problems, pages 49–52

1. (a) $y = 17$.   (b) $f(6) = 17; f(-10) = -7$.

(c)

x	-10	-6	-2	0	2	6	10	14	18
y	-7	5	13	$15\frac{1}{2}$	17	17	13	5	-7

(d) See Figure 7.   (e) $f(12) = 9\frac{1}{2}$.   (f) $x \approx -7.8$ and $x \approx 15.8$.

(g) $x = 4$.   (h) $f(4) = 17\frac{1}{2}$.   (i) $\Delta y = 2$.   (j) $\Delta y = -9$.

3. (a) $\dfrac{\Delta s}{\Delta t} = \dfrac{78 - 66}{6 - 2} = \dfrac{12}{4} = 3$ points per hr.

(b) $\dfrac{\Delta s}{\Delta t} = \dfrac{84 - 66}{14 - 2} = \dfrac{18}{12} = 1.5$ points per hr.

(c) $\dfrac{\Delta s}{\Delta t} = \dfrac{84 - 78}{14 - 6} = \dfrac{6}{8} = \dfrac{3}{4}$ points per hr.

5. (a) $\dfrac{\Delta h}{\Delta t} = \dfrac{112 - 0}{1 - 0} = 112$ ft. per sec.   (b) $\dfrac{\Delta h}{\Delta t} = \dfrac{192 - 112}{2 - 1} = 80$ ft. per sec.

(c) $\dfrac{\Delta h}{\Delta t} = \dfrac{192 - 0}{2 - 0} = 96$ ft. per sec.   (d) $\dfrac{\Delta h}{\Delta t} = \dfrac{256 - 240}{4 - 3} = 16$ ft. per sec.

7. $\dfrac{\Delta v}{\Delta t} = \dfrac{40 - 0}{2 - 0} = 20$ ft. per sec. per sec.;

$\dfrac{\Delta v}{\Delta t} = \dfrac{288 - 40}{4 - 2} = 124$ ft. per sec. per sec.

9. (a) $\dfrac{\Delta B}{\Delta t} = \dfrac{61 - 13}{6 - 3} = 16$ bacteria per hr.

(b) $\dfrac{\Delta B}{\Delta t} = \dfrac{373 - 61}{12 - 6} = 52$ bacteria per hr.

11. $\dfrac{\Delta p}{\Delta l} = \dfrac{80,000 - 45,000}{40 - 30} = 3500$, or \$3500 per ft.   **13.** $l = \frac{3}{2}x$.

15. $\Delta r / \Delta n = 0.0172, r = 0.522 + 0.0172n$.

Chapter 2, Section 1, page 58

1. (a) -3.   (b) 0.   (c) 1.   (d) $\frac{1}{2}$.   (e) 0.   (f) -2.

3. 4.   **5.** 2.

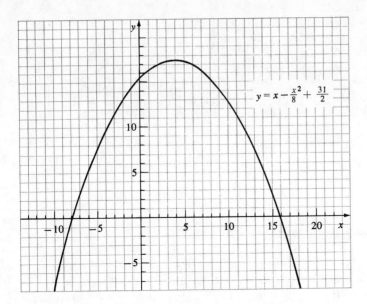

Figure 7

Chapter 2, Section 2, pages 64–65

1. (a) 1. (b) 1.6. (c) 0. (d) $\frac{1}{2}$.
3. (a) 1 g per day. (b) 0 g per day. (c) -1 g per day.
(d) -3 g per day. (e) It is shrinking at a rate of 1 g per day (the rate of change is -1 g per day).
5. (a) 2 million people per yr. (b) 1 million people per yr.
7. (a) About 1.5 sq. in. per day. (b) About 0.25 sq. in. per day.

Chapter 2, Section 3, pages 72–73

1. $dy/dx\,|_{x=x_0} = 2x_0;\ dy/dx\,|_{x=2} = 4.$
3. $dy/dx\,|_{x=x_0} = 10x_0;\ dy/dx\,|_{x=2} = 20.$
5. $dy/dx\,|_{x=x_0} = 10x_0;\ dy/dx\,|_{x=2} = 20.$
7. $dy/dx\,|_{x=x_0} = -2x_0;\ dy/dx\,|_{x=2} = -4.$
9. $dy/dx\,|_{x=x_0} = -10x_0;\ dy/dx\,|_{x=2} = -20.$
11. $dy/dx\,|_{x=x_0} = 3;\ dy/dx\,|_{x=2} = 3.$

13. $dy/dx\big|_{x=x_0} = 3;\ dy/dx\big|_{x=2} = 3.$

15. $dy/dx\big|_{x=x_0} = 16/x_0^2;\ dy/dx\big|_{x=2} = 4.$

17. $dy/dx\big|_{x=x_0} = -3;\ dy/dx\big|_{x=2} = -3.$

19. $dy/dx\big|_{x=x_0} = 0;\ dy/dx\big|_{x=2} = 0.$

21. $dy/dx\big|_{x=x_0} = 0;\ dy/dx\big|_{x=2} = 0.$

23. The slope is $dy/dx\big|_{x=2};\ dy/dx\big|_{x=x_0} = 3x_0^2$, so $dy/dx\big|_{x=2} = 3(2)^2 = 12.$

25. $\dfrac{\Delta q}{\Delta p} = \dfrac{[100 - (p_0 + \Delta p)^2] - [100 - p_0^2]}{\Delta p} = \dfrac{-2p_0\Delta p + (\Delta p)^2}{\Delta p} = -2p_0 + \Delta p,$

so $dq/dp\big|_{p=p_0} = -2p_0.$ (a) $-2p_0.$ (b) $-2(5) = -10.$

Chapter 2, Section 4, pages 79–80.

1. (a) $dy/dx = 35x^{34}.$ (b) $dy/dx = 45x^8.$ (c) $dy/dx = 30x^5 - 12x^3 + 2.$
(d) $dP/dt = 2.64t + 4.06.$ (e) $dC/du = 2u^3 - u^2.$ (f) $dC/dr = 2\pi.$
(g) $dA/dr = 2\pi r.$ (h) $dy/dx = (6x^2 + 1)/3.$ (i) $dy/dt = (10/3)t + (2/5).$
(j) $ds/du = 9u^{26}.$

3. $21(5^2) + 2 = 527.$ **5.** 25.   **7.** $3(0.6)(4^2) - 0.8 = 28.$

9. $f'(x) = 27x^2 - 12, f'(6) = 27 \cdot 6^2 - 12 = 960.$

11. $h'(t) = 36t^2 - 1$ so $h'(1/2) = 36(1/4) - 1 = 8.$

13. $g'(u) = 3u^8 - 2$ so $g'(1) = 1.$

15. (a) $B'(8) = 2500 - 250 \cdot 8 = 500.$ (b) $B'(11) = 2500 - 250 \cdot 11 = -250.$

17. When $x = 2$, we find that $y = 6$ and $dy/dx = 7$. The line through $(2, 6)$ with slope 7 is $y - 6 = 7(x - 2)$ or $y = 7x - 8.$

Chapter 2, Section 5, pages 90–92

1. 4.   **3.** 10.   **5.** 2.   **7.** 6.   **9.** 3.

11. $\lim\limits_{x \to 3} F(x) = \lim\limits_{x \to 3} \dfrac{(x - 3)(x + 3)}{x - 3} = \lim\limits_{x \to 3} [x + 3] = 6 = F(3).$

13. $e/8.$ **15.** (a) 2.   (b) 2.   (c) $+\infty.$ (d) $-\infty.$

17. $+\infty; b.$

Chapter 2, Miscellaneous Problems, pages 92–94

1. (a) $-4.$ (b) $-2.$ (c) $0.$ (d) $2.$ (e) $6.$

3. (a) $\Delta u = \dfrac{1}{x_0 + \Delta x} - \dfrac{1}{x_0}$, and $\dfrac{\Delta u}{\Delta x} = \dfrac{-1}{x_0(x_0 + \Delta x)}$ after simplifying.
Let $\Delta x \to 0$ and obtain $du/dx = -1/x^2$.   (b) $dy/dx = -5/x^2.$

5. (a) $35x^{34}.$ (b) $70x^{34}.$ (c) $15x^4.$ (d) $3x^5 - 3.$ (e) $6.28x - 2.5.$
(f) $4x^5 + \frac{5}{3}.$

7. (a) $dy/dx = 12x^2 - 9x^{14}$, so $dy/dx = 3$ when $y = 1.$
(b) $dC/dp = 8.2p - 9.1$, so $dC/dp = 15.5$ when $p = 3.$ (c) $f'(x) = 32x^7.$
(d) $f'(x) = 15x^4 - 4x^3$, so $f'(2) = 208.$ (e) $G'(t) = \frac{6}{27}t^5 - 4$, so $G'(3) = 50.$

9. (a) $dy/dx = 4x - 1.$ (b) $dy/dx = \frac{40}{7}x^3 + \frac{1}{2}.$ (c) $f'(t) = 8t^3 + 3t^2.$
(d) $g'(t) = 8t - 12.$ (e) $dp/dq = 24q^2 + 24q + 6.$ (f) $dw/dv = 2.$

Chapter 3, Section 1, pages 101–102

1. It is the speed of the car in mi. per hr. 4.5 hours after the beginning of the trip.
3. (a) The velocity v at time t is $v = dh/dt = 0.6t + 0.6$; the velocity at $t = 9$ is
$0.6(9) + 0.6 = 6$ ft. per sec.
(b) Since the velocity is $v = 0.6t + 0.6$, the acceleration is $a = dv/dt = 0.6$. Thus
the acceleration (at any time) is 0.6 ft. per sec. per sec.
5. $C'(x) = 0.04x + 10$, so when $x = 1000$, the marginal cost is $C'(1000) = 40 + 10$
$= \$50$. Since the mfg. can make an extra tire for \$50 and sell it for \$65, he or she
should make at least one extra tire a month and thereby increase the profit.
7. $R(x) = x \cdot p(x) = x(75 - 0.01x) = 75x - 0.01x^2$, so the marginal revenue is
$R'(x) = 75 - 0.02x$. When $x = 1000$, this marginal revenue is $75 - 20$ or \$55.

Chapter 3, Section 2, page 106

1. (a) When $x = 2$, $dy/dx = 6(2) = 12$; y increasing.
(b) When $x = 2$, $dy/dx = 6(2) - 20 = -8$; y decreasing.
(c) When $x = 5$, $dy/dx = 6(5) - 20 = 10$; y increasing.
(d) When $u = 1$, $dw/du = 6(1)^5 = 6$; w increasing.
(e) $P'(q) = 2q - 3$, so $P'(1) = -1$; P decreasing.
(f) $P'(q) = 2q - 3$, so $P'(2) = 1$; P increasing.
(g) $f'(t) = -5$ so $f'(6) = -5$; f decreasing.
(h) $R = 5x - 2x^2$, $dR/dx = 5 - 4x$; when $x = 2$, $dR/dx = 5 - 8 = -3$;
R decreasing.

3. $dM/dx = 2x - 100$, and this is negative if $x < 50$; it is positive when $x > 50$. Therefore M is decreasing if $x < 50$ and increasing if $x > 50$.
5. $x'(t) = 0.16 - 0.02t$, so $x'(6) = 0.04$ and the concentration is increasing; $x'(10) = -0.04$ and the concentration is decreasing.

Chapter 3, Section 3, pages 115–116

1. $x = 3$. **3.** $x = 2$. **5.** $x = 5$. **7.** $x = -1$ and $x = 1$.
9. $q = 0$ and $q = 12$. **11.** Relative maximum of $y = 9$ when $x = 3$.
13. Relative minimum of $y = 26$ when $x = 2$.
15. Relative maximum of $y = 25$ when $x = 5$.
17. Relative maximum of $y = 2$ when $x = -1$; relative minimum of $y = -2$ when $x = 1$.
19. Relative minimum of $y = 0$ when $x = 0$; relative maximum of $y = 96$ when $x = 12$.
21. From $y = 8$ when $x = 2$, $y = 9$ when $x = 3$, and $y = 5$ when $x = 5$, we conclude that the absolute minimum is $y = 5$ and the absolute maximum is $y = 9$.
23. From $y = 46$ when $x = 0$ and $y = 31$ when $x = 1$, we conclude that the absolute minimum is $y = 31$ and the absolute maximum is $y = 46$.
25. From $y = 0$ when $x = 0$, $y = 25$ when $x = 5$, and $y = 24$ when $x = 6$, we conclude that the absolute minimum is $y = 0$ and the absolute maximum is $y = 25$.
27. From $y = 2$ when $x = -1$ and $y = -2$ when $x = 1$, we conclude that the absolute minimum is $y = -2$ and the absolute maximum is $y = 2$.
29. From $P = 0$ when $q = 0$ and $P = 96$ when $q = 12$, we conclude that the absolute minimum is $P = 0$ and the absolute maximum is $P = 96$.
31. He should charge 23¢ a glass. (P attains an absolute maximum when $x = 23$.)

Chapter 3, Section 4, pages 123–124

1. $f''(x) = 12x^2 - 12x; f''(2) = 12(2)^2 - 12(2) = 24$.
3. $F''(s) = 12s - 20; F''(3) = 12(3) = -20 = 16$.
5. When $x = 5$, $d^2y/dx^2 = 6(5) + 8 = 38$.
7. When $t = 6$, $d^2s/dt^2 = 6 - 6(6) = -30$.
9. Vel $= 32t$ and acc $= 32$, so when $t = 4$, the velocity is 128 ft. per sec. and the acceleration is 32 ft. per sec. per sec.
11. Vel $= 6t^2 + 12t + 2$ and acc $= 12t + 12$, so when $t = 4$, the velocity is 146 ft. per sec. and the acceleration is 60 ft. per sec. per sec.
13. When $x = 2$, $d^2y/dx^2 = 12(2)^2 - 12(2) - 12 = 12$; concave upward.
15. $y = x^3 - x^4$; $dy/dx = 3x^2 - 4x^3$; $d^2y/dx^2 = 6x - 12x^2$, so when $x = 3$, $d^2y/dx^2 = 6(3) - 12(3)^2 = -90$; concave downward.

17. When $x = 8$, $d^2y/dx^2 = 24 - 6(8) = -24$; relative maximum.

19. $d^2y/dx^2 = -2$; relative maximum.

21. Let M be the marginal cost for q units. Let the minimum of the U-shaped graph of M occur at $q = q_0$. Then M decreases for $q < q_0$ (so $dM/dq < 0$ there), M has a minimum when $q = q_0$ (so $dM/dq = 0$ there), and M increases for $q > q_0$ (so $dM/dq > 0$ there). Since $M = dC/dq$, the above parenthetical statements can be rephrased as: $d^2C/dq^2 < 0$ when $q < q_0$; $d^2C/dq^2 = 0$ when $q = q_0$; and $d^2C/dq^2 > 0$ when $q > q_0$. Thus C has a point of inflection when $q = q_0$.

Chapter 3, Section 5, pages 131–133

1. The largest corral has an area of 15,000 square yards. It is obtained from a rectangle of sides 100 yards by 150 yards, with a dividing fence of length 100 yards.

3. The length should be $R\sqrt{2}$ and the width should be $R/\sqrt{2}$.

5. The maximum area of $80,000/(4 + \pi)$ square yards occurs when the rectangular part of the garden has dimensions $400/(4 + \pi)$ yards by $800/(4 + \pi)$ yards, and the diameter of the semicircular part lies along the side of length $800/(4 + \pi)$.

7. Sell 600 hamburgers each day at $1.50 each.

9. $1.25. [If tickets are $50 + x$ cents, the total receipts will be

$$r = (50 + x)\left(20,000 - \frac{x}{10} \cdot 1000\right); \ r \text{ has a minimum when } x = 75.]$$

Chapter 3, Miscellaneous Problems, pages 134–135

1. $ds/dt = 3t^2 - 6t + 3$ and $d^2s/dt^2 = 6t - 6$, so when $t = 2$, the velocity is $3(2)^2 - 6(2) + 3 = 3$ ft. per sec. and the acceleration is $6(2) - 6 = 6$ ft. per sec. per sec.

3. $R = (6.4/0.3)q = (1/0.3)q^2$; $dR/dq = 6.4/0.3 - 2q/0.3$; so when $q = 3$, the marginal revenue dR/dq is $(6.4 - 6)/0.3 = 4/3$.   (b)   Increasing, since $dR/dq = 4/3$ is positive.   (c)   Decreasing, since when $q = 4$, $dR/dq = (6.4 - 8)/0.3 = -1.6/0.3$ is negative.

5. Absolute maximum (of $y = 0.333$) when $x = 1$; absolute minimum (of $y = 3.2$) when $x = 0$.

7. $x = 900$ when $t = 30$.

9. Absolute minimum $y = -37$ when $x = -2$; absolute maximum $y = 3$ when $x = 0$.

11. Absolute maximum of 208,000 when $T = 60°C$; absolute minimum of 0 when $T = 100°C$.

13. See Figure 8.

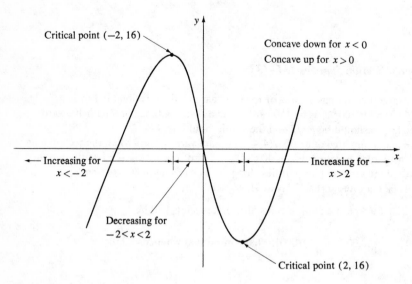

Critical point (−2, 16)

Concave down for $x < 0$
Concave up for $x > 0$

Increasing for $x < -2$

Decreasing for $-2 < x < 2$

Increasing for $x > 2$

Critical point (2, 16)

Figure 8

Chapter 4, Section 1, pages 141–142

1. $dy/dx = x^3(5) + (5x - 1)(3x^2)$.

3. $ds/dt = (t^3 - 5t^2 + 1)(4t^3) + (t^4 + 3)(3t^2 - 10t)$.

5. $\dfrac{dy}{dx} = \dfrac{(1 + x)(2x) - x^2}{(1 + x)^2}$.

7. $f'(x) = \dfrac{(2x + 1)(2x) - x^2 \cdot 2}{(2x + 1)^2}$.

9. $\dfrac{dP}{dt} = \dfrac{(1 + 2t)3 - (3t) \cdot 2}{(1 + 2t)^2}$.

11. $f'(x) = (3x)(2x) + (x^2 - 1) \cdot 3$.

13. $H'(u) = -2/(2u + 1)^2$. **15.** 285. **17.** 1.

19. Decreasing, because $dy/dx = -2/(x - 1)^2$ is negative when $x = 5$.

21. (a) Decreasing because $dy/dx = -299/(3t + 1)^2$ is negative for all $t \geq 0$.
(b) $-299/100^2$ or -0.0299.

23. 1. (*Hint:* Find the absolute minimum of $y = x + (1/x)$ for $x > 0$.)

25. $f'(1) = 4$ (because $f'(x) = xL'(x) + L(x) \cdot 1 = 1 + L(x)$).

Chapter 4, Section 2, pages 148–150

1. $dy/dx = -3x^{-4}$.   **3.** $f'(s) = \frac{1}{3}s^{-2/3}$.   **5.** $C'(x) = -12x^{-2} + 5$.
7. $dp/dq = 15.33q^{-3.1}$.   **9.** $dy/dx = -35x^{-8}$.
11. $C'(x) = -4.8x^{1.4}(x^{2.4} + 1)^{-2}$.   **13.** $\frac{1}{12}$ $(dp/dx = \frac{1}{3}x^{-2/3})$.
15. 3 $(dp/dx = -16q^{-3}$ and $d^2p/dx^2 = 48q^{-4})$.   **17.** 2.
19. (a) $dq/dp = -3.615p^{-1.3}$.   (b) $R = 12.05p^{0.7}$ and the marginal revenue
is $dR/dp = 8.435p^{-0.3}$.
21. 5 in. wide and 10 in. long.

Chapter 4, Section 3, pages 155–156

1. $dy/dx = 5(2x + 1)^4(2) = 10(2x + 1)^4$.
3. $dw/du = 6(u^2 + 1)^5(2u) = 12u(u^2 + 1)^5$.
5. $ds/dt = 7(4t - 1)^6(4) = 28(4t - 1)^6$.
7. $dw/dv = (1/3)(2v - 1)^{-2/3}(2) = (2/3)(2v - 1)^{-2/3}$.
9. $f'(x) = -3(x^4 - x)^{-4}(4x^3 - 1)$.
11. $H'(r) = -4(5 - r^2)^{-5}(-2r) = 8r(5 - r^2)^{-5}$.
13. $dy/dx = -(5x/2)(5x + 1)^{-3/2}$.
15. $dy/dx = (-3)(x^3 + 1)^{-4}(3x^2) = -9x^2(x^3 + 1)^{-4}$.
17. $dy/dx = (5)(-1)(1 + x^2)^{-2}(2x) = -10x(1 + x^2)^{-2}$.
19. $dy/dx = (\frac{15}{2})(-4)(x^3 + 1)^{-5}(3x^2) = -90x^2(x^3 + 1)^{-5}$.
21. $dy/dx = -x(1 - x^2)^{-1/2}$.   **23.** $dy/dx = 10(x^3 - x)^9(3x^2 - 1)$.
25. $dy/dx = -5(1 - x)^4$; $d^2y/dx^2 = 20(1 - x)^3$; y is decreasing. Concave up
when $x = 0$; concave down when $x = 2$. There is a point of inflection when $x = 1$.

Chapter 4, Section 4, page 159

1. $dy/dx = (1 - y)/x$.   **3.** $dy/dx = (3x^2 - y^2)/(2xy - 6)$.
5. $dy/dx = (y - 2xy^2)/(2x^2y - x)$, or $-y/x$.   **7.** $dy/dx = (1 - 3)/2 = -1$.
9. $dy/dx = (3.4 - 1)/(-4 - 6) = -\frac{11}{10}$.
11. By implicit differentiation, $y'(3x^3y^2 + 4) = (6x - 3x^2y^3)$, so $y' = 0$ when
$x = 2$ and $y = 1$.

Chapter 4, Section 5, pages 164–165

1. For $y = \sqrt{x}$, $dy/dx = 1/2\sqrt{x} = 1/8$ when $x = 16$; $\Delta y = \sqrt{16.1} - \sqrt{16}$
$\approx (1/8)(0.1) = 0.0125$, so $\sqrt{16.1} \approx 4.0125$.
3. $\Delta y = \sqrt{101} - \sqrt{100} \approx (1/2\sqrt{100})(1) = 0.05$, so $\sqrt{101} \approx 10.05$.
5. $\Delta y = \sqrt{26.5} - \sqrt{25} \approx (1/2\sqrt{25})(1.5) = 0.15$, so $\sqrt{26.5} \approx 5.15$.
7. For $y = x^{1/5}$, $dy/dx = \frac{1}{5}x^{-4/5} = 1/80$ when $x = 32$; $\Delta y = 33^{1/5} - 32^{1/5}$
$\approx (1/80)(1) = 0.0125$, so $33^{1/5} \approx 2.0125$.
9. For $y = x^{0.351}$, $dy/dx = 0.351$ when $x = 1$; $\Delta y = 1.004^{0.351} - 1 \approx (0.351)(0.004)$
$= 0.001404$, so $1.004^{0.351} \approx 1.001404$.
11. For $y = \sqrt{x}$, $dy/dx = \frac{1}{6}$ when $x = 9$; $\Delta y = \sqrt{9.6} - \sqrt{9} \approx (1/6)(0.6) = 0.1$,
so $\sqrt{9.6} \approx 3.1$.
13. $dq/dp = -84.77$ when $p = 1$; $\Delta q \approx (dq/dp)\Delta p = (-84.77)(0.02) = -1.6954$,
so q decreases by about 1.6954.
15. $\Delta V \approx (4\pi r^2)\Delta r = (36\pi)(-0.02) = -0.72\pi \approx -2.26$ and $\Delta A \approx (8\pi r)\Delta r$
$= (24\pi)(-0.02) = -0.48\pi \approx -1.51$.
17. $\Delta R \approx (300x - 3x^2)\Delta x = (1425)(0.5) = 712.5$.
19. $I = kr^{-4}$, so $\Delta I \approx -4kr^{-5}\Delta r$, or $\Delta I/I \approx -4\Delta r/r$. If $\Delta r/r = 0.01$, then
$\Delta I/I \approx -4(0.01) = -0.04 = -4\%$; a 4% decrease.

Chapter 4, Miscellaneous Problems, pages 167–168

1. $dy/dx = (x^2 + 3x + 1)(6x) + (3x^2 + 5)(2x + 3) = 12x^3 + 27x^2 + 16x + 15$.
3. $dw/dt = 1/(4t + 1)^2$.   **5.** $dw/dt = 30(5t + 1)^5$.
7. $f'(x) = (1/5)x^{-4/5}$.
9. $g'(x) = 12x^2(3x + 1)^3 + 2x(3x + 1)^4 = 2x(9x + 1)(3x + 1)^3$.
11. $h'(t) = 12t^{1.4} - 4.5t^{-2.5}$.
13. Implicit differentiation gives $2y(dy/dx) + 2(dy/dx) = 1$. Let $y = 1$ and solve
for dy/dx; $dy/dx = \frac{1}{4}$.
15. Implicit differentiation gives $2y(dy/dx) + x(dy/dx) + y = 3$. Let $x = 3$ and
$y = 2$, and solve for dy/dx; $dy/dx = \frac{1}{7}$.
17. When $r = 0.2$, $R = \frac{1}{3}$ and $dR/dr = 2/(1.2)^2 \approx 1.39$.   **19.** $L = 2b/3av$.
21. $\Delta F \approx -2Gr^{-3}\Delta r$ so $\Delta F/F \approx -2\Delta r/r$. When $r = 4000$, $\Delta r = 4$; this gives
$\Delta F/F \approx -2(4/4000) = -0.002$. Thus F decreases by about 0.2%.

Chapter 5, Section 1, pages 178–179

1. (a) $x = 2$.   (b) $x = 5$.   (c) $x = 1/2$.   (d) $x = 2^{\sqrt{2}-1}$.
(e) $x = -3/2$.   (f) $x = 1/2$.

3. See Figure 9.   **5.** See Figure 10.   **7.** $h \approx 0.69$.

9. Because of the independence on when the time interval occurred, $g(t + h)/g(t) = g(0 + h)/g(0)$. If we define $f(t) = g(t)/g(0)$, then this equation can be written $f(t + h) = f(t)f(h)$. Therefore $f(t) = a^t$ and so $g(t) = g(0)a^t$, which is the desired answer when $C = g(0)$.

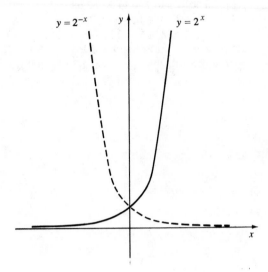

Figure 9

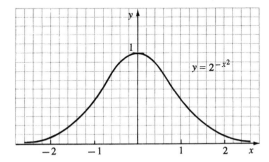

Figure 10

1. See Figure 11.

3. (a) Let S be the strength of the magnitude 6 quake and let T be the strength of the magnitude 3 quake. Then $\log S = 6$ and $\log T = 3$. Therefore $S = 10^6$, $T = 10^3$, and $S/T = 10^3 = 1000$, as asserted. (b) Let J be the strength of the Japan quake and S the strength of the San Francisco quake. $J = 4S$ and so $\log J = \log 4 + \log S$. Since $\log J = 8.9$, we obtain $\log S = 8.9 - 0.6 = 8.3$ as the magnitude of the San Francisco quake.

5. (a) $\ln 15 = \ln 3 + \ln 5 = 2.71$. (b) $\ln 40 = 3 \ln 2 + \ln 5 = 3.68$.
(c) $\ln \sqrt[3]{5} = \frac{1}{3} \ln 5 = 0.54$. (d) $\ln 0.4 = \ln \frac{2}{5} = \ln 2 - \ln 5 = -0.92$.
(e) $\ln 30 = \ln 2 + \ln 3 + \ln 5 = 3.4$. (f) $\ln \frac{9}{25} = 2 \ln 3 - 2 \ln 5 = -1.02$.

7. (a) $\log_2 6 = \log_2 2 + \log_2 3 = 1 + 1.58 = 2.58$. (b) $\log_2 9 = 2 \log_2 3 = 3.16$.
(c) $\log_2 \frac{3}{2} = \log_2 3 - \log_2 2 = 1.58 - 1 = 0.58$. (d) $\log_2 \sqrt{3} = \frac{1}{2} \log_2 3 = \frac{1}{2}(1.58) = 0.79$.

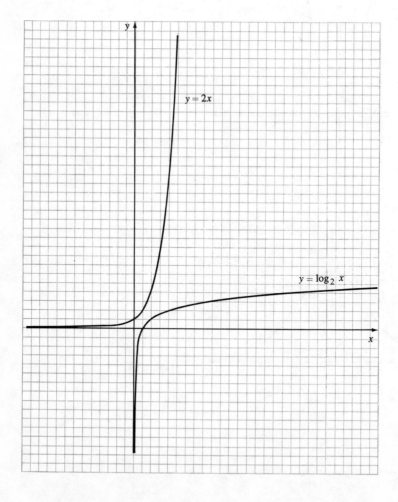

Figure 11

9. $p = 30e^{-2}$, which is approximately 4.06 in. of mercury.

11. $N = 2^{63}$, $\log N = \log 2^{63} = 63 \log 2 = 63(0.3) = 18.9$; $N = 10^{18.9}$, which is between 10^{18} and 10^{19}. It would require 18 digits to write the number N. This is more rice than the earth has produced to date.

13. (a) As t increases $e^{-.06t}$ decreases. Thus the denominator of y decreases, so y increases. (b) $t = -\dfrac{1}{0.06} \ln \left[\dfrac{1}{232} \left(\dfrac{573}{y} - 1 \right) \right]$.

Chapter 5, Section 3, pages 194–196

1. $dy/dx = 3e^x$.   **3.** $dy/dx = 2x + 5e^x$.   **5.** $dy/dx = 20e^{5x}$.
7. $dy/dx = -45e^{-3x}$.   **9.** $dP/dq = 2qe^{-3q} - 3q^2 e^{-3q}$.
11. $M'(t) = 8e^{2t}(1 + e^{2t})^3$.   **13.** $f'(x) = \frac{1}{2}(x + e^x)^{-1/2}(1 + e^x)$.
15. $dy/dx = 2e^{2x}$ or $2(e^x)^2$.   **17.** $\dfrac{dy}{dx} = \dfrac{-xe^{-x^2/2\sigma^2}}{\sigma^3 \sqrt{2\pi}}$.
19. $dy/dx = 220e^{-t/10}(1 + \frac{1}{2}e^{-t/10})^{-2}$.
21. $y = e^{x\ln 2}$, so $dy/dx = (\ln 2)e^{x\ln 2}$ or $2^x \ln 2$.

Chapter 5, Section 4, pages 201–202

1. (a) $dy/dx = 10/x$.   (b) $dw/dt = 10t - (1/t)$.   (c) $ds/dr = 2/(2r + 1)$.
(d) $dy/dx = \ln(3x - 1) + 3x/(3x - 1)$.
3. From $T = a \ln x - a \ln(k - x) + b$, we obtain $dT/dx = (a/x) + a/(k - x)$
$= ak/x(k - x)$.
5. (a) $dy/dx = x(3x + 1)^2(x + 6)^{10}\{(1/x) + [6/(3x + 1)] + [10/(x + 6)]\}$.
(b) $dT/dx = 10^x \ln 10$.   (c) $\dfrac{du}{ds} = \dfrac{53.1}{s(s - 1)\sqrt{s - 2}}\left(-\dfrac{1}{s} - \dfrac{1}{s - 1} - \dfrac{1}{2(s - 2)} \right)$.
(d) $dy/dx = 2x^{2x}(1 + \ln x)$.   (e) $dy/dx = (1/x)(2x^{\ln x} \ln x)$.
(f) $f'(x) = x^{x^3}(x^2 + 3x^2 \ln x)$, or $x^{2 + x^3}(1 + 3 \ln x)$.
7. From $n = u - a + \ln u - \ln a$, we obtain $dn/du = 1 + (1/u) = (u + 1)/u$.

Chapter 5, Section 5, pages 210–211

1. $100e^{8 \ln 10/5.4} \approx 3030$.
3. $(\ln 3)/0.035 \approx 31.39$ yr.; $(\ln 3)/0.012 \approx 91.55$ yr.; $(\ln 3)/0.005 \approx 219.72$ yr.
5. $10,000[1 + (0.075/2)]^{18} \approx \$19,399.29$.   **7.** $(1/18) \ln 3 \approx 6.1\%$.

Chapter 5, Miscellaneous Problems, pages 212–214

1. See Figure 12.
3. (a) $\ln 2^2 \cdot 3^2 = 2 \ln 2 + 2 \ln 3 = 1.38 + 2.20 = 3.58$.
(b) $\ln(\frac{4}{3}) = 2 \ln 2 - \ln 3 = 1.38 - 1.10 = 0.28$.
(c) $\ln(\frac{4}{3})^{1/7} = (\frac{1}{7})\ln(\frac{4}{3}) = (\frac{1}{7})(0.28) = 0.04$.
(d) $e^{1.79} = e^{0.69+1.10} = e^{\ln 2 + \ln 3} = e^{\ln 2}e^{\ln 3} = 2 \cdot 3 = 6$.
5. (a) $dy/dx = (\frac{2}{3})x^{-2/3} - 3e^{3x} + (1/x)$. (b) $ds/dt = 0.06e^{0.06t}$.
(c) $f'(x) = -4016e^{-3.2x}$. (d) $dy/dx = (\frac{1}{2})(e^x - e^{-x})$.
(e) From $w = e^{(2 \ln 6)t}$, we obtain $dw/dt = 2 \ln 6 \, e^{(2 \ln 6)t}$ or $dw/dt = (2 \ln 6)6^{2t}$.
(f) $dq/dp = 2/(2p + 1)$.
7. $P = \lambda^4 e^{-\lambda}/24$ so $dP/d\lambda = (4\lambda^3 e^{-\lambda} - \lambda^4 e^{-\lambda})/24$.
9. $800e^{10 \ln(11/8)/6.4} \approx 1316$.
11. $M = 100e^{-0.038t}$ (M in mg, t in centuries), so when $t = 100$ (that is, 10,000 years), $M = 100e^{-3.8} \approx 2.24$ mg.
13. $D = 2e^{-3.2t}$.

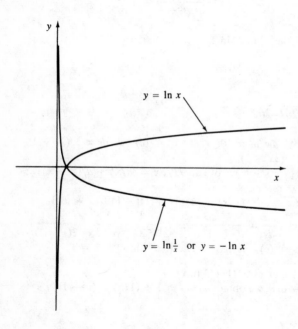

Figure 12

Chapter 6, Section 1, pages 222–223

1. (a) $(x^7/7) + C$. (b) $(q^{2.3}/2.3) + C$. (c) $(-1/2t^2) + C$.
(d) $(-2x^{-1/2}) + C$.
3. (a) 2. (b) 4. (c) 1. (d) 3.
5. (a) $y = (x^7/7) + (\frac{6}{7})$. (b) $p = (q^{2.3}/2.3) + 8.17$.
(c) $s = (-1/2t^2) + 10$. (d) $C = (-2x^{-1/2}) + 1$.
7. (a) $u^2v^2 + C$. (b) $(2/3)uv^3 + C$.

Chapter 6, Section 2, pages 228–230

1. $x^2 + x + C.$ **3.** $3u^2 - 2u + C.$ **5.** $3x^4 + (x/2) + C.$
7. $(x^3/3) - (3x^2/2) + 5x + C.$ **9.** $0.3x^5 + 10x^{-0.3} + C.$
11. $(2x^3/3) - 2x^2 + (\frac{1}{2}) \ln x + C.$ **13.** $x^2 - x - \ln x + C.$
15. $(e^{2x}/2) + 2e^{-3x} + C.$ **17.** $u + 3 \ln u + C.$ **19.** $(\frac{5}{3}) \ln x + C.$
21. $(\frac{1}{16})(2x + 1)^8 + C.$ **23.** $(\frac{1}{12})(x^2 + 1)^6 + C.$
25. $x^2 - (x^4/4) + C.$ **27.** $(\frac{1}{8})(e^{2x} + 1)^4 + C.$
29. $(\frac{2}{15})(5t + 1)^{3/2} + C.$ **31.** $y = 1 + 3 \ln x.$ **33.** $R = e^{2t} + t + 1.$
35. $y = (2x + 1)^5 - 1.$

Chapter 6, Section 3, page 238

1. $\frac{2}{9}(3x + 1)^{3/2} + C.$ **3.** $\frac{1}{3}(x^2 - 1)^{3/2} + C.$ **5.** $\ln(x + 1) + C.$
7. $-\frac{1}{2}\ln(5 - 2w) + C.$ **9.** $\frac{1}{2}e^{2x+1} + C.$ **11.** $x + \ln(x - 1) + C.$
13. $x \ln x - x + C$ (try $u = \ln x, dv = dx$).
15. $\frac{1}{2}x^2(\ln x)^2 - \frac{1}{2}x^2 \ln x + \frac{1}{4}x^2 + C$ (try $u = (\log x)^2, dv = xdx$ and reduce to Problem 14).
17. $(1/k)x^2e^{kx} - (2/k^2)xe^{kx} + (2/k^3)e^{kx} + C$ (try $u = x^2, dv = e^{kx}dx$ and reduce to Problem 16).
19. $\frac{2}{5}(x + 2)^{5/2} - \frac{4}{3}(x + 2)^{3/2} + C$ (try the substitution $u = x + 2$).
21. $\frac{1}{15}(3x - 1)^5 + C$ (guess, or try substitution $u = 3x - 1$).
23. $-\frac{1}{3}xe^{-3x} - \frac{1}{9}e^{-3x} + C$ (guess, or use integration by parts).

Chapter 6, Section 4, pages 242–243

1. (a) $y = 2x^4 - 2x^2 + 3x^{-1} - 3.$ (b) $y = \frac{2}{9}(3x + 1)^{3/2} + \frac{2}{9}.$
(c) $P = -\frac{1}{2}e^{1-t^2} + \frac{1}{2}.$ (d) $w = s \ln s - s + \pi.$
3. $C = 0.125x + 0.00439x^2 + 16.68.$
5. 256 ft. $(ds/dt = 32t, s = 16t^2;$ when $t = 4,$ the distance s is $16 \cdot 4^2 = 256.)$
7. 37 tn. $(W =$ tons evaporated up to time $t; dW/dt = \frac{1}{3}T + 2$
$= \frac{1}{3}(20 - \frac{1}{4}t^2 + 8t) + 2;$ integration gives $W = \frac{1}{3}(20t - \frac{1}{9}t^3 + 4t^2) + 2t;$ when
$t = 3, W = 37).$
9. 168 l. $(y = l$ pumped in t minutes; $dy/dt = 0.07(150 - 3t), y = 0.07(150t - \frac{3}{2}t^2)$: when $t = 20, y = 168.)$

1. $[x^5]_0^2 = 32.$ **3.** $\frac{1}{4}[e^{4x}]_0^1 = \frac{1}{4}(e^4 - 1).$ **5.** $\frac{1}{3}[\ln(x+2)]_{-1}^e{}^{-2} = \frac{1}{3}.$

7. $\frac{1}{22}[(2t+1)^{11}]_{-1/2}^0 = \frac{1}{22}.$ **9.** $[10x^3 - 3x^2 + 5x]_1^3 = 246.$

11. $[2x^3 + \frac{1}{2}x^2]_1^3 = 56.$

13. See Figure 13; Area $I = \int_0^1 (x^3 - 3x^2 + 2x)dx = [\frac{1}{4}x^4 - x^3 + x^2]_0^1 = \frac{1}{4};$

$-\text{Area II} = \int_1^2 (x^3 - 3x^2 + 2x)dx = [\frac{1}{4}x^4 - x^3 + x^2]_1^2 = -\frac{1}{4};$ so Area II $= \frac{1}{4}.$

15. Area $\approx 18,440 \approx \int_0^{20,000} (dC/dx)dx = C(20,000) - C(0),$ so the total cost is

$C(20,000) = 18,440 + C(0) = \$19,440.$

17. 2360 beats; $2360 \times 0.07 = 165.2 \, l.$

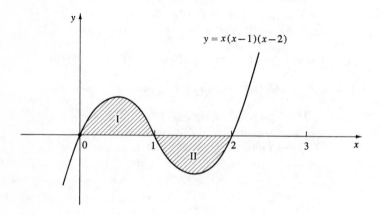

$y = x(x-1)(x-2)$

Figure 13

Chapter 6, Section 6, pages 258–259

1. $V = \int_0^{20} A(x)dx = \int_0^{20} \frac{1}{2}\pi x \, dx = 100\pi$ cu. ft.

3. $V = \int_0^1 A(x)dx = \int_0^1 \pi(e^x - 1)^2 dx = \pi \int_0^1 (e^{2x} - 2e^x + 1)dx$

$= \pi(\frac{1}{2}e^2 - 2e + \frac{5}{2})$ (or ≈ 2.38) cu. m.

5. A cross section parallel to the base and x units from the top is a circle of radius

$xr/h.$ Therefore $A(x) = \pi(xr/h)^2.$ $V = \int_0^h \frac{\pi x^2 r^2}{h^2}dx = \frac{\pi r^2}{h^2}\left[\frac{x^3}{3}\right]_0^h = \frac{\pi r^2 h}{3}.$

7. $V = \int_0^5 \pi \cdot \frac{9}{25}(25 - x^2)dx = 60\pi$ cu. in.

Chapter 6, Miscellaneous Problems, pages 260–264

1. (a) $-x^{-1} + C.$ (b) $[-x^{-1}]^1_{1/2} = -1 + 2 = 1.$ (c) $\frac{1}{18}(2x - 1)^9 + C.$
(d) $[\frac{1}{18}(2x - 1)^9]^1_0 = \frac{1}{18} + \frac{1}{18} = \frac{1}{9}.$ (e) $\frac{1}{3}t^3 - t + C.$ (f) $[\frac{1}{3}t^3 - t]^3_0 = 6.$
(g) $\frac{1}{3}(u^2 + 16)^{3/2} + C.$ (h) $[\ ]^3_0 = \frac{1}{3}(25)^{3/2} - \frac{1}{3}(16)^{3/2} = \frac{61}{3}.$
(i) $x + \ln x + C.$ (j) $[\ ]^e_1 = e.$ (k) $\frac{1}{3}e^{3x} + C.$ (l) $[\ ]^1_0 = \frac{1}{3}(e^3 - 1).$
(m) $xe^x - e^x + C.$ (n) $[\ ]^1_0 = 1.$
3. About 23,900 million barrels.   5. About 12.65 micrograms.

7. $c = 0.672y + 113.10.$ 9. $\int_0^{10} (10x - x^2)dx = [5x^2 - \frac{1}{3}x^3]^{10}_0 = 166\frac{2}{3}.$

11. $\int_0^6 (49 - q^2)dq - 6 \cdot 13 = 222 - 78 = 144.$

13. The volume of water in the figure can be calculated by slicing with vertical cross
sections; it is $w \cdot \int_a^b f(x)dx,$ where w is the width of the box. When the water is calm
the volume is $w \cdot h \cdot (b - a),$ where h is the water level. Since these volumes are equal,
$wh(b - a) = w\int_a^b f(x)dx,$ so $h = \bar{y}.$
15. $\bar{R} = \frac{1}{70}$ (area under graph from 0 to 70) $\approx \frac{1}{70} \cdot 23{,}900 \approx 341.5$ (million barrels
per year).
17. Above average. (In this city, the average consumption was 7 gal. per person per
hour.)
19. $\bar{y} = 50/3.$ 21. About 3870 ft.

Chapter 7, Section 1, pages 272–274

1. (a) 17.   (b) 19.   (c) $p(5, 0).$ (d) $(0, 0), (1, -\frac{2}{5}), (2, -\frac{4}{9}), (3, -\frac{6}{13})$
(*Note:* In general, $(x, -2x/(1 + 4x))$ will work for any value of x except $-\frac{1}{4}.$)
3. $Q = (1.14)(32)^{4/5}(243)^{1/5} = (1.14)(2^5)^{4/5}(3^5)^{1/5} = (1.14)(2^4)(3)$
$= (1.14)(16)(3) = 54.72.$
5. $g(12, 8, 64) = 0.9[(12\sqrt{64}/12\sqrt[3]{8}) + (12 + \sqrt{64})/4] = 0.9(4 + 5) = 8.1.$
7. $f(s, t) = 2000s + 60t.$ 9. $f(x, y, z) = 0.24xyz.$

11. $f(r_1, r_2) = \left(\dfrac{1}{r_1} + \dfrac{1}{r_2}\right)^{-1}.$

Chapter 7, Section 2, pages 279–281

1. $\partial z/\partial x = 8xy^4,$ $\partial z/\partial y = 16x^2y^3.$ 3. $\partial w/\partial u = 3ve^u,$ $\partial w/\partial v = e^{3u}.$
5. $\partial Q/\partial r = 2rt - 4s^2r^{-5},$ $\partial Q/\partial s = 2sr^{-4},$ $\partial Q/\partial t = r^2.$
7. $\partial S/\partial m = 24(12m + 5b);$ $\partial S/\partial b = 10(12m + 5b).$
9. $\partial P/\partial L = 5(2 + 3C);$ $\partial P/\partial C = 15L.$
11. $\partial w/\partial x = e^x \ln y;$ $\partial w/\partial y = e^x/y.$ 13. $g_s = 2.3/s;$ $g_t = 0.4/t.$
15. $F_x = 1;$ $F_y = -z/\sqrt{3 - 2y}.$ 17. 96.
19. $\partial w/\partial u = 6e^3$ and $\partial w/\partial v = e^3$ when $(u, v) = (1, 2).$

21. $\partial v/\partial p = r^4/L$; $\partial v/\partial r = 4pr^3/L$; $\partial v/\partial L = -pr^4L^{-2}$.

23. $f_L = 0.76L^{-0.24}C^{0.24}$ and $f_C = 0.24L^{0.76}C^{-0.76}$. When $(L, C) = (80, 20)$, we find $f_L(80, 20) = (0.76)(0.35)(2.05) = 0.55$ and $f_C(80, 20) = (0.24)(27.95)(0.10) = 0.67$, so an increase in C will produce the greater change.

25. $\dfrac{\partial D_x}{\partial P_z}\dfrac{P_z}{D_x} = (0.5)\dfrac{6}{63.3 - (1.9)(10) + (0.2)(5) + (0.5)(6)} \approx 0.06$.

Chapter 7, Section 3, pages 285–286

1. $\Delta z \approx 40\Delta x - 192\Delta y = 40(-0.1) - 192(0.3) = -61.6$; z decreases by about 61.6.
3. $\Delta f \approx \Delta u + 200\Delta v + 10\Delta w = (-10) + 200(\tfrac{1}{2}) + 10(1) = 100$.
5. $\Delta Q \approx 4\Delta x + 3\Delta y = 4(0.5) + 3(-0.1) = 1.7$.
7. $\Delta z \approx 80\Delta x + 300\Delta y = 80(-0.3) + 300(0) = -24.0$; z decreases by about 24.
9. $\Delta F \approx \Delta Q - 0.3\Delta E = (1) - 0.3(-2) = 1.6$.

Chapter 7, Section 4, pages 290–292

1. $\partial^2 z/\partial x^2 = 80x^3y^2$; $\partial^2 z/\partial x\partial y = \partial^2 z/\partial y\partial x = 40x^4y$; $\partial^2 z/\partial y^2 = 8x^5$.
3. $\partial^2 w/\partial r\partial s = 30re^{5s}$; $\partial^2 w/\partial r^2 = 6e^{5s}$; $\partial^2 w/\partial s^2 = 75r^2e^{5s}$.
5. $g_{xy} = 0$; $g_{yy} = 12y$; $g_{xz} = 0$; $g_{xy}(2, -1, 3) = 0$; $g_{yy}(2, -1, 3) = -12$; $g_{xz}(2, -1, 3) = 0$.
7. $P_{xx} = 3x^{-0.5}y^3$; $P_{xy} = 18x^{0.5}y^2$; $P_{yy} = 24x^{1.5}y$; $P_{xy}(4, 3) = 324$.
9. $\partial^2 Q/\partial x\partial y = 6xy^2$; $\partial^2 Q/\partial y^2 = 6(1)^2(5) = 30$ when $x = 1$ and $y = 5$.
11. -18.
13. $\partial^2 w/\partial x^2 = 6$; $\partial^2 w/\partial x\partial y = 0$; $\partial^2 w/\partial x\partial z = 6z^2$; $\partial^2 w/\partial y\partial z = 0$; when $(x, y, z) = (3, 2, 4)$, $\partial^2 w/\partial x\partial z = 6 \cdot 4^2 = 96$.
15. $9e^{3x-2y}$; $-6e^{3x-2y}$; $4e^{3x-2y}$; $\partial^2 s/\partial x^2 = 9e$ when $(x, y) = (1, 1)$; $\partial^2 s/\partial x\partial y = -6$ when $(x, y) = (4, 6)$; $\partial^2 s/\partial y^2 = 4$ when $(x, y) = (0, 0)$.
17. $Q_{xx} = e^x \ln y$; $Q_{xy} = e^x/y$; $Q_{yy} = -e^x/y^2$; $Q_{xx}(2, e) = e^2$; $Q_{xy}(3, 5) = e^3/5$; $Q_{yy}(1, 2) = -e/4$.
19. $f_t = e^{-x^2/t_t - 1/2}[(x/t)^2 - (1/2t)]$; $f_{xx} = e^{-x^2/t_t - 1/2}[4(x/t)^2 - (2/t)]$; therefore $f_t = \tfrac{1}{4}f_{xx}$.
21. $\partial^2 z/\partial x\partial y = 2xt^3 = \partial^2 z/\partial y\partial x$; $\partial^2 z/\partial x\partial t = 6xyt^2 - 3x^2 = \partial^2 z/\partial t\partial x$; $\partial^2 z/\partial y\partial t = 3x^2t^2 = \partial^2 z/\partial t\partial y$.

Chapter 7, Section 5, pages 297–299

1. $(x, y) = (4, 3)$.   **3.** $(x, y) = (1, 5)$.   **5.** $(x, y) = (3, 1)$.
7. $(x, y) = (4, 3)$ gives a relative minimum.
9. $(x, y) = (1, 5)$ gives a relative maximum.

BASIC CALCULUS WITH APPLICATIONS

11. There are no relative maxima nor minima (Case 2 in the second derivative test occurs).

13. (a) $R = (50 - 5r)r + (200 - s)s - [2r + s + 500]$, or
$R = 48r + 199s - 5r^2 - s^2 - 500$.

(b) R attains a relative (actually, an absolute) maximum when $r = \frac{48}{10}, s = \frac{199}{2}$.

Chapter 7, Section 6, pages 304–305

1. $(x, y) = (9, 3)$. **3.** $(x, y) = (5, 6)$. **5.** $(x, y) = (\frac{1}{2}, 0)$.
7. $C = 60$ and $L = 240$.

Chapter 7, Section 7, pages 314–316

1. $\hat{y}_1 = 1.65, \hat{y}_2 = 2.72, \hat{y}_3 = 4.95, \hat{y}_4 = 12.18$; $E \approx 1.82 + 0.61 + 0.90 + 1.39$ ≈ 4.7; see Figure 14.

3. $m = \dfrac{(11.2)(21.5) - 4(77.8)}{(11.2)^2 - 4(40.24)} = 1.98$ and $b = \dfrac{(21.5) - (11.2)(1.98)}{4} = -0.17$,

so $y = 1.98x - 0.17$ is the regression line; when $x = 6$, this predicts $y = 11.71$.

5. $m = \dfrac{(15)(122.3) - 6(337.1)}{(15)^2 - 6(55)} = 1.79$ and $b = \dfrac{(122.3) - (15)(1.79)}{6} = 15.91$,

so $y = 1.79x + 15.91$ is the regression line; when $x = 7$, this predicts $y = 28.44$.

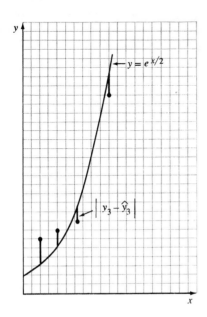

Figure 14

Chapter 7, Miscellaneous Problems, pages 316–318

1. If $f(L, C) = aL^bC^{1-b}$, then $f(kL, kC) = a(kL)^b(kC)^{1-b} = ak^bL^bk^{1-b}C^{1-b}$
$= kaL^bC^{1-b} = kf(L, C)$; if $f(L, C) = (a_1L^b + a_2C^b)^{1/b}$ then $f(kL, kC)$
$= (a_1k^bL^b + a_2k^bC^b)^{1/b} = [k^b(a_1L^b + a_2C^b)]^{1/b} = k(a_1L^b + a_2C^b)^{1/b} = kf(L, C)$.
3. $f_u = (1 + v^2)^5$; $f_v = 10uv(1 + v^2)^4$; $f_{uu} = 0$; $f_{vv} = 10u(1 + v^2)^4$
$+ 80uv^2(1 + v^2)^3$; $f_{uv} = 10v(1 + v^2)^4$; $f_v(2, 0) = 0$; $f_{vv}(2, 0) = 20$.
5. $\Delta Q \approx (0.64)(-2) + (0.36)(3) = -0.2$, so production would decrease by approximately 0.2 units.
7. Critical points are $(0, 0)$ and $(2\sqrt{2}, \sqrt{2})$. The first one $(0, 0)$ gives neither a relative maximum nor minimum. The second one $(2\sqrt{2}, \sqrt{2})$ gives a relative minimum.
9. Relative maximum at $u = 4, v = 1$.   **11.** $C = 25, L = 75$.
13. The least squares regression line is $p = 0.5t + 3.83$, so $P = e^{3.83}e^{0.5t} = 46.06e^{0.5t}$.
When $t = 20$, this predicts $P \approx 1,014,539$ (computations by hand calculator).
15. $D = 8 - 1.2P$, which predicts $D = 0.8$ when $P = 6$.

Chapter 8, Section 1, pages 325–326

1. (a) Left side (L.S.) of differential equation (D.E.) is $(x/3)(3Cx^2)$; right side
(R.S.) of D.E. is $(Cx^3 + 1) - 1$. Therefore L.S. = R.S.   (b) The initial conditions
imply that $C = -1$, so the desired particular solution is $y = 1 - x^3$.
3. (a) L.S. of D.E. is $(2x)(\frac{3}{2}Cx^{1/2})$. R.S. of D.E. is $3(Cx^{3/2})$. Therefore L.S. = R.S.
(b) $y = 2x^{3/2} - 1$.
5. (b) $y = x$.   **7.** (b) $y^4 = 2x^2$.   **9.** (b) $w = x \ln x - x$.
11. (a) Differentiate implicitly and obtain $3y^2\dfrac{dy}{dx} - x\dfrac{dy}{dx} - y = 0$. This equation is
equivalent to the given D.E.   (b) $y^3 - xy = 8$.
13. (b) $tP = 1$.

Chapter 8, Section 2, pages 332–333

1. $\int y^3dy = \int x\,dx$; $\frac{1}{4}y^4 = \frac{1}{2}x^2 + C_1$, or $y^4 = 2x^2 + C$.
3. $\ln s = \frac{1}{4}t^4 + C$.   **5.** $-q^{-1} = p^2 + 3p + C$.   **7.** $y^2 = t^2 - e^{-2t} + C$.
9. $\ln(1 - 3U) = 3s + C$.
11. $\int dy/y = \int \ln x$; $\ln y = x \ln x - x + C$ (see Formula 66 in the Table
of Integrals).
13. $y^4 = 2x^2$.   **15.** $\ln s = \frac{1}{4}t^4$.   **17.** $-q^{-1} = p^2 + 3p - 5$.
19. $y^2 = t^2 - e^{2t} + 1$.   **21.** $\ln(1 - 3U) = 3s - 1$.
23. $\ln y = x \ln x - x + 1$.   **25.** $s = e^Ce^{t^4/4}$, or $s = Ke^{t^4/4}$.
27. $U = \frac{1}{3}(1 - Ke^{3s})$.   **29.** $y = x^xe^{1-x}$.

1. $\int dr = \int \dfrac{kds}{s}$; $r = k \ln s + C$.

3. (a) $\int \dfrac{dP}{Pr + 36{,}500} = \int dt$; $\dfrac{1}{r} \ln(Pr + 36{,}500) = t + C_1$.

(b) Solve the preceding implicit solution for P: $\ln(Pr + 36{,}500) = rt + C_2$;

$e^{\ln(Pr+36,500)} = e^{rt+C_2}$; $Pr + 36{,}500 = e^{rt}e^{C_2}$, $P = Ce^{rt} - \dfrac{36{,}500}{r}$. Set $t = 0$ and $P = 0$

to obtain $0 = Ce^{r\cdot 0} - \dfrac{36{,}500}{r}$, or $C = \dfrac{36{,}500}{r}$. With this value for C, the solution

becomes $P = \dfrac{36{,}500}{r}(e^{rt} - 1)$.

5. (a) $\int \dfrac{dt}{m} = \int \dfrac{dv}{gm - kv}$, $\dfrac{t}{m} = -\dfrac{1}{k} \ln(gm - kv) + C_1$, for $gm - kv > 0$.

(b) Solve the preceding equation for v: $e^{-kt/m} = e^{\ln(gm-kv)+C_2}$,

$e^{-kt/m} = (gm - kv)C_3$, or $kv = gm - C_4 e^{-kt/M}$; $v = \dfrac{gm}{k} - C_5 e^{-kt/m}$. Set $t = 0$,

$v = 0$ and obtain $C_5 = \dfrac{gm}{k}$. With this value of C_5, the solution becomes

$v = \dfrac{gm}{k} - \dfrac{gm}{k} e^{-kt/m}$.

7. (a) For Function 23, we have (*logarithmic differentiation*):

$\ln y = \ln M - \ln(1 + Ce^{-kMt})$, $\dfrac{y'}{y} = \dfrac{kCMe^{-kMt}}{1 + Ce^{-kMt}}$. We also have

$M - y = M - \dfrac{M}{1 + Ce^{-kMt}} = \dfrac{MCe^{-kMt}}{1 + Ce^{-kMt}}$. We see, therefore, that $\dfrac{y'}{y} = k(M - y)$;

thus Equation 24 is true. (b) Put $y = y_0$ and $t = 0$ into Function 23; the result is

$y_0 = \dfrac{M}{1 + C}$, so $C = \dfrac{M}{y_0} - 1 = \dfrac{M - y_0}{y_0}$. With this value for C, Function 23 becomes

Equation 24.

9. $\int k \, dt = \int \dfrac{1}{(a - y)(b - y)} dy$; $kt + C_1 = \dfrac{1}{b - a} \int \left(\dfrac{1}{a - y} - \dfrac{1}{b - y} \right) dy$;

$kt + C_1 = \dfrac{1}{b - a}(-\ln(a - y) + \ln(b - y))$, $kt + C_1 = \dfrac{1}{b - a} \ln \dfrac{b - y}{a - y}$;

$e^{(b-a)kt}e^{(b-a)C_1} = \dfrac{b - y}{a - y}$. Set $y = 0$, $t = 0$ and obtain $e^{(b-a)C_1} = b/a$. Thus

$\dfrac{be^{(b-a)kt}}{a} = \dfrac{b - y}{a - y}$. This is a solution in implicit form. It can be solved for y explicitly

as follows: $(a - y)be^{(b-a)kt} = ab - ay$, $ay - ybe^{(b-a)kt} = ab - abe^{(b-a)kt}$,

$y = a\dfrac{b - be^{(b-a)kt}}{a - be^{(b-a)kt}}$. This explicit solution can be transformed into Equation 26 as

follows: $a\dfrac{b - be^{(b-a)kt}}{a - be^{(b-a)kt}} = a\dfrac{[a - be^{(b-a)kt}] + [b - a]}{a - be^{(b-a)kt}} = a\left(1 + \dfrac{b - a}{a - be^{(b-a)kt}}\right)$.

11. (a) At time t, the balance is P. In a subsequent short time interval Δt, the interest accruing causes the balance to grow by $0.05P\Delta t$; on the other hand, $600\Delta t$ dollars are withdrawn. Thus we have the (approximate) equation $\Delta P = 0.05P\Delta t - 600\Delta t$. Divide by Δt and let $\Delta t \to 0$ to obtain the differential equation $dP/dt = 0.05P - 600$. After separating variables, one obtains
$$\int \frac{dP}{0.05P - 600} = \int dt.$$ At this stage, we must exercise care to avoid the pitfall of writing $\ln(0.05P - 600)$; note that $0.05P - 600$ is negative and hence has no logarithm defined. Therefore use the lower equation in Formula 13 on page 389 and proceed as follows: $\dfrac{1}{0.05}\ln(600 - 0.05P) = t + C_1,\ e^{\ln(600-0.05P)} = e^{0.05t}e^{0.05C_1},$

$600 - 0.05P = Ce^{0.05t}$. Evaluate C by setting $t = 0$, $P = 10{,}000$; the result is $C = 100$. Thus $600 - 0.05P = 100e^{0.05t}$ or $P = 12{,}000 - 2{,}000e^{0.05t}$.
(b) Set $P = 0$ and solve for t. The result is $e^{0.05t} = 6$; $0.05t = \ln 6$; $t = 20 \ln 6$, or (use a calculator or a table of logarithms) $t = 35.84$ yr.
(c) The amount withdrawn is $(35.84)(600) = \$21{,}504$.
13. As $t \to +\infty$, $e^{-kt/V}$ tends to zero. Therefore $(A - y_0)e^{-kt/V}$ tends to zero. Therefore $\lim_{t \to +\infty} y = A$. The concentration inside the cell will approach the extreme concentration A as time goes on.
15. Since k and M are positive, e^{-kMt} tends to zero as $t \to +\infty$. Hence the denominator in Equation 24 tends to y_0. Therefore the fraction tends to y_0M/y_0 or M. Thus $\lim_{t \to +\infty} y = M$. The population will get closer and closer to M as time goes on.
17. $d^2y/dt^2 = k(-yy' + y'(M - y))$. At a point of inflection, $d^2y/dt^2 = 0$, so $-yy' + y'(M - y) = 0$; $y'M = 2yy'$; or $M = 2y$ (*Note:* $y' \neq 0$ for the logistic function). Thus $y = M/2$ at the point of inflection.

Chapter 8, Section 4, pages 354–356

1. See Figure 15.

3. x	0	0.1	0.2	0.3	0.4	0.5	0.6	0.7	0.8	0.9	1.0
y	1	1.1	1.22	1.38	1.57	1.84	2.2	2.72	3.51	4.8	7.19
5. x	0.5	1		1.5	2.0	2.5	3.0				
y	0.3	0.45		0.45	0.38	0.28	0.20				

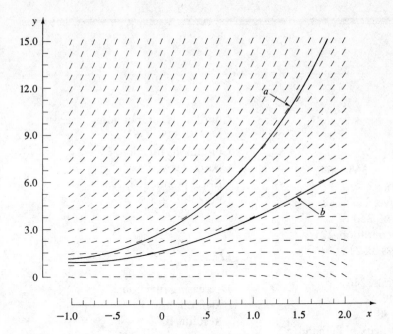

Figure 15

Chapter 8, Miscellaneous Problems, pages 356–358

1. $y = 10e^{-x} - x$.   **3.** $x^2 + y^2 = 2x$.   **5.** $y^2/2 = x - (x^2/2) + C$.
7. $\ln(s - 2) = \frac{2}{3}t^{3/2} + C$.   **9.** $(2P + 1)^5 = -10e^{-q} + C$.
11. Separation of variables is not applicable.
13. $p = 2 + te^{t-1}$. $\left(\text{Solution: } \int \dfrac{dP}{p - 2} = \int \left(1 + \dfrac{1}{t}\right) dt, \ln(p - 2) = t + \ln t + C_1,\right.$
$e^{\ln(p-2)} = e^{(t + \ln t + C_1)}, p - 2 = e^t e^{\ln t} e^{C_1}, p - 2 = Cte^t, p = 2 + Cte^t, \text{ set } p = 3 \text{ and}$

$\left. t = 1 \text{ and solve for } C, \text{ find } C = e^{-1} \text{ so } p = 2 + te^{-1}e^t \text{ or } p = 2 + te^{t-1}\right).$

15. The differential equation is $dP/dt = 0.08P - 5000$. The solution is
$P = (1/0.08)(5000 + Ce^{0.08t})$ where $C = -1000$. Set $P = 0$ and solve for t to obtain
$t = (1/0.08)\ln 5 = 20.12$; he can cruise for over 20 years!

17.

x	0	0.2	0.4	0.6	0.8	1.0
y	0	0	0.04	0.12	0.24	0.41

Index

Maximum-minimum in several
   variables: 292–296
   with constraints, 299
   Lagrange multiplier method, 301
   second derivative test, 296
Monopoly, 127

N

Napier, John, 183
Newton, Issac, 68
Nondifferentiability, 114
Normal distribution, 177, 195–196

O

Ordinate, 11

P

Parabola, 375
Paraboloid, 270
Parallel lines, 375
Partial derivative, 274
Particular solution, 323
Perpendicular lines, 375
Point-slope formula, 45
Poisson probability distribution, 213
Polynomial: 366
   differentiation of, 77
Production functions, 159, 272–273
Propensity to consume, 32
Psychophysical laws, 333–334

Q

Quadrant, 13
Quadratic equation, 368
Quadratic formula, 369

R

Rashevsky, N., 199, 243
Rate of change:
   average, 27
   graphical interpretation, 38
   instantaneous, 59–60, 66
   units for, 29
Rate of reaction, 106, 341–342
Rectangular coordinates, 11
Revenue, total and marginal, 149
Richter scale for earthquakes, 187
Robinson, Abraham, 218
Rolls Royce, 252, 254

S

Saddle point, 294
Saddle surface, 294
Samuelson, P. A., 100, 101
Scatter diagram, 306
Schultz, Henry, 31, 100–101, 134,
   149, 165, 168, 272
Secant line, 55
Shannon, C. E., 213
Slope, 34, 374
Slope-intercept formula, 44
Smith, Adam, 101
Smith, J. Maynard, 101, 150, 168, 202
Speed, 61–64
Standard deviation, 195
Square-sum error, 308
Supply curve, 8–9
Surface, 269

T

Tangent line, 54
Taylor series, 161
Total differential, 281
Transcendental function, 169
Trigonometric functions: 237
   graphs of, 382–383

U

Utility theory, 299

V

Variable, dependent and
    independent, 13

Velocity, 32, 61–64
Volume as an integral, 256

W

Weber, Ernest, 334
Weber's law, 334
Weierstrass, Karl, 68

applications index (cont'd)

*These applications are accompanied by references to original source material or to suggested further readings.